Chemical Principles in Practices

Chemical Principles in Practices

Contributors

Ruilin Yang, Yonghua Han et al.

www.aurisreference.com

Chemical Principles in Practices

Contributors: Ruilin Yang, Yonghua Han et al.

Published by Auris Reference Limited

www.aurisreference.com

United Kingdom

Chemical Principles in Practices

ISBN: 978-1-78154-859-2

British Library Cataloguing in Publication Data
A CIP record for this book is available from the British Library

Printed in the United Kingdom

Exclusively distributed by CBS Publishers & Distributors Pvt. Ltd.

Sales & Distribution Rights only for India, Pakistan, Bangladesh, Sri Lanka, Nepal and Bhutan. This book is not to be sold outside these territories.

Contents

List of Abbreviations

APC	Anaphase-Promoting Complex
BOP	Bond overlap population
CRNT	Chemical reaction network theory
CRO	Chemical reaction optimization
CCPs	Constrained critical paths
CEFT	Constrained earliest finish time
CCPDAG	Constrained-critical-path directed acyclic graph
DAG	Directed acyclic graph
EA	Electron affinity
ELNES	Electron energy loss near-edge structure
EELS	Electron energy loss spectroscopy
EEL	Electron energy-loss
KEGG	Encyclopedia of Genes and Genomes
GRN	Gene regulatory network
HSAB	Hard and Soft Acids and Bases Principle
IE	Ionization energy
LMT	Levelized-min time
LCR	ligase chain reaction
LSHF	Linear surface heat flux
LST	Linear surface temperature
MCP	Modified critical path
NET	Network-embedded thermodynamic
ODE	Ordinary differential equation
OLCAO	Orthogonalized linear combination of atomic orbital
PDOS	Partial density of states
PCR	Polymerase chain reaction
SCF	Self-Consistent Field
SAC	Spindle assembly checkpoint
SVR	Support vector regression
TF	Transcription factor
TEM	Transmission electron microscopy
UTI	Urinary Tract Infection

List of Contributors

Ruilin Yang
Guangdong and Shenzhen Key Laboratory of Male Reproductive Medicine and Genetics, Institute of Urology, Peking University Shenzhen Hospital, Shenzhen PKUH-KUST Medical Center, Shenzhen, China,
Guangzhou Panyu Central Hospital, Guangzhou, China,
Shantou University Medical College, Shantou, China,

Yonghua Han
Guangdong and Shenzhen Key Laboratory of Male Reproductive Medicine and Genetics, Institute of Urology, Peking University Shenzhen Hospital, Shenzhen PKUH-KUST Medical Center, Shenzhen, China,

Yiwang Ye
Guangdong and Shenzhen Key Laboratory of Male Reproductive Medicine and Genetics, Institute of Urology, Peking University Shenzhen Hospital, Shenzhen PKUH-KUST Medical Center, Shenzhen, China,

Yuchen Liu
Guangdong and Shenzhen Key Laboratory of Male Reproductive Medicine and Genetics, Institute of Urology, Peking University Shenzhen Hospital, Shenzhen PKUH-KUST Medical Center, Shenzhen, China,
Shantou University Medical College, Shantou, China,

Zhimao Jiang
Guangdong and Shenzhen Key Laboratory of Male Reproductive Medicine and Genetics, Institute of
Urology, Peking University Shenzhen Hospital, Shenzhen PKUHKUST Medical Center, Shenzhen, China,

Yaoting Gui
Guangdong and Shenzhen Key Laboratory of Male Reproductive Medicine and Genetics, Institute of
Urology, Peking University Shenzhen Hospital, Shenzhen PKUHKUST Medical Center, Shenzhen, China,

Zhiming Cai
Guangdong and Shenzhen Key Laboratory of Male Reproductive Medicine and Genetics, Institute of Urology, Peking University Shenzhen Hospital, Shenzhen PKUH-KUST Medical Center, Shenzhen, China,
Urogenital Institute of Shenzhen University, Shenzhen Second People's Hospital, Shenzhen, China,
Shenzhen Second People's Hospital, Shenzhen, China

Qiang Zhu
National Key Laboratory of Crop Genetic Improvement, College of Life Science and Technology, Huazhong Agricultural University, Wuhan, China,
Center for Bioinformatics, Huazhong Agricultural University, Wuhan, China,

Tao Qin
Shandong Provincial Research Center for Bioinformatic Engineering and Technique, School of Life
Sciences, Shandong University of Technology, Zibo, China

Ying-Ying Jiang
Shandong Provincial Research Center for Bioinformatic Engineering and Technique, School of Life
Sciences, Shandong University of Technology, Zibo, China

Cong Ji
Center for Bioinformatics, Huazhong Agricultural University, Wuhan, China,

De-Xin Kong
Center for Bioinformatics, Huazhong Agricultural University, Wuhan, China,

Bin-Guang Ma
National Key Laboratory of Crop Genetic Improvement, College of Life Science and Technology, Huazhong Agricultural University, Wuhan, China,
Center for Bioinformatics, Huazhong Agricultural University, Wuhan, China,

Hong-Yu Zhang
National Key Laboratory of Crop Genetic Improvement, College of Life Science and Technology, Huazhong Agricultural University, Wuhan, China,
Center for Bioinformatics, Huazhong Agricultural University, Wuhan, China,

Kazuyoshi Tatsumi
Department of Materials, Physics and Energy Engineering, Nagoya University, Chikusa, Nagoya 464-8603, Japan

Shunsuke Muto
Department of Materials, Physics and Energy Engineering, Nagoya University, Chikusa, Nagoya 464-8603, Japan

Kazutaka Ikeda
Institute of Materials Structure Science, High Energy Accelerator Research Organization (KEK), 1-1 Oho, Tsukuba, Ibaraki 305-0801, Japan

Shin-Ichi Orimo
Institute for Materials Research, Tohoku University, Sendai 980-8577, Japan

P. K. Chattaraj
Department of Chemistry, Indian Institute of Technology, Kharagpur, 721302, In-

dia time dependent Schrödinger equation (TDSE) minimum polarizability principle (MPP).

B. Maiti
Department of Chemistry, Indian Institute of Technology, Kharagpur, 721302, India time dependent Schrödinger equation (TDSE) minimum polarizability principle (MPP).

P. Geerlings
Eenheid Algemene Chemie, Free University of Brussels (VUB), Pleinlaan 2,1050 Brussels, Belgium

F. De Proft
Eenheid Algemene Chemie, Free University of Brussels (VUB), Pleinlaan 2, 1050 Brussels, Belgium

Stefano Zambelli
University of Padova, Italy

Dennis Go¨rlich
Bio Systems Analysis Group, Institute of Computer Science, Jena Centre for Bioinformatics and Friedrich Schiller University Jena, Jena, Germany,
Institute of Biostatistics and Clinical Research, University of Muenster, Muenster, Germany

Peter Dittrich
Bio Systems Analysis Group, Institute of Computer Science, Jena Centre for Bioinformatics and Friedrich Schiller University Jena, Jena, Germany,

Bapuji Pullepu
Department of Mathematics, SRM University, Kattankulathur, Tamil Nadu 603203, India

P. Sambath
Department of Mathematics, SRM University, Kattankulathur, Tamil Nadu 603203, India

K. K. Viswanathan
UTM Centre for Industrial and Applied Mathematics and Department of Mathematical Sciences, Faculty of Science, Universiti Teknologi Malaysia, 81310 Johor Bahru, Johor, Malaysia

Swetha Ravi
Department of Mathematics, Gudlavalleru Engineering College, Gudlavalleru, India

Jagdish Prakash
Department of Mathematics, University of Botswana, Gaborone, Botswana

Viswanatha Reddy Gottam
Department of Mathematics, S. V. University, Tirupati, India

Vijaya Kumar Varma Sibyala
Department of Mathematics, S. V. University, Tirupati, India

Yuyi Jiang
College of Information Science and Engineering, East China University of Science and Technology, Shanghai 200237, China

Zhiqing Shao
College of Information Science and Engineering, East China University of Science and Technology, Shanghai 200237, China

Yi Guo
College of Information Science and Engineering, East China University of Science and Technology, Shanghai 200237, China

Thomas Wilhlm
Theoretical Systems Biology, Institute of Food Research, Norwich Research Park, Colney Lane

Preface

The text *Chemical Principles in Practices* helps students to develop chemical insight by showing the connections between fundamental chemical ideas and their applications. The idea of chemical principles developed out of the classical elements. The purpose of first chapter is to carry out chemical synthesis of the bacteriophage G4 and the study of its infectivity. Second chapter focuses on chemical basis of metabolic network organization. In third chapter, we systematically examine differences in the chemical bonding states of Al-containing compounds (including AlH_3) by comparing their $Al\text{-}L_{2,3}$ EEL spectra. The interaction of a hydrogen atom in its ground electronic state and an excited electronic state with laser fields of different colors has been studied in fourth chapter. Fifth chapter deals with chemical reactivity as described by quantum chemical methods. An historical introduction of chemical kinetics has been presented in sixth chapter. Molecular codes in biological and chemical reaction networks have been described in seventh chapter. Eighth chapter focuses on effects of chemical reactions on unsteady free convective and mass transfer flow from a vertical cone with heat generation/absorption in the presence of VWT/VWC. The effects of thermal radiation and radiation absorption on flow past an impulsively started infinite vertical plate with Newtonian heating and chemical reaction have been investigated in ninth chapter. Tenth chapter proposes a tuple molecular structure-based chemical reaction optimization (TMSCRO) method for DAG scheduling on heterogeneous computing systems, based on a very recently proposed metaheuristic method, chemical reaction optimization (CRO). The smallest chemical reaction system with bistability has been presented in last chapter.

Chapter 1

CHEMICAL SYNTHESIS OF BACTERIOPHAGE G4

Ruilin Yang[1,2,3], Yonghua Han[1]., Yiwang Ye[1], Yuchen Liu[1,3], Zhimao Jiang[1], Yaoting Gui[1], Zhiming Cai[1,4,5]

[1] Guangdong and Shenzhen Key Laboratory of Male Reproductive Medicine and Genetics, Institute of Urology, Peking University Shenzhen Hospital, Shenzhen PKUHKUST Medical Center, Shenzhen, China,

[2] Guangzhou Panyu Central Hospital, Guangzhou, China,

[3] Shantou University Medical College, Shantou, China,

[4] Urogenital Institute of Shenzhen University, Shenzhen Second People's Hospital, Shenzhen, China,

[5] Shenzhen Second People's Hospital, Shenzhen, China

ABSTRACT

Background

Due to recent leaps forward in DNA synthesis and sequencing technology, DNA manipulation has been extended to the level of whole-genome synthesis. Bacteriophages occupy a special niche in the micro-organic ecosystem and have potential as a tool for therapeutic agent. The purpose of this study was to carry out chemical synthesis of the bacteriophage G4 and the study of its infectivity.

Methodology/Principal Findings

Full-sized genomes of bacteriophage G4 molecules were completed from short overlapping synthetic oligonucleotides by direct assembly polymerase chain reaction and ligase chain reaction followed by fusion polymerase chain reaction with flanking primers. Three novel restriction endonuclease sites were introduced to distinguish the synthetic G4 from the wild type. G4 particles were recovered after electroporation into Escherichia coli and were efficient enough to infect another strain. The phage was validated by electron microscope. Specific polymerase chain reaction assay and restriction analyses of the plaques verified the accuracy of the chemical synthetic genomes.

Conclusions

Our results showed that the bacteriophage G4 obtained is synthetic rather than a wild type. Our study demonstrated that a phage can be synthesized and manipulated genetically according to the sequences, and can be efficient enough to infect the Escherichia coli, showing the potential use of synthetic biology in medical application.

INTRODUCTION

Bacteriophages, being viruses that infect bacteria, have played an important role in underpinning the development and advancement of the biosciences since the dawn of molecular biology. Bacteriophages, first discovered around 1915, occupy a special place in viral biology. They are perhaps the best understood viruses. Their genome being less than 10,000 bases long is particularly amenable to genetic alterations at the level of whole genome synthesis thanks to the recent advance made in DNA synthesis and sequencing technology. We may now modify genetic information to an extent not possible before. Due to the advances in synthetic biology, it is now possible to obtain large segments of synthetic DNA, assemble them into entire genomes of infectious agents, and boot them to life. Cello J et al [1] achieved the first chemical synthesis of a DNA (7,500 bp) corresponding to the entire genome of poliovirus. The DNA of the poliovirus was the largest DNA sequence ever synthesized in 2002. By establishing conditions for assembly of the genome, Venter improved upon the methodology and completed the infectious genome of bacteriophage ΦX174 (5,386 bp) within 2 weeks [2]. Researchers further made a distinct improvement in the scale of synthesis which was the 582,970 bp genome of Mycoplasma genitalium [3]. Once again, Venter and his colleagues succeeded in creating the first living organism with a completely synthetic genome[4]. A few medical applications of chemical genome synthesis also have been described [5],[6]. The common goal of this new strategy is to further our understanding of an organism's properties and to make use of this new information to prevent or treat human disease.

Bacteriophage G4, first isolated in 1973 [7], is a genus of Escherichia coli (E. coli) phages of the family Microviridae. G4 or ΦX174-like, is an icosahedra particle with almost the same density as bacteriophage ΦX174 which has been used in many landmark experiments, and contains single-stranded circular DNA [8]. The 5,577 nucleotide long sequence of bacteriophage G4

DNA has been determined, with an average of 66.9% nucleotide sequence similarity compared to that of bacteriophage ΦX174 [9]. These, along with the successful whole genome assembly of bacteriophage ΦX174, can validate G4 as an attractive target for chemical synthesis in medical use.

G4 can only multiply within E. coli and kill the cells by lysis. E. coli are the most common etiologic agent clinically associated with Urinary Tract Infection (UTI) which is the most frequently diagnosed kidney and urologic disease [10], [11]. However, there is documentation of increases multidrug-resistant pathogens [12]. For antibiotic-resistant infection, phage therapy may have potential uses in human medicine [13]. Chemical synthesis of G4 genomes will provide a new perspective of medical application.

All these facts induced us to carry out the synthesis of the ΦX -like phage and the study of its infectivity. We set out to generate a synthetic G4 genome (syn-G4) by whole genome assembly. To tell the difference from the wild type of the phage G4, we genetically alter 3 restriction endonuclease sites during the synthesis of the genome and subsequently generate a mutated G4 genome (m-G4). Two of these infectious circlular genomes produced were electropotated into DH5α and the plaques of syn-G4 and m-G4 were observed after incubation. Both of the phages were confirmed and were able to infect E. coli strain BL21. Our report demonstrates that an infectious phage can be synthesized and manipulated genetically according to the sequence published, and efficient enough to infect the E. coli, showing the potential use of synthetic biology for medical application.

RESULTS

Chemical synthesis of syn-G4 and m-G4 genomes

The oligonucleotides were designed to synthesize a G4 genome with exactly the same sequence reported by Godson et al. in 1978 [9] [NCBI Reference Sequence: NC_001420.2]. The sequence was designed to make a molecule that could be cleaved to size with *PstI* and then circularized to produce a circular molecule (Fig. 1).Three restriction endonuclease sites (*NcoI, KpnI* and *EcoRI*, respectively) were introduced to generate a m-G4 genome. Since the mutation sites are located at the third base of the corresponding codon, minimal structural and functional impact was expected. The oligonculeotides were pooled and gel-purified by Sangon Biotech as described in Materials and methods. The sequences of the short custom-made segments were consistent with the results from both sequencing and agarose gel. Full-sized genomes of syn-G4 and m-G4 molecules were completed and provided by Shanghai Sangon Biotech. Co., Ltd (data shown in Material S1).

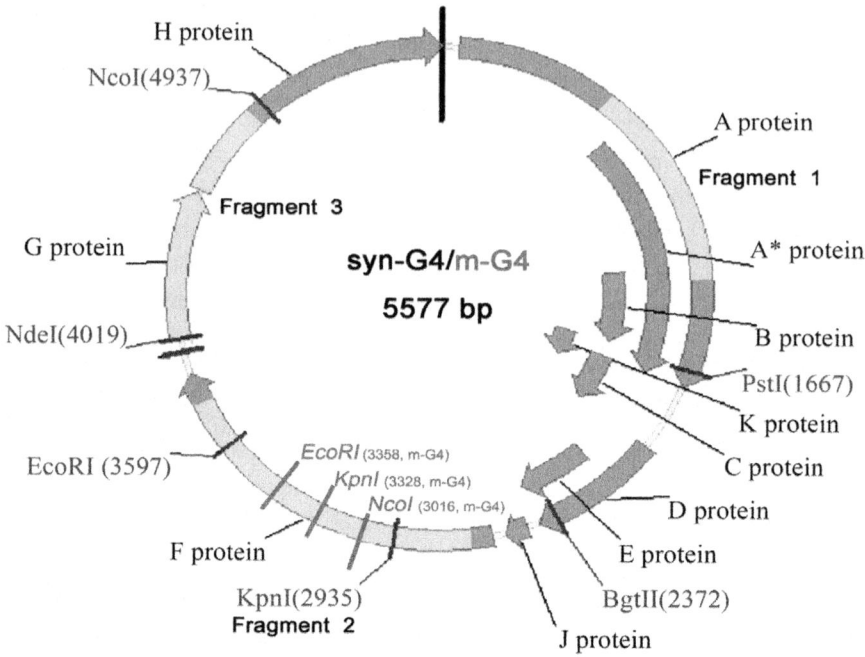

Figure 1: Structure drawing of single-stranded circular syn-G4/m-G4 genome (5577 bp) and sites of restriction endonucleases. New NcoI, KpnI and EcoRI sites marked in red slash were added into the m-G4 genome. Three specific PCR fragments were shown in yellow.

Circularization of full-length genome and infectivity testing

The polymerase chain reaction (PCR) productions of full-length genomes were cleaved with *PstI* and then circularized by ligation with T4 ligase using recommended conditions to generate products of syn-G4 and m-G4 circular molecules, then each 10 μl of ligation product was electroporated into DH5α cells, immediately diluted with 600 μl of liquid Luria-Bertani (LB) medium, and then divided into two screwcapped glass culture tubes. The tubes were rotated at 37°C for 60 min and then 200 μl of culture fluid was stained to LB solid culture medium to stay overnight at 37°C. Phage plaques were visualized after 6–12 h of incubation at 37°C (Fig. 2, A, syn-G4; B, m-G4). The phage formed small, clear, round plaques and some merged or ran together to form a mass on the LB lawn. 10 μl of pure water underwent the same procedures for control group. No plaque was observed after incubation at 37°C (Fig. 2, C, control). Phage from the plate underwent plaque purification and 100 μl of phage suspension with an estimated concentration of 10^{-5}/L was added to

100 μl of log phase BL21. Phage plaques were visualized incubation of 6–10 h but formed smaller plaques compared to which generated by with a same concentration.

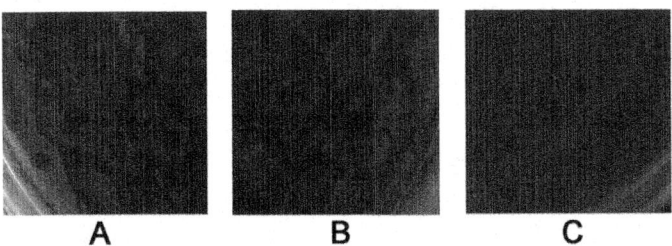

A **B** **C**

Figure 2: Close up images of the plaques (A, syn-G4; B, m-G4) and the bacterial lawns without plaques (C, control). The plaques appeared small, clear and round. Some merged or ran together to form a mass. The appearance of the plaques of both syn-G4 and m-G4 showed no difference in A and B (A, syn-G4; B, m-G4). doi:10.1371/journal.pone.0027062.g002

Morphology by electron microscope

Synthetic G4 samples were positive stained with 0.2% (w/v) phosphotungstic acid and examined by transmission electron microscopy, as seen in Fig. 3. The phage looked like an icosahedral protein shell and the shape was seen approximately 50 nm in diameter (Average of anteroposterior diameter and left-right diameter).

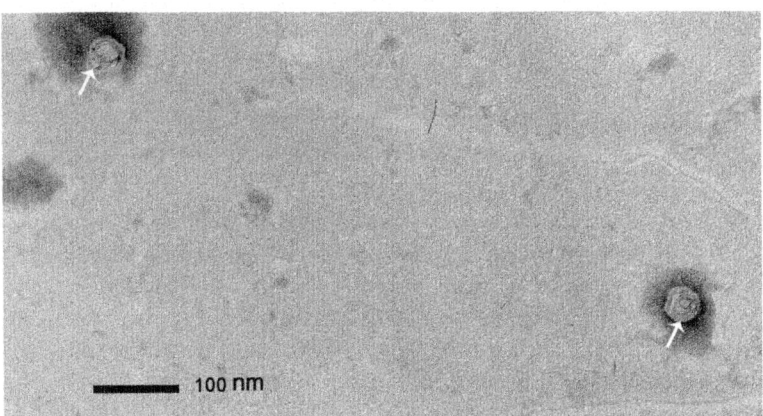

100 nm

Figure 3: Synthetic G4 morphology by transmission electron microscope (Arrow). Magnification: 800,000×. Bar: 100 nm. doi:10.1371/journal.pone.0027062.g003

PCR assay and restriction analyses

Three different parts of the genome were targeted for PCR assay, considering each part coding important structural proteins. The three specific PCR fragments, shown in yellow in Figure 1, of both of syn-G4 and m-G4 were successfully amplified from plaques purification using the specific primer pair (Tab. 1) and confirmed with the results of sequencing. The resulting PCR product was subjected to agarose gel electrophoresis (Fig. 4). The 645 bp PCR amplification products of both syn-G4 and m-G4 were validated by sequencing to ensure the sequences without mutations and including the previous 3 restriction endonuclease sites (data shown inFile S1). The 645 bp sequence of syn-G4 contained only 1 *KpnI* site. These sequences of m-G4 contained 2 *KpnI* sites, 1 *NcoI* site and 1 *EcoRI* site after genetic alteration shown in red inFigure 1 with slash at the position of 3016, 3328 and 3358 respectively. The 645 bp DNA products were subjected to restriction enzyme digestion with *EcoRI*, *KpnI* and *NcoI* to excise different segments according to the supplier's recommendations. Subsequently, m-G4 could be digested by *EcoRI*, *KpnI* and *NcoI* respectively, but syn-G4 just could be digested by *KpnI*(Fig. 5). Consistent with the sequencing results, the genetic alteration sites were confirmed by restriction enzyme digestion.

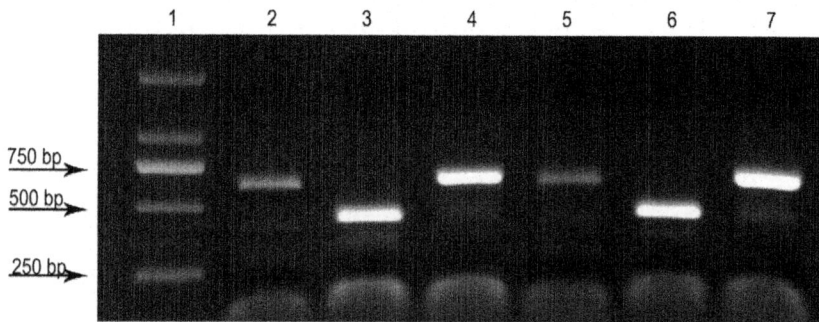

Figure 4: PCR of specific sequences. Agarose gel showing the PCR results obtained when specific primers were used. Lane 1, D2000 marker; Lane 2, 3 and 4, PCR amplification of the three specific products 603 bp, 443 bp and 645 bp, respectively, of syn-G4 (Location shown in Fragment 3, 1 and 2 in yellow in Fig. 1, respectively); Lane 4, 5 and 6, PCR amplification of the three specific products 603 bp, 443 bp and 645 bp, respectively, of m-G4 (Location shown in Fragment 3, 1 and 2 in yellow in Fig. 1, respectively). doi:10.1371/journal.pone.0027062.g004

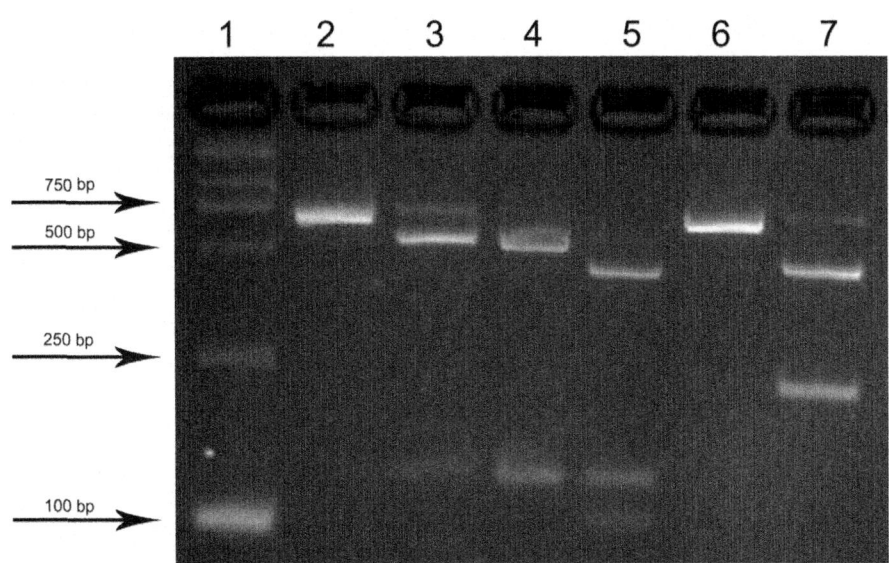

Figure 5: Digestion of restriction endonuclease. Lane 1, D2000marker. Agarose gels (Lane 2, 4, 6) showing the products (shown in Fragment 2 in Fig. 1) of syn-G4 digested by restriction endonucleases and generations of restriction endonuclease digestion fragments of 645 bp (EcoRI), 125 bp+520 bp (KpnI) and 645 bp (NcoI), respectively. The 3 modified restriction endonuclease sites of m-G4 were confirmed by agarose gels and different digestion fragments of 518 bp+127 bp (EcoRI), 125 bp+423 bp+97 bp (KpnI), and 209 bp+436 bp (NcoI) (Lane 3, 5 and 7 respectively), were observed. doi:10.1371/journal.pone.0027062.g005

In this report, the natural DNA was unavailable to the experimenter. A chemical synthesis of a circular G4 genome was transfected into E. coli, which produced viable bacteriophages and caused E. coli lysis. To distinguish from the wild type G4, m-G4 was genetically modified with 3 restriction endonuclease sites, which subsequently turned out to be an infectious phage after transfection into E. coli. Both of the plaques could not be identified in morphology, but there were obvious differences observed after digestion of genomes with 3 restriction endonucleases. These results demonstrate that the phages causing lysis of E. coli. were originally from our designed syn-G4 and m-G4 ones. The difference between G4 and M-G4 showed that this isolate is synthetic rather than a wild type. The chemical synthesis of genome was efficient enough to form infectious molecules and subsequently infect the E. coli.

DISCUSSION

The result in this report proves the correctness of Godson's sequence meaning that is accurate enough to produce an infectious phage. The genetic modification of syn-G4 subsequently generating m-G4 validates the capability of manual manipulation. The development of synthetic approach is still at an early stage, but this field of research has attracted much attention [14], [15]. The methodology of synthetic biology has been improved, from forming an infectious agent to a cellular genome with high efficiency and yield. Apart from the resurrection of the 1918 influenza virus and the generation of codon- and codon pair-deoptimized polioviruses, researchers have reengineered bacteriophages to combat antibiotic-resistant bacteria by endowing them with genetic mechanisms that destroy bacterial mechanisms for evading antibiotic action [16], [17]. Therefore, phage therapy, especially using synthetic phages, might be an alternative method to control and combat pathogenic bacteria in human medicine. From a practical perspective, UTI mainly caused by uropathogenic E. coli (UPEC) is easy to recur. For such a stubborn UTI infection, a phage designed to destroy UPEC is one potential treatment. We propose to assemble larger genomes required for an entire phage that can be programmed to target specific pathogenic agents and pathological mechanisms, without significant impairment of the phage's ability to replicate. Chemical synthesis of G4 is the first attempt but there are still many challenges. Meanwhile, a lot of work still needs to be done to test the relevant theories and principles. However, the oligonucleotide synthesis combined with the modifications of a phage genome described here will serve as a basis for understanding the procedure. Given the synthetic approaches available and the present knowledge of phage biology, this new strategy may have potential use in antimicrobial therapies. Whole G4 genome assembly will further enhance the role of a synthetic approach for medical application. This new field of research needs to be explored thoroughly in order to meet the present and future medical demands.

MATERIALS AND METHODS

Chemical synthesis of genomes

The oligonucleotides were designed to generate a syn-G4 genome according to the sequence reported by Godson [9]. According to the reference sequence, we mutated T-A, T-G and T-C at the position of 3016 bp 3244 bp and 3328 bp respectively to produce three novel restriction endonuclease sties (*NcoI*, *KpnI* and *EcoRI*, respectively) in order to distinguish from the wild type G4 genome. The full-sized genomes of bacteriophage G4 and m-G4 were synthesized and provided by Sangon Biotech Co., Ltd (Shanghai). Basically, for synthesis of the full-sized genomic DNA, each of syn-G4 and m-G4 genomes was divided into 4 segments, including A,B1,B2,and B3(syn-G4: 5996A: 925 bp, 5996B1:1410 bp, 5996B2: 1250 bp, 5996B3 2091 bp;m-G4: 5996A: 925 bp, 5997B1:1410 bp, 5996B2: 1250 bp, 5996B3 2091 bp). All oligonucleotides were chemically synthesized by Sangon Biotech. The resultant fragments of various lengths with overlap were added together and finally assembled into full-sized genome by direct assembly PCR and ligase chain reaction (LCR) followed by fusion PCR with flanking primers from short overlapping synthetic oligonucleotides. The linear full-sized genomic DNA was circularized by means of enzymatic ligation and the use of preexisted *PstI* restriction sites for preparation of sticky ends using recommended conditions. Transfection of cells with the circular synthetic genome resulted in appearance of infective plaques.

Circularization of linear genome DNA molecules

PCR products of full-sized genomes were pooled and confirmed by sequencing. The linear DNA products were cleaved with *PstI* and then circularized by ligation with T4 ligase using recommended conditions. The ligation mixture was stayed overnight at 22°C in preparation for infectivity testing.

Assay of G4 plaques

One microliter of syn-G4 ligation product was electroporated into DH5α cells (Biononatech Co., Ltd, Shenzhen, China), immediately diluted with 600 µl of liquid LB medium, and then divided into two screwcapped glass culture tubes. The two tubes were rotated at 37°C for 60 min and then 200 µl of culture fluid was stained to LB medium A to stay overnight at 37°C. Another microliter of m-G4 and pure water underwent the same procedure and each of 200 µl of culture fluid was stained to LB medium B and C respectively, to stay overnight at 37°C. Phage plaques in A and B were visualized after 6–12 h of incubation at 37°C and recorded by photography at 8 h. No plaque was observed in C.

Electron microscopy

Several plaques were picked up to 1.5 ml Eppendorf and 0.5 ml of 2.0% (w/v) glutaraldehydewas added for fixation before examination. A concentrated phage sample was negatively stained with 0.2% (w/v) phosphotungstic acid on a carbon-coated grid and examined by JEM100CXII (Japan) transmission electron microscope (Electron Microscope Laboratory, School of Life Sciences, Sun Yat sen University, Guangzhou) at an accelerating voltage of

80 kV. Electron micrographs were taken at a magnification of 800,000×. The phage size was determined from the average of two independent measurements (Anteroposterior diameter and left-right diameter).

Plaques purification and assay of infectivity

A confluent plaque was picked up and covered with 50 μl SM buffer (50 ml 1 M tris-cl(PH 7.5), 2 g MgSO4, 5.8 g NaCl, 5 ml 2% gelatin and ddH$_2$O up to 1 L), gently shaken and maintained in the refrigerator at 4°C for 3 min. After adding a few drops of chloroform, the plate is manually shaken. The phage suspension was centrifuged (5000×g for 10 min) to remove DH5α and debris. The 200 μl of the supernatant was retrieved and diluted to the proper concentration (10^{-7}/L~10^{-5}/L) for infectivity test. E. coli strain BL21 (Biononatech Co., Ltd, Shenzhen, China) colony was picked up to 5 ml LB culture medium to prepare log phase of BL21. BL21 were grown in LB broth without antibiotic selection at 37°C until an OD600 of 0.4 was reached. Bacterial cells were diluted in LB broth to a final density of 10^6 CFU/ml. An aliquot of cells (10^5CFU, 100 μl) was added with 100 μl phage suspension with a diluted concentration of 10^{-5}/L and mixed with 0.7% soft agar, slightly shaken, then the mixture was plated on LB plates, following overnight incubation at 37°C. Phage plaques were visualized after 6~18 h of incubation at 37°C.

PCR assay and DNA sequencing

Specific primers for syn-G4 and m-G4 were synthesized by Jierui Biotech co., Ltd. (Guangzhou), and their sequences were listed in Table 1. The expected length of the amplification product was 443 bp, 603 bp and 645 bp respectively. Plaques purified were picked directly into 50 μl PCRs respectively. Each reaction mixture contained 2.5 of phage suspension, 1.5 μl of 25 mmol/L MgCl$_2$, 2 μl of 2.5 mmol/L dNTP, 1.0 μl of each 20 pmol/L primer, 0.4 μl of 5 U/L Taq polymerase (Takara), 25 of primerstar, and ddH$_2$O up to total 50 μl. PCR of NLE-pCMV10 for NLE gene was also performed as the control group (date not shown). w The conditions for PCR were as follows: 99°C for 2 min, followed by 30 cycles of 98°C, 10 s, 68°C, 40 s, 72°C, 10 min and a final extension of 72°C for 7 min. The PCR was performed as previously described. The PCR products were analyzed using a Rapid Agarose Gel Electrophoresis System (Wealtec Corp, Sparks, Nev) in 1.0% agarose gels in 5.0 μl loading buffer (10 min at 90 V). The genome was gel-purified and sequenced on a 3730 XL sequencer (3730 XL sequencer, ABI, Sangon) with depth of 2-fold. Sequencing of the cloned amplification product confirmed that it was identical to part of the syn-G4 and m-G4.

Table 1: Primers for specific PCR. doi:10.1371/journal.pone.0027062.t001

Target	Sequences	PCR Product (bp)
Syn-G4	5'-CAAAAATCTTGGAGGAGTCAACTATGAAGTCTC-3'	443
	5'-GGAGTACCGGACTGCGATGGGCATAGAGTAAC-3'	
Syn-G4	5'-GCCTACGGGAGATACTCGAGTCTCCGATAC-3'	603
	5'-GGTTGAAGGACGGTTGCTTCACGGTTTAC-3'	
Syn-G4	5'-TCTATATCCCACACCGTCATATCTACGGTC-3'	645
	5'-TAATTACGCGATGCTCAGGAACATAGAAG-3'	
m-G4	5'-CAAAAATCTTGGAGGAGTCAACTATGAAGTCTC-3'	443
	5'-GGAGTACCGGACTGCGATGGGCATAGAGTAAC-3'	
m-G4	5'-GCCTACGGAGATACTCGAGTCTCCGATAC-3'	603
	5'-GGTTGAAGGACGGTTGCTTCACGGTTTAC-3'	
m-G4	5'-TCTATATCCCACACCGTCATATCTACGGTC-3'	645
	5'-TAATTACGCGATGCTCAGGAACATAGAAG-3'	

Restriction analyses

For restriction analyses, 10 µl of PCR products with expected length of 645 bp were digested with restriction endonucleases (*EcoRI*, *KpnI* and *NcoI*) according to the supplier's recommendations (Takara, Dalian, China). Digestion of syn-G4 and m-G4 with 2.0 µl of *EcoRI* restriction endonuclease (Takara, Dalian) was done at 37°C in 10 m M Tris-HCl, 50 mM NaCl, 15 mM MgCl2, pH 7.5. DNA was digested with 2.0 µl of *KpnI* endonuclease (Takara, Dalian) at 37°C in 10 mM Tris-HCl, 10 mM MgCl2, 0.02% Triton X-100, 0.1 mg/ml BSA. Digestion of DNA with 2.0 µl of *NcoI* restriction endonuclease (Takara, Dalian) was done at 37°C 10 mM Tris-HCl, 50 mM KCl, 1 mM DTT, 0.1 mM EDTA, 0.2 mg/ml BSA and 50% glycerol. The digested DNA was examined by gel electrophoresis after 10 hours digestion. The resultant fragments were separated and purified as described above.

Gel Electrophoresis

DNA was analyzed using a 6.0% (w/v) agarose gel in TAE buffer (40 mmol l⁻¹ Tris acetate; 2 mmol L⁻¹ EDTA, pH 8.3) with 0.5 µmg ml⁻¹ ethidium bromide. Electrophoresis was performed at constant voltage 90 V) for 30 min and visualized by UV light (300 nm) after staining with ethidium bromide (1 Ag/ml). This procedure was followed for all the experiments except where stated differently. The relative sizes of the DNA fragments were estimated by comparing their electrophoretic mobility with that of the standards run with the samples on each gel (D2000marker).

Theoretical Considerations

In order to accomplish the synthesis of infectious G4 genome, we had considered the replication mechanism of G4 [18] and analyzed the methodology of chemical synthesis of genome [19]–[21]. When the sequence of phage G4 was determined by Godson, its structure was almost understood. During the stage of phage G4 replication, there was a circular double-stranded DNA called replication form (RF form) [22]. It was not yet possible for us to synthesize entire genome as long continuous strands of DNA. In designing the oligonclotiedes sets, we adopted the basic methods available for assembling long DNA sequences and the strategy of dividing the genome into 4 segments. Each segment was also divided into 2 or 3 shorter pieces (data not shown). These entire short custom-made single-stranded DNA with overlap were assembled from the oligonucleotides. These segments were finally assembled into full-sized genome by direct assembly PCR and ligase chain reaction (LCR) [21]. These synthetic phage G4 was referred to the G series of X-like phages. They were subsequently grown and handled by all of the methods commonly used for ΦX174 [7]. The linear G4 molecules were then circularized to form a circular RF form after gel-purified using recommended conditions [2].

ACKNOWLEDGMENTS

We are indebted to the faculty and staff of the Peking University Shenzhen Hospital, whose names were not included in the author list, but who contributed to this work. And we would like to acknowledge the useful discussion with Changye Hui.

AUTHOR CONTRIBUTIONS

Conceived and designed the experiments: RY ZC. Performed the experiments: RY. Analyzed the data: RY. Contributed reagents/materials/analysis tools: RY YH YY ZJ YL YG. Wrote the paper: RY. Obtained permission for use of cell line: ZJ.

REFERENCES

1.   Cello J, Paul AV, Wimmer E (2002) Chemical synthesis of poliovirus cDNA: generation of infectious virus in the absence of natural template. Science 297(5583): 1016–1018.

2.   Smith HO, Hutchison CA 3rd, Pfannkoch C, Venter JC (2003) Generating a synthetic genome by whole genome assembly: φX174 bacteriophage from synthetic oligonucleotides. Proc Natl Acad Sci USA 100(26): 15440–15445.

3.   Gibson DG, Benders GA, Andrews-Pfannkoch C, Denisova EA, Baden-Tillson H, et al. (2008) Complete chemical synthesis, assembly, and cloning of a Mycoplasma genitalium genome. Science 319(5867): 1215–1220.

4.   Gibson DG, Glass JI, Lartigue C, Noskov VN, Chuang RY, et al. (2010) Creation of a bacterial cell controlled by a chemically synthesized genome. Science 2;329(5987): 38–9.

5.   Kobasa D, Takada A, Shinya K, Hatta M, Halfmann P, et al. (2004) Enhanced virulence of influenza A viruses with the haemagglutinin of the 1918 pandemic virus. Nature 431(7009): 703–707.

6.   Coleman JR, Papamichail D, Skiena S, Futcher B, Wimmer E, et al. (2008) Virus attenuation by genome-scale changes in codon pair bias. Science 320: 1784–1787.

7.   Godson GN (1974) Evolution of phi-chi 174. Isolation of four new phi-chi-like phages and comparison with phi-chi 174. Virology 58(1): 272–89.

8.   Sanger F, Coulson AR, Friedmann T, Air GM, Barrell BG, et al. (1978) The nucleotide sequence of bacteriophage phiX174. J Mol Biol 25;125(2): 225–46.

9.   Godson GN, Barrell BG, Staden R, Fiddes JC (1978) Nucleotide sequence of bacteriophage G4 DNA. Nature 16;276(5685): 236–47.

10.   Kahlmeter G, ECO.SENS (2003) An international survey of the antimicrobial susceptibility of pathogens from uncomplicated urinary tract infections: the ECO.SENS Project. J Antimicrob Chemother 51(1): 69–76.

11.   Ronald A (2003) The etiology of urinary tract infection: traditional and emerging pathogens. Dis Mon 49(2): 71–82.

12.   Swaminathan S, Alangaden GJ (2010) Treatment of resistant enterococcal urinary tract infections. Curr Infect Dis Rep 12(6): 455–64.

13.   Levskaya A, Chevalier AA, Tabor JJ, Simpson ZB, Lavery LA, et al. (2005) Synthetic biology: engineering Escherichia coli to see light. Nature 438(7067): 441–442.

14.   Rad'ko SP, Il'ina AP, Bodoev NV, Archakov AI (2007) The synthesis of artificial genome as the basis of synthetic biology. Biomed Khim 53(3): 237–48.

15.   Chan LY, Kosuri S, Endy D (2005) Refactoring bacteriophage T7. Mol Syst Biol 1: 2005.0018.

16. Lu TK, Collins JJ (2007) Dispersing biofilms with engineered enzymatic bacteriophage. Proc Natl Acad Sci 104(27): 11197–11202.

17. Lu TK, Collins JJ (2009) Engineered bacteriophage targeting gene networks as adjuvants for antibiotic therapy. Proc Natl Acad Sci 106(12): 4629–4634.

18. Zechel K, Bouché JP, Kornberg A (1975) Replication of phage G4. A novel and simple system for the initiation of deoxyribonucleic acid synthesis. J Biol Chem 25;250(12): 4684–4689.

19. Wimmer E, Mueller S, Tumpey TM, Taubenberger JK (2009) Synthetic viruses: a new opportunity to understand and prevent viral disease. Nat Biotechnol 27(12): 1163–72.

20. Rad'ko SP, Il'ina AP, Bodoev NV, Archakov AI (2007) The synthesis of artificial genome as the basis of synthetic biology. Biomed Khim 53(3): 237–48.

21. Mueller S, Coleman JR, Wimmer E (2009) Putting synthesis into biology: a viral view of genetic engineering through de novo gene and genome synthesis. Chem Biol 16(3): 337–47.

22. Godson GN, Boyer H (1974) Susceptibility of the phiX-like phages G4 and G14 to R-EcoRi endonuclease. Virology 62(1): 270–5.

Chapter 2

CHEMICAL BASIS OF METABOLIC NETWORK ORGANIZATION

Qiang Zhu[1,2]., Tao Qin[3]., Ying-Ying Jiang[3].a , Cong Ji[2].b, De-Xin Kong[2]., Bin-Guang Ma[1,2]., Hong-Yu Zhang[1,2]

[1] National Key Laboratory of Crop Genetic Improvement, College of Life Science and Technology, Huazhong Agricultural University, Wuhan, China,

[2] Center for Bioinformatics, Huazhong Agricultural University, Wuhan, China,

[3] Shandong Provincial Research Center for Bioinformatic Engineering and Technique, School of Life Sciences, Shandong University of Technology, Zibo, China

ABSTRACT

Although the metabolic networks of the three domains of life consist of different constituents and metabolic pathways, they exhibit the same scale-free organization. This phenomenon has been hypothetically explained by preferential attachment principle that the new-recruited metabolites attach preferentially to those that are already well connected. However, since metabolites are usually small molecules and metabolic processes are basically chemical reactions, we speculate that the metabolic network organization may have a chemical basis. In this paper, chemoinformatic analyses on metabolic networks of Kyoto Encyclopedia of Genes and Genomes (KEGG), *Escherichia coli* and *Saccharomyces cerevisiae* were performed. It was found that there exist qualitative and quantitative correlations between network topology and chemical properties of metabolites. The metabolites with larger degrees of connectivity (hubs) are of relatively stronger polarity. This suggests that metabolic networks are chemically organized to a certain extent, which was further elucidated in terms of high concentrations required by metabolic hubs to drive a variety of reactions. This finding not only provides a chemical explanation to the preferential attachment principle for metabolic network expansion, but also has important implications for metabolic network design and metabolite concentration prediction.

Author Summary

The metabolic networks of the three domains of life exhibit the same scale-free organization, which has been hypothetically explained in terms of preferential attachment principle. Here we reveal that the scale-free organization of metabolic networks may have a chemical basis. Through a chemoinformatic analysis on metabolic networks of Kyoto Encyclopedia of Genes and Genomes (KEGG), *Escherichia coli* and *Saccharomyces cerevisiae*, it was found that the metabolites with higher degrees of connectivity (hubs) are of relatively stronger polarity. The reason underlying this phenomenon is that to drive a variety of reactions, metabolic hubs have to be highly concentrated. Since the intracellular environments are hydrophilic, metabolic hubs have to be strong-polar to reach high concentrations. This finding has direct implications for metabolic network design and provides a chemical explanation to the preferential attachment principle, which has been validated by numerical simulations of metabolic network expansion. In addition, the correlations between metabolite concentrations, metabolic network topology and metabolite chemical properties also suggest that we can use chemical and topological properties of metabolites to predict their intracellular concentrations. A support vector regression model has been successfully established to predict the metabolite concentrations for *Escherichia coli*.

INTRODUCTION

One of the most intriguing findings in systems biology is that despite the varied constituents and metabolic pathways of three domains of life, their metabolic networks exhibit the same scale-free organization. That is, a small part of metabolites participate in a large number of reactions (which are also termed hubs), while others are involved in a few reactions [1]. As the scale-free architectures are robust and error-tolerant, this finding provides meaningful insights into the design principle of metabolic networks.

The scale-free organization of metabolic networks has been hypothetically explained in terms of evolution that the new-recruited metabolite members attach preferentially to those that are already well connected (rich get richer, also known as preferential attachment principle) [2]–[4]. This implies that the metabolic network hubs originated relatively earlier than others in evolutionary history [5]. However, several issues about this evolutionary explanation remain elusive. First, the molecular basis of preferential attachment principle has not been fully elucidated, as it is inexplicable how the new metabolites "know" which metabolites are well connected. Second, the evolutionary explanation to the metabolic network organization has little implications for network design, because we do not know how to choose metabolites as hubs to construct a new metabolic network. Since most metabolites are small molecules and metabolic processes are basically chemical reactions, we speculate that the metabolic network organization may have a chemical basis, which stimulated our interest to address these issues by combining bioinformatics and chemoinformatics. The latter is a discipline devoted to encoding, storing, managing, searching and analyzing all kinds of chemical data by information technology [6], [7].

RESULTS/DISCUSSION

Correlations between Network Topology and Chemical Properties

Primarily, we explored the relationships between network topology and chemical properties for the metabolites recorded in Kyoto Encyclopedia of Genes and Genomes (KEGG). As illustrated in Figure S1, the metabolic network of KEGG is scale-free. There are 154 metabolites with degrees (defined as the number of edges linked to the metabolites) higher than 10, while 1180 are connected with only one metabolite. As shown in Table 1 and Figure 1, there exist qualitative and even quantitative correlations between degree and some chemical properties. In particular, molecular polarity, characterized by partition coefficients (ClogP, AlogP and LogD), ratio of atomic charge weighted partial positive surface area on total molecular surface area (FPSA3) and water solubility, rises with the increase of degree. Similar correlations can be observed for the metabolic networks of *Escherichia coli* (*E. coli*) (Figure 2) and *Saccharomyces cerevisiae* (*S. cerevisiae*) (Table 2). Therefore, it seems that metabolites get more polar and thus more water-soluble with the rise of degrees, which implies that the organization of the metabolic networks has a chemical basis. It is of apparent interest to explore the reasons underlying these correlations.

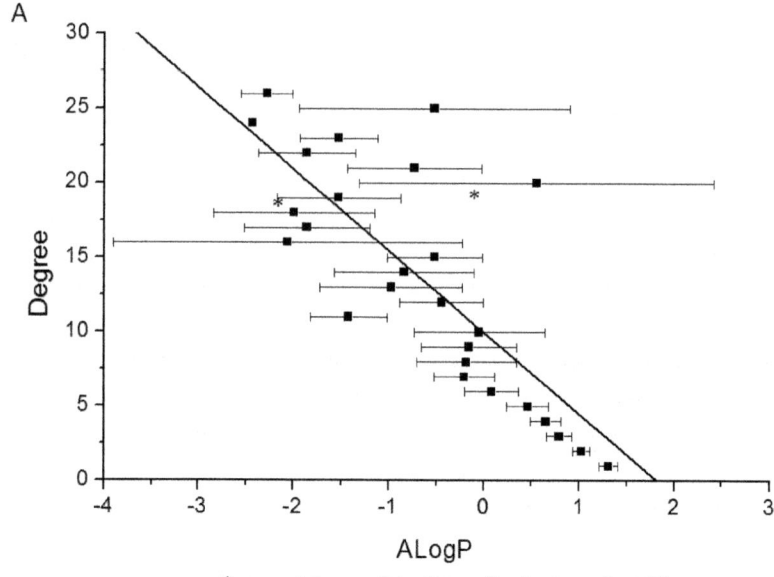

*comprising metabolites with degree of >= 26

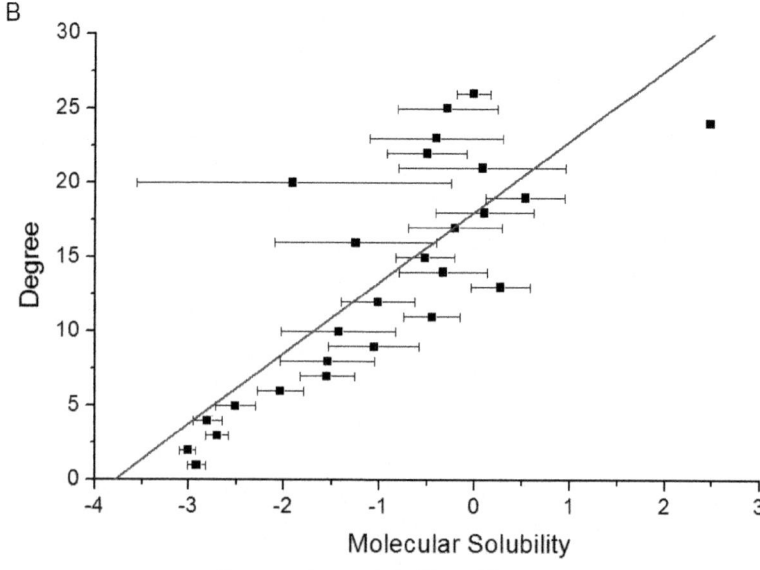

*comprising metabolites with degree of >= 26

Figure 1: Correlations between topological and chemical properties of KEGG metabolites. (**A**) Degree-ALogP (mean ± SE) correlation for KEGG metabolites ($R=-0.778$, $P<0.001$). (**B**) Degree-Molecular Solubility (mean ± SE) correlation for KEGG metabolites ($R=0.795$, $P<0.001$). doi:10.1371/journal.pcbi.1002214.g001

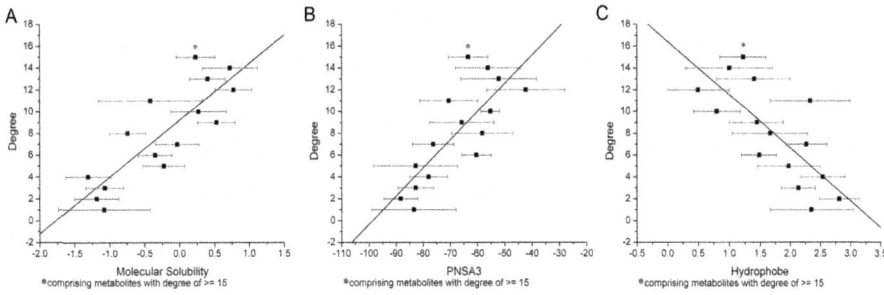

Figure 2: Correlations between topological and chemical properties of *E. coli* metabolites. (**A**) Degree-Molecular Solubility (mean ± SE) correlation (*R*=0.835, P<0.001). (**B**) Degree-PNSA3 (mean ± SE) correlation (*R*=0.796, *P*<0.001). (**C**) Degree-Hydrophobe (mean ± SE) correlation (*R*=−0.743, *P*<0.005). PNSA3 is defined as atomic charge weighted partial negative surface area. Hydrophobe is the number of hydrophobe. doi:10.1371/journal.pcbi.1002214.g002

Table 1: Mean values of some chemical descriptors for KEGG-recorded metabolites. doi:10.1371/journal.pcbi.1002214.t001

Descriptors	Characterization	Mean values		
		Degree 1 (n = 1180)	Degree 2-6 (n = 3327)	Degree > 6 (n = 368)
ClogP[a]	Partition coefficient octanol/water	1.30[d]	0.70[d]	−1.10[d]
FPSA3[b]	Ratio of atomic charge weighted partial positive surface area on total molecular surface area	0.062[d]	0.067[d]	0.079[d]
LogD[c]	Octanol-water partition coefficient calculated taking into account the ionization states of the molecule	0.43[d]	−0.53[d]	−2.31[d]
Molecular Solubility[c]	Water solubility, expressed as logS, where S is the solubility in mol/L	−2.91[d]	−2.82[d]	−0.98[d]

Table 2: Mean values of some chemical descriptors for *S. cerevisiae* metabolites. doi:10.1371/journal.pcbi.1002214.t002

Descriptors	Characterization	Mean values		
		Degree 1-3 (n = 301)	Degree 4-15 (n = 285)	Degree > 15 (n = 26)
ClogP[a]	Partition coefficient octanol/water	0.46[d]	−0.54[d]	−3.05[d]
FPSA3[b]	Ratio of atomic charge weighted partial positive surface area on total molecular surface area	0.066[d]	0.068[d]	0.080[d]
LogD[c]	Octanol-water partition coefficient calculated taking into account the ionization states of the molecule	−0.89[e]	−1.94[e]	−3.88[e]
Molecular Solubility[c]	Water solubility, expressed as logS, where S is the solubility in mol/L	−2.47[e]	−1.99[e]	0.11[e]

Explanation to the Correlations between Network Topology and Chemical Properties

As metabolic reactions are basically chemical reactions, it is natural to resort to chemical principles to explain the correlations. It is well known that the precondition for a chemical reaction to occur is $\Delta G = \Delta G^{0 + RT} \ln Q < 0$, where Q is the reaction quotient and is determined by the relative concentrations of reactants and products. Thus, for metabolites that participate in a large number of reactions as reactants (which usually have large degrees, as shown in Table S4), they must reserve high concentrations (quantities) to drive the reactions. Since metabolic reactions mainly occur in non-membrane systems which are hydrophilic environments, the metabolic network hubs must be highly water-soluble to reach high concentrations, which means that the hubs tend to be strong-polar. Therefore, the observed correlations between degree and chemical properties could be basically explained in terms of chemical property requirements of metabolic hubs. This explanation is supported by the correlations between degree and metabolite concentration and between metabolite concentration and chemical properties.

Recently, the absolute concentrations for over 100 metabolites of *E. coli*, exponentially growing in aerobic environment, were determined by Bennett and co-workers [8]. The concentrations of the measured metabolites are strongly biased. The top 10 abundant compounds account for 77% of the total concentration, while the less abundant half comprise only 1.3%, reminiscent of the topological structures of metabolic networks. As shown in Figure 3, there exists a correlation between the concentration and degree for *E. coli* metabolites. The metabolites with larger degrees have relatively higher concentrations and the degrees decline gradually with the drop of concentrations. However, one may argue that the metabolite concentrations oscillate during different phases of life, so how the concentrations of metabolites can correlate with degrees of connectivity–a static property? The answer resides in the fact that the amplitude of metabolite oscillation is rather low. For instance, during the life cycle of a yeast cell the amplitude of metabolite oscillation is usually within 10-fold, with a median of ~2.4-fold [9]. Therefore, it is reasonable to consider that the observed correlation between degree and metabolite concentration (at the level of order of magnitude) is robust.

A stepwise multiple linear regression analysis was conducted by SPSS (Version 15.0. SPSS Inc. Chicago, IL.) to select the most meaningful chemical properties from 83 descriptors to correlate with negative logarithm of *E. coli* metabolite concentrations ($-$LogC). The final regression equation is: $-$LogC = 6.105 + 0.431 × "ClogP" + 15.595 × "FNSA3" + 16.727 × "FPSA3" $-$ 5.333 × "RPCG", in which ClogP, FNSA3 (ratio of atomic charge weighted partial negative surface area on total molecular surface area), FPSA3 and RPCG (ratio of most positive charge on sum total positive charge) are all descriptors characterizing molecular polarity. The fitted concentrations by the chemical properties correlate well with the experimental values (Figure 4), indicating that the metabolite concentrations (at least for *E. coli*) are determined to a certain extent by their polarity and solubility, namely, strong-polar metabolites have relatively high concentrations.

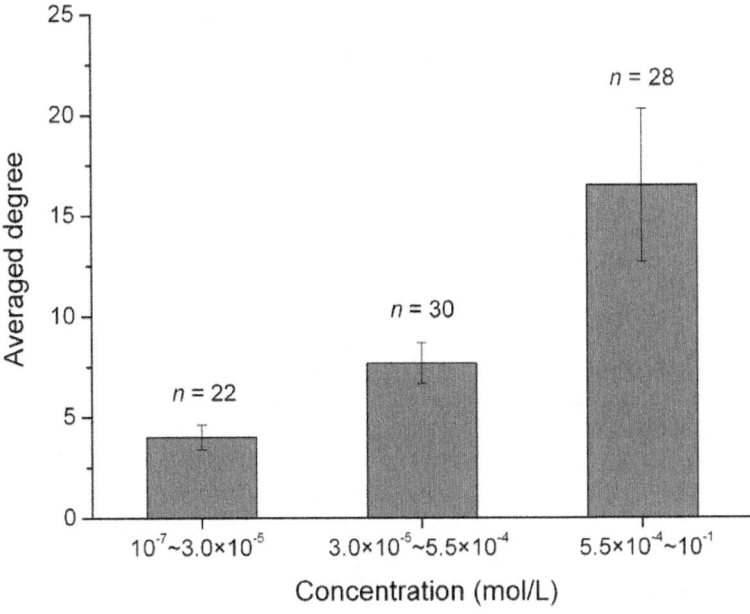

Figure 3: Degree-concentration correlation for ***E. coli*** metabolites (***P***<0.01, Kruskal-Wallis test). doi:10.1371/journal.pcbi.1002214.g003

This finding is similar to the observation about protein abundance of *E. coli* that highly abundant proteins are on average more hydrophilic than those with low copy numbers [10]. However, in protein-protein interaction (PPI) networks, protein degree is negatively correlated with concentration [11], just contrary to the observation on metabolic networks. The underlying reason was suggested as that the hub proteins of PPI networks tend to use hydrophobic residues at surface to bind diverse partners through nonspecific hydrophobic interactions [11]. The cellular concentrations of hub proteins are thus constrained by their hydrophobicity. Therefore, the different behaviors of PPI and metabolic network hubs can be well understood by basic chemical rules.

Taken together, the above observations offer an explanation to the correlation between topology and chemistry of metabolic networks. This finding also provides new clues to understanding the molecular basis of preferential attachment principle underlying the evolution of metabolic networks.

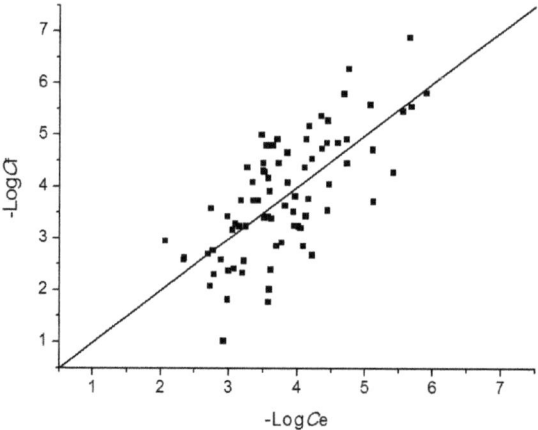

Figure 4: Theoretical fitting of *E. coli* metabolite concentrations by chemical properties. A stepwise multiple linear regression analysis was conducted to select the most
meaningful chemical properties that correlate with concentration (*C*). The final regression equation is: $-LogC=6.105 + 0.431 \times$ "ClogP" + 15.595 $\times$ "FNSA3" + 16.727 $\times$
"FPSA3" $- 5.333 \times$ "RPCG". The negative logarithm of fitted concentrations ($-LogC_f$)
for 80 *E. coli* metabolites correlates well with that of experimental values ($-LogC_e$)
($R=0.704$, $P<0.0001$). doi:10.1371/journal.pcbi.1002214.g004

Chemical Basis for the Preferential Attachment Principle

Since life originated from water environments, the primordial metabolites
must be highly hydrophilic. With the evolution of organisms, more and more
complex membrane systems evolved, which required hydrophobic metabolites
to perform intercellular and intracellular communications [12]. As a result,
the evolutionary direction of metabolites is from hydrophilic to hydrophobic,
which is clearly shown in the chemical evolution of *S. cerevisiae* metabolomes
(Table 3). According to the correlation between metabolite concentration and
chemical properties (Figure 4), it is reasonable to infer that the early-originated
metabolites have relatively higher concentrations than the late-recruited
counterparts in water environments. Since high-concentrated metabolites
have more potential to drive new reactions, it is understandable why the new-
recruited metabolites prefer to select old members as initial reactants (because
they are more abundant and thus more accessible). Taken together, the present
analysis reveals that metabolite concentration is a key factor to govern the
metabolic network expansion. Although the late metabolites can not "know"
which counterpart is well connected, they can "sense" which member is
abundant, which provides a self-consistent explanation to the preferential
attachment principle in terms of chemistry.

Descriptors	Characterization	Mean values	
		Early metabolites (n = 243)	Late metabolites (n = 369)
ClogP[a]	Partition coefficient octanol/water	−1.98[d]	0.98[d]
FPSA3[b]	Ratio of atomic charge weighted partial positive surface area on total molecular surface area	0.079[d]	0.061[d]
LogD[c]	Octanol-water partition coefficient calculated taking into account the ionization states of the molecule	−3.12[d]	−0.44[d]
Molecular Solubility[e]	Water solubility, expressed as logS, where S is the solubility in mol/L	−0.74[d]	−3.06[d]

Table 3: Mean values of some chemical descriptors for early and late metabolites of *S. cerevisiae*. doi:10.1371/journal.pcbi.1002214.t003

This explanation was validated by numerical simulations that were based on three rules. First, the network expands continuously by adding new metabolites (vertices) with a constant rate, namely, *n* metabolites are added in each step (*n* = 1 in the present simulations).

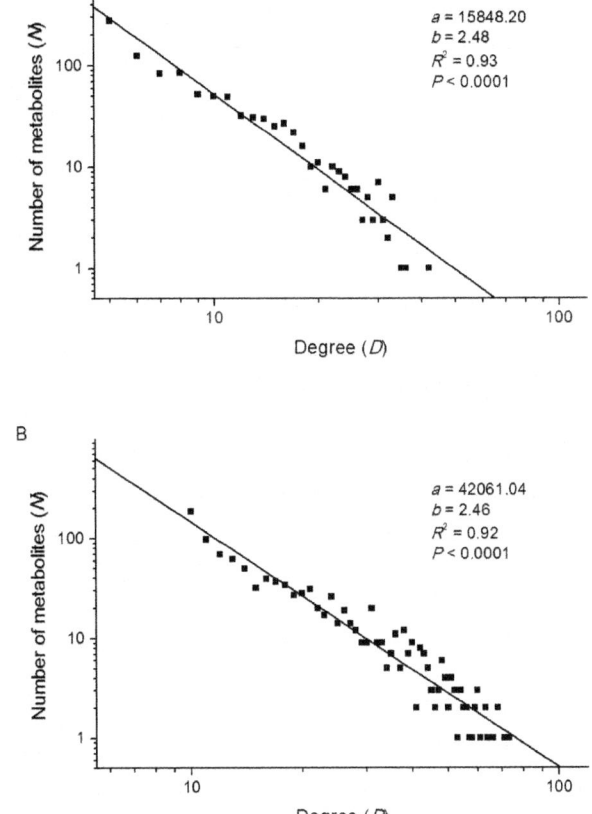

Figure 5: Numerical simulations of metabolic network expansion. The simulations were based on three rules: i) *n* metabolites are added in each expansion step (*n*=1 in

the present simulations); ii) the newly added metabolites have lower concentrations compared to the old ones; iii) the metabolites of higher concentrations have higher probability to be involved in the emerging reactions (edges). The simulations start with 1 metabolite with the initial concentration (C_i) of 1,000,000 and terminate when a metabolite reaches a concentration (C_f) of ≤ 10. The concentration decline (d) in each step is 1,000, with a random fluctuation (f) of 1,500. (**A**) The number of reactions (edges) added in each step is 5; (**B**) The number of reactions (edges) added in each step is 10. In both simulations, the number of metabolites (N) decays with the increase of degrees (D) and follows the equation $N=aD^{-b}$. doi:10.1371/journal.pcbi.1002214.g005

Second, the newly added metabolites have lower concentrations compared to the old ones, *i.e.*, there is a declining trend for the concentrations of emerging metabolites. Third, the metabolites of higher concentrations have higher probability to be involved in the emerging reactions (edges). The present simulations start with 1 metabolite with the initial concentration (C_i) of 1,000,000 and terminate when a metabolite reaches a concentration (C_f) of $\leq$ 10. This concentration range spans five orders of magnitude, which coincides with the variation range of metabolite concentrations in *E. coli* (from $\sim 10^{-7}$ to $\sim 10^{-2}$ mol/L) [8]. The concentration decline (d) in each step is 1,000, with a random fluctuation (f) of 1,500. As a result, the total number of generated metabolites reaches around 1,000, which is close to the real number of metabolites of organisms. The numbers of reactions (edges) added in each step are 5 or 10. As shown inFigure 5, the simulations with different parameters exhibit similar power-law distributions of node degrees, which suggests that the concentration-governed model provides a viable explanation to the scale-free organization of metabolic networks.

Implications for Metabolic Network Design

The above finding implies a chemical criterion in metabolic network design that the polarity of hubs should be compatible with the working environments to guarantee the high concentrations of these critical metabolites. If the environments are polar (*e.g.*, water), one should use hydrophilic molecules as hubs, while if the environments are non-polar (*e.g.*, hydrocarbon solutions) [13], hydrophobic molecules should be selected as hubs. This opinion is preliminarily supported by the fact that the "core" of organic chemical network (*i.e.*, a small set of strongly connected, chemically diverse substances) identified by Bishop *et al.* [14] are really much less polar than the hubs of metabolic networks (Table 4), well reflecting the fact that organic chemical reactions are mainly performed in organic solvents which are less polar than water. Thus, this chemical criterion is of apparent value in metabolic network design.

Table 4: Mean values of some chemical descriptors for hubs of KEGG-based network and cores of organic chemical network. doi:10.1371/journal.pcbi.1002214.t004

Descriptors	Characterization	Mean values	
		KEGG hubs (n = 279)	Chemical cores (n = 300)
ClogP[a]	Partition coefficient octanol/water	−1.26[d]	2.11[d]
FNSA3[b]	Ratio of atomic charge weighted partial negative surface area on total molecular surface area	−0.110[d]	−0.060[d]
FPSA3[b]	Ratio of atomic charge weighted partial positive surface area on total molecular surface area	0.080[d]	0.040[d]
LogD[c]	Octanol-water partition coefficient calculated taking into account the ionization states of the molecule	−2.56[d]	2.08[d]
Molecular Solubility[c]	Water solubility, expressed as logS, where S is the solubility in mol/L	−0.80[d]	−2.61[d]
RPCG[b]	Ratio of most positive charge on sum total positive charge (Relative positive charge)	0.158[d]	0.233[d]

Implications for Metabolite Concentration Prediction

A primary goal of systems biology is to quantitatively characterize cellular behaviors, which requires the information about the absolute concentrations of metabolites. As the intracellular content of metabolites is quite low [15], it is a big challenge to determine their concentrations experimentally. Thus, it is of great significance to use theoretical methods to do predictions. In a pioneering study, Kümmel *et al* established a network-embedded thermodynamic (NET) method to predict intracellular metabolite concentrations [16]. However, this method depends largely on Gibbs energies of formation for metabolites, so its use is restricted to a small part of metabolites. The correlations between metabolite concentration and their topological/chemical properties revealed in this study suggest that intracellular metabolite concentrations may be predicted by their topological and chemical properties.

By using the support vector regression (SVR) [17] method in R (version 2.11.1), a SVR model was established to predict *E. coli* metabolite concentrations by their topological and chemical properties. This model was evaluated by leave-one-out cross validation. The squared correlation coefficient is 0.5906 and the total mean squared error is 0.5316. The fitted metabolite concentrations by this model correlate well with the original experimental values (Figure 6). To evaluate the relative contribution of each descriptor to the performance of SVR model, we constructed SVR models by deleting one parameter each time and calculated the squared correlation coefficients of leave-one-out cross validation by using grid search over supplied parameter ranges. The smaller the squared correlation coefficient becomes, the more important the deleted descriptor is to the SVR model. As shown in Table 5, the deletion of degree results in the lowest squared correlation coefficient, followed by the deletion of ClogP, which means that degree and ClogP make most important contributions to the performance of SVR model.

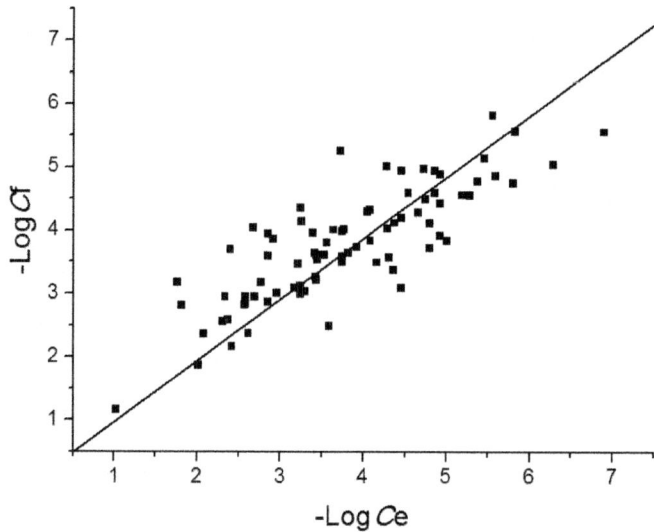

Figure 6: Theoretical fitting of *E. coli* metabolite concentrations by the SVR model. The negative logarithm of fitted concentrations ($-LogC_f$) for 80 *E. coli* metabolites correlates well with that of experimental values ($-LogC_e$): $-LogC_f=0.9678 \times -LogC_e$ ($R=0.827$, $P<0.0001$, regression without intercept). doi:10.1371/journal.pcbi.1002214.g006

Table 5: Performance of SVR models evaluated by descriptor deletion. doi:10.1371/journal.pcbi.1002214.t005

Deleted descriptor	Characterization	Squared correlation coefficient[f]	Total mean squared error[f]
Degree[a]	Number of edges linked to the node of network	0.4547	0.7094
Clogp[b]	Partition coefficient octanol/water	0.5185	0.6304
Amide Molecules[c]	Number of amide	0.5489	0.5952
N Count[c]	Number of Nitrogen atoms	0.5674	0.5963
6mem rings Molecules[c]	Number of 6 membered rings	0.5680	0.5628
FNSA3[d]	Ratio of atomic charge weighted partial negative surface area on total molecular surface area	0.5691	0.5594
HBD Count[e]	Number of hydrogen bond donating groups in the molecule	0.5717	0.5744
FPSA3[d]	Ratio of atomic charge weighted partial positive surface area on total molecular surface area	0.5778	0.5482
ALogP[c]	The Ghose and Crippen octanol-water partition coefficient	0.5806	0.5449
LScore Molecules[c]	Floating point Lipinski measure	0.5860	0.5373
RPCG[d]	Ratio of most positive charge on sum total positive charge (Relative positive charge)	0.6045	0.5134

The *E. coli* metabolite concentrations that have been predicted by the NET method [16] were also estimated by the SVR model. The SVR predictions agree well with the NET results and those determined by prior experiments (at the level of order of magnitude) (Table 6). By the SVR method, the intracellular concentrations for other *E. coli* metabolites were also predicted and presented in Table S6, which can be used as initial data in *E. coli* metabolic network simulation. As the SVR model only depends on very basic (topological and chemical) properties of metabolites, it is expected to be applicable in metabolite concentration prediction for other bacteria.

Metabolite[a]	Predicted concentration[b]	Predicted concentration[c]		Experimental concentration[d]
		Lower limit	Upper limit	
13DPG	n.a.[e]	3.237	3.959	n.d.[j]
2PG	3.347	3.292	3.770	2.394
3PG	3.260	2.387	2.495	2.394
3PHP	2.906	5.046	7.000	n.d.[j]
DHAP	3.221	3.155	3.252	3.174
F6P	3.416	3.796	6.000	3.319
G1P	3.935[f]	3.959	6.000	n.d.[j]
G6P	3.577[g]	3.301	3.523	3.319
G3P	3.170	4.301	5.046	3.174
R5P	3.341	3.959	4.699	3.824
RU5P	3.617[h]	3.824	4.699	3.824
X5P	3.594[i]	3.959	6.000	3.824

Table 6: Comparison of predicted and experimental concentrations for some E. coli metabolites. doi:10.1371/journal.pcbi.1002214.t006

In summary, the present analysis indicates that the organization of metabolic networks has a chemical basis. That is, metabolic hubs prefer to select relatively strong-polar metabolites. This basis can be explained in terms of high concentrations required by metabolic hubs to drive a variety of reactions. The present finding not only provides a molecular-level explanation to the preferential attachment principle for metabolic network expansion but also has direct implications for metabolic network design and metabolite concentration prediction.

MATERIALS AND METHODS

Metabolic Network Reconstruction and Topological Parameter Calculation

The KEGG-based metabolic network was reconstructed by manually screening the 8100 small-molecule reactions recorded in KEGG Ligand Database (http://www.genome.jp/kegg/ligand.html) (up to Sep 2009) [18]. The screening criteria are as follows: i) The reactions involving macromolecules (*e.g.*, polymers, proteins and nucleic acids) and metabolites with unspecified residues (denoted by R group) were deleted; ii) Currency metabolites, including gases, metal ions and cofactors were discarded, except that they directly participate in metabolic reactions [19], [20]. The resulting small-molecule metabolic network consists of 4875 nodes (compounds) and 9263 undirectional edges (substrate-product relations).

The metabolic network of *E. coli* was reconstructed by manually screening the 1317 small-molecule reactions for *E. coli* K-12 recorded in EcoCyc Database (http://www.ecocyc.org) [21]. The screening criteria are the same as above described. The resulting small-molecule metabolic network consists of 601 nodes (compounds) and 1538 undirectional edges (substrate-product relations).

The metabolic network of *S. cerevisiae* was reconstructed by manually screening the 1923 small-molecule reactions recorded in YEASTNET (http://www.comp-sys-bio.org/yeastnet) [22]. The screening criteria are the same as above described. The resulting small-molecule metabolic network consists of 612 nodes (compounds) and 2654 undirectional edges (substrate-product relations).

The parameters describing the network topology were calculated by Network Analyzer Plugin in Cytoscape-2.7.0 [23], [24]. The node degree of a node *n* is defined as the number of edges linked to *n*. The basic information for KEGG, *E. coli* and *S. cerevisiae* metabolites that are involved in the metabolic networks are presented in Tables S1-S5.

Identification of Early and Late Members of *S. Cerevisiae* Metabolome

To elucidate the molecular basis of preferential attachment principle underlying the evolution of metabolic networks, we identified the early and late members from *S. cerevisiae* metabolome. Recently, Prachumwat and Li classified yeast proteins into five age groups, according to the occurring patterns of their orthologs in other species [25]. The oldest age group, consisting of 1806 members, includes proteins that can be traced back to eubacterial genomes. Among these proteins, 972 are enzymes. According to the KEGG records, 633 metabolites associated with these ancient enzymes were collected, 12 of which are aerobic metabolites (according to the aerobic metabolite information provided by Raymond and Segrè [26]) and thus are not early metabolites. The remained 621 metabolites constitute the set of early metabolites of *S. cerevisiae*, in which 243 members are involved in the metabolic network of *S. cerevisiae*. The other 369 (= 612−243) metabolites of *S. cerevisiae* metabolic network were thus regarded as late members.

Chemical Property Calculation, Network Expansion Simulation and Statistical Analysis

83 commonly used property descriptors were calculated with Cerius2 (Version 4.11L. Accelrys Inc. San Diego, CA.), Sybyl (Version 7.0. Tripos Associates

Inc. St. Louis, MO.), Pipeline Pilot (Student Edition. Version 6.1.5. SciTegic Accelrys Inc. San Diego, CA.) and Tripos Benchware DataMiner (Version 1.6. Tripos Associates Inc. St. Louis, MO.). Stepwise multiple linear regression analysis was performed by Cerius2 (Version 4.11L. Accelrys Inc. San Diego, CA.). The numerical simulations of metabolic network expansion were performed based on python package "networkx" (version 1.2). All of the statistical analyses were performed with SPSS (Version 15.0. SPSS Inc. Chicago, IL.).

Support Vector Regression Model Construction

By a trial-and-deletion procedure, 11 properties that have largest contributions to the support vector regression (SVR) model were selected, which include degree and 10 chemical properties, *i.e.*, 6mem rings Molecules (number of 6 membered rings), Amide Molecules (number of amide), ALogP (the Ghose and Crippen octanol-water partition coefficient), ClogP (partition coefficient octanol/water), FNSA3 (ratio of atomic charge weighted partial negative surface area on total molecular surface area), FPSA3 (ratio of atomic charge weighted partial positive surface area on total molecular surface area), HBD Count (number of hydrogen bond donating groups in the molecule), N Count (number of Nitrogen atoms), LScore Molecules (floating point Lipinski measure) and RPCG (ratio of most positive charge on sum total positive charge (Relative positive charge)). Radial basis kernel function $e^{-\gamma|u-v|^2}$ was chosen to construct a ε-SVR model. The parameters were trained by using grid search over supplied parameter ranges and the best parameters were obtained as follows: gamma =0.01, epsilon =0.22, cost =7.9. The SVR algorithm for metabolite concentration prediction is available on request.

AUTHOR CONTRIBUTIONS

Conceived and designed the experiments: HYZ. Performed the experiments: QZ TQ YYJ CJ DXK BGM. Analyzed the data: QZ TQ YYJ CJ DXK BGM HYZ. Contributed reagents/materials/analysis tools: QZ TQ YYJ CJ DXK BGM. Wrote the paper: HYZ.

REFERENCES

1.   Jeong H, Tombor B, Albert R, Oltvai ZN, Barabási AL (2000) The large-scale organization of metabolic networks. Nature 407: 651–654.

2.   Barabási AL, Albert R (1999) Emergence of scaling in random networks. Science 286: 509–512.

3.   Barabási AL, Oltvai ZN (2004) Network biology: understanding the cell's functional organization. Nat Rev Genet 5: 101–113.

4.   Light S, Kraulis P, Elofsson A (2005) Preferential attachment in the evolution of metabolic networks. BMC Genomics 6: 159.

5.   Fell DA, Wagner A (2000) The small world of metabolism. Nat Biotechnol 18: 1121–1122.

6.   Chen WL (2006) Chemoinformatics: past, present, and future. J Chem Inf Model 46: 2230–2255.

7.   Engel T (2006) Basic overview of chemoinformatics. J Chem Inf Model 46: 2267–2277.

8.   Bennett BD, Kimball EH, Gao M, Osterhout R, Van Dien SJ, et al. (2009) Absolute metabolite concentrations and implied enzyme active site occupancy in Escherichia coli. Nat Chem Biol 5: 593–599.

9.   Tu BP, Mohler RE, Liu JC, Dombek KM, Young ET, et al. (2007) Cyclic changes in metabolic state during the life of a yeast cell. Proc Natl Acad Sci U S A 104: 16886–16891.

10.   Ishihama Y, Schmidt T, Rappsilber J, Mann M, Hartl FU, et al. (2008) Protein abundance profiling of the Escherichia coli cytosol. BMC Genomics 9: 102.

11.   Heo M, Maslov S, Shakhnovich E (2011) Topology of protein interaction network shapes protein abundances and strengths of their functional and nonspecific interactions. Proc Natl Acad Sci U S A 108: 4258–4263.

12.   Jiang YY, Kong DX, Qin T, Zhang HY (2010) How does oxygen rise drive evolution? Clues from oxygen-dependent biosynthesis of nuclear receptor ligands. Biochem Biophys Res Commun 391: 1158–1160.

13.   Ball P (2005) Seeking the solution. Nature 436: 1084–1085.

14.   Bishop KJ, Klajn R, Grzybowski BA (2006) The core and most useful molecules in organic chemistry. Angew Chem Int Ed Engl 45: 5348–5354.

15.   Thiele I, Palsson BØ (2010) A protocol for generating a high-quality genome-scale metabolic reconstruction. Nature Protoc 5: 93–121.

16.   Kümmel A, Panke S, Heinemann M (2006) Putative regulatory sites unraveled by network-embedded thermodynamic analysis of metabolome data. Mol Syst Biol 2: 2006.0034.

17.   Smola AJ, Schölkopf B (2004) A tutorial on support vector regression. Stat Comput 14: 199–222.

18. Goto S, Okuno Y, Hattori M, Nishioka T, Kanehisa M (2002) LIGAND: database of chemical compounds and reactions in biological pathways. Nucleic Acids Res 30: 402–404.

19. Huss M, Holme P (2007) Currency and commodity metabolites: their identification and relation to the modularity of metabolic networks. IET Syst Biol 1: 280–285.

20. Ma H, Zeng AP (2003) Reconstruction of metabolic networks from genome data and analysis of their global structure for various organisms. Bioinformatics 19: 270–277.

21. Keseler IM, Bonavides-Martínez C, Collado-Vides J, Gama-Castro S, Gunsalus RP, et al. (2009) EcoCyc: A comprehensive view of Escherichia coli biology. Nucleic Acids Res 37: D464–D470.

22. Herrgård MJ, Swainston N, Dobson P, Dunn WB, Arga KY, et al. (2008) A consensus yeast metabolic network reconstruction obtained from a community approach to systems biology. Nat Biotechnol 26: 1155–1160.

23. Shannon P, Markiel A, Ozier O, Baliga NS, Wang JT, et al. (2003) Cytoscape: a software environment for integrated models of biomolecular interaction networks. Genome Res 13: 2498–2504.

24. Assenov Y, Ramírez F, Schelhorn SE, Lengauer T, Albrecht M (2008) Computing topological parameters of biological networks. Bioinformatics 24: 282–284.

25. Prachumwat A, Li WH (2006) Protein function, connectivity, and duplicability in yeast. Mol Biol Evol 23: 30–39.

26. Raymond J, Segrè D (2006) The effect of oxygen on biochemical networks and the evolution of complex life. Science 311: 1764–1767.

Chapter 3

CHEMICAL BONDING OF ALH3 HYDRIDE BY AL-L2,3 ELECTRON ENERGY-LOSS SPECTRA AND FIRST-PRINCIPLES CALCULATIONS

Kazuyoshi Tatsumi[1,], Shunsuke Muto[1], Kazutaka Ikeda[2] and Shin-Ichi Orimo[3]

[1]Department of Materials, Physics and Energy Engineering, Nagoya University, Chikusa, Nagoya 464-8603, Japan

[2]Institute of Materials Structure Science, High Energy Accelerator Research Organization (KEK), 1-1 Oho, Tsukuba, Ibaraki 305-0801, Japan

[3]Institute for Materials Research, Tohoku University, Sendai 980-8577, Japan

ABSTRACT

In a previous study, we used transmission electron microscopy and electron energy-loss (EEL) spectroscopy to investigate dehydrogenation of AlH_3 particles. In the present study, we systematically examine differences in the chemical bonding states of Al-containing compounds (including AlH_3) by comparing their Al-$L_{2,3}$ EEL spectra. The spectral chemical shift and the fine peak structure of the spectra were consistent with the degree of covalent bonding of Al. This finding will be useful for future nanoscale analysis of AlH_3 dehydrogenation toward the cell.

INTRODUCTION

Aluminum trihydride (AlH_3, alane) has high gravimetric and volumetric hydrogen densities (10 wt % and 149 kg·H_2/m³, respectively). It has been investigated for hydrogen storage applications [1,2,3] after Sandrock *et al*. reported that ball-milling with small amounts of LiH sufficiently accelerated its dehydrogenation kinetics to enable it to be used as an onboard power supply for vehicles [1]. To reveal the mechanism for this accelerated dehydrogenation, it is desirable to analyze the chemical bonding changes of the system on at least sub-micron order.

After the report of Sandrock *et al.*, aluminum trihydride has been intensively studied both experimentally and theoretically. In particular, its high pressure phases and phase stabilities have been investigated theoretically [4,5,6] and experimentally [3,6,7]. A theoretical search was conducted for AlH_3 crystal structures and two crystal structures were proposed: the β and γ-AlH_3 phases [4]. The crystal structures of the β [8] and γ [9] phases were subsequently analyzed experimentally. The calculated electronic structures (including chemical bonding) of the polymorphs and/or other hydrides have been compared [4,10]. However, apart from reference [11], there have been no experimental studies of their electronic structures. Chemical bonding around H atoms in solid materials cannot be directly investigated by X-ray or electron spectroscopy due to the low scattering powers of H for both x-rays and electrons. In contrast, chemical bonding in compounds in the Al–H–Na system have been theoretically investigated [4]. Alternatively, as has been done only in reference [11], spectroscopic information about the counter element Al in several Al-containing compounds can be systematically investigated to determine the basic chemical bonding of AlH_3 with the aid of theoretical electronic structure calculations.

We recently investigated dehydrogenation of AlH_3 by transmission electron microscopy (TEM) and electron energy loss spectroscopy (EELS) [12,13]. We obtained TEM images and electron diffraction patterns during dehydrogenation. Moreover, EELS (including Al core-electron excitation spectra) revealed that single hydride crystals were coated with a thin amorphous alumina layer.

The present study extends this earlier study by investigating the relationship between Al EEL spectra and the chemical bonding of Al in the hydride and other Al-containing compounds. Because the electron energy loss near-edge structure (ELNES) reflects the local electronic structure around the excited atom in the illuminated area, the relationship obtained will provide basic information for future EELS analysis of the change in the local chemical bonding that is responsible for accelerating dehydrogenation.

RESULTS AND DISCUSSION

Al-$L_{2,3}$ EELS

Figure 1 shows experimental and theoretical Al-$L_{2,3}$ spectra of AlH_3. For reference, it also shows experimental spectra of Al_2O_3 and Al to evaluate how well the theoretical calculations reproduce the spectra. Theoretical spectra were calculated for four different AlH_3 phases (α, α', β, and γ), typical aluminum compounds (Al_2O_3 and Al), and β-AlF_3, which has the same crystal structure as α'-AlH_3.

The peak profiles, positions, and chemical shifts of the threshold in the experimental and theoretical ELNES of metallic Al and Al_2O_3 are in reasonable agreement, although the intensities on the lower energy side (*i.e.*, from the onset of ionization to 10 eV) of the theoretical spectra tend to be smaller than those of the experimental spectra. This underestimation may be a result of excitonic effects beyond the one-electron approximation due to the relatively shallow core-shell excitation [14,15,16]. The characteristics of the experimental spectrum of the hydride are qualitatively consistent with those of the theoretical spectrum. We were unable to identify the phase of the hydride sample from the spectral data because the pre-peak intensities are not very reliable in the theoretical spectra. A higher experimental energy resolution and improved theoretical reproducibility are desirable for phase identification.

To clarify the most characteristic features of the spectral data and the chemical shifts of the Al compounds, the vertical green, red and black lines in Figure 1 respectively indicate the energy positions of the onset for the metallic Al spectrum, and the first peaks in the hydride and Al_2O_3 spectra. The calculations correctly reproduce their experimental order. The energy of the first peak in the theoretical β-AlF_3 spectrum (80.2 eV) is higher than that of Al_2O_3 (79.0 eV) by half the energy difference between AlH_3 and Al_2O_3 (2.5 eV).

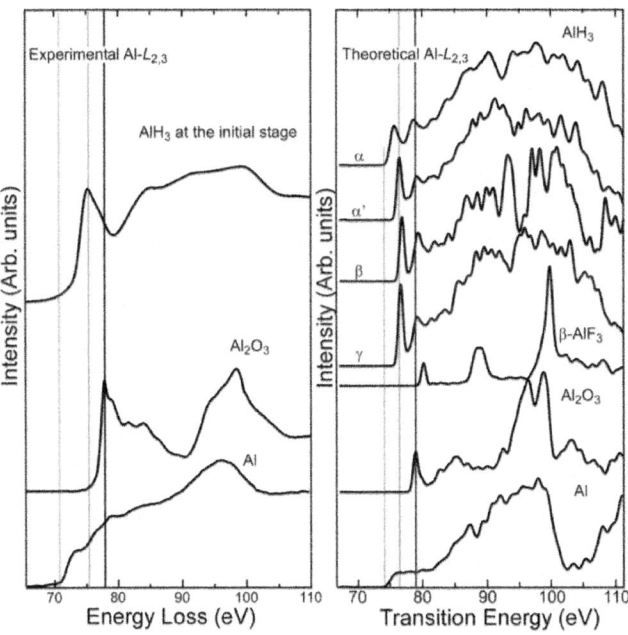

Figure 1: Experimental (**a**) and theoretical (**b**) Al-$L_{2,3}$ energy loss near-edge structure (ELNES) of AlH_3 and reference compounds. The vertical green, red, and black lines

indicate the energy positions of the onset of metallic Al spectrum, the first peak of AlH$_3$ and the first peak of Al$_2$O$_3$, respectively.

Electronic Structure near the Band Gap

The hydride and other compounds are expected to have different theoretical ground-state electronic structures near the band gap and different chemical bonding: these differences are thought to give rise to the observed chemical shifts in the EELS Al-$L_{2,3}$ spectra. Figure 2 shows the partial density of states (PDOS) near the band-gap for α, α'-AlH$_3$, and Al$_2$O$_3$.

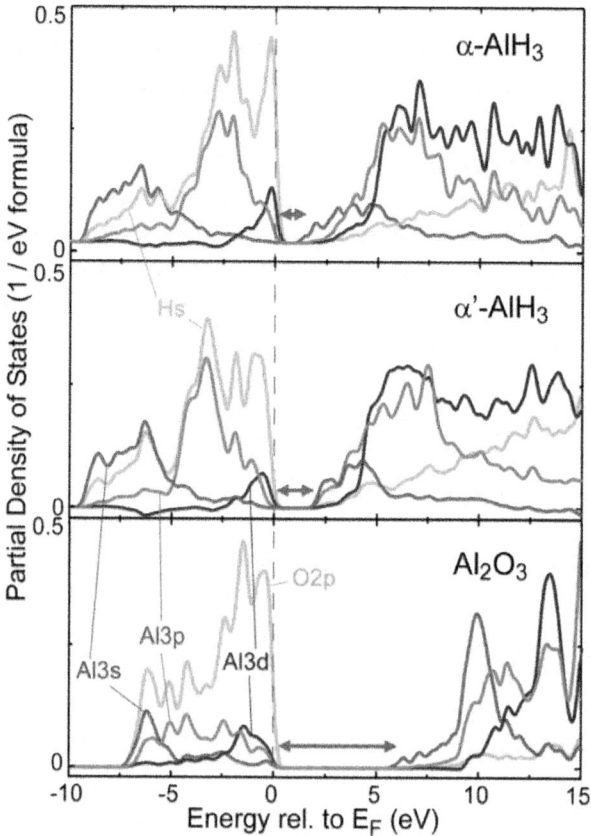

Figure 2: Calculated partial density of states (PDOS) for α-AlH$_3$ and Al$_2$O$_3$. The energy is relative to the Fermi level. The double-headed arrows indicate the band gaps of the electronic structures.

The band gap widths indicated by the double-headed arrows are respectively 1.4, 2.1, and 5.9 eV for α, α'-AlH$_3$, and Al$_2$O$_3$, indicating that the chemical bonding of AlH$_3$ is less ionic than that of Al$_2$O$_3$. In contrast, the experimentally observed band gap is about 9 eV for Al_2O_3. Thus, the present calculation significantly underestimates the band-gap energy, which is common for calculations based on the local density approximation. Improved techniques for calculating the electronic exchange and correlation potential such as the self-interaction-corrected local density approximation [17] are more promising for accurately reproducing the chemical shifts between metallic and insulating compounds; we intend to use such a method in a future study. In AlH_3 PDOS, the valence bands mainly consist of hydrogen orbitals, indicating that H is anionic. Nevertheless, the contribution of the Al orbitals to the valence bands is greater in AlH_3 than in Al_2O_3. In contrast, the conduction bands are mainly formed by Al orbitals, which are more hybridized with anion (H) orbitals in the hydride than the anion (O) orbitals in the oxide. These compositional differences in the valence and conduction bands confirm that AlH_3 is less ionic than Al_2O_3.

The measured Al-$L_{2,3}$ spectra reflect the unoccupied PDOS of Al s and d symmetry orbitals under the electric dipole approximation. For both AlH_3 and Al_2O_3, the Al $3s$ orbital is dominant over the Al $3d$ orbital at the bottom of the conduction bands. Therefore, the first peaks in the spectra in Figure 1 are roughly assigned to the lower energy states of Al $3s$ bands. The relative energy of the bottom of the conduction bands of α and α'-AlH_3 with respect to the Al $2p$ inner level is 2.7 (2.1) eV lower than that of the oxide. Since this energy difference is comparable to the experimental chemical shift between these compounds, the difference in the Al ionicity of the theoretical ground electronic structure can well explain the experimental chemical shift. Chemical bonding is discussed in detail in the next section.

The differences between the two hydride polymorphs are rather small. The other hydride phases, which are not shown inFigure 2, had similar PDOS to that of the α› phase. At the conduction band bottom, Al $3p$ is less hybridized with Al $3s$ in the α phase than in the α' and the other hydride phases. This might be responsible for the differences in the pre-peak structures of the polymorphs in the theoretical Al $L_{2,3}$ ELNES, although the incorporated excitonic effects should be more rigorously considered.

Covalent Bond Strength

To investigate the covalent bond strength between the Al and its neighboring atoms, Figure 3 plots the bond overlap population (BOP) between each atom pair with respect to the interatomic distance. A positive BOP indicates that the covalent bond charge accumulates between the atom pair. On the other hand, a negative BOP indicates that the electron charge between the atoms is deficient relative to that of the superposed atomic electron densities, which is regarded as an antibonding mechanism between the atoms. The hydride contains a significant covalent bond charge for both Al–Al and Al–H. In contrast, Al_2O_3 has large negative BOPs for Al–Al, resulting in a negative total BOP of -0.2 per Al atom. This result implies that an alternative mechanism, namely an ionic bonding mechanism, is dominant for Al_2O_3. β–AlF_3 has a similar interatomic distance distribution to that of α-AlH_3, which is consistent with β–AlF_3 having an identical crystal structure to α'-AlH_3. However, unlike for the hydride, all Al–Al pairs have negative BOPs.

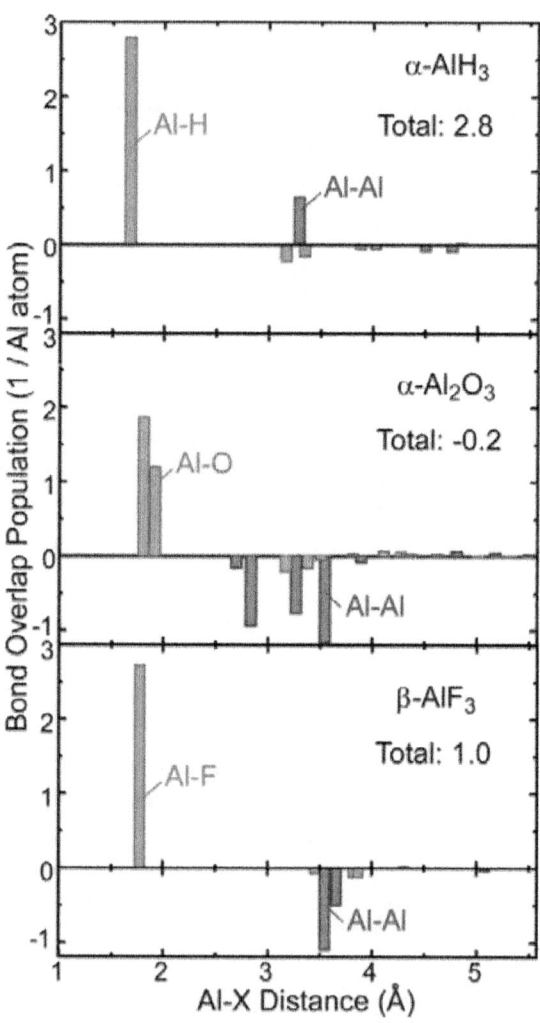

Figure 3: Plots showing the calculated bond overlap population (BOP) and the corresponding atomic distances between an Al atom and its near neighbors. The number in each graph indicates the sum of the BOPs.

Comparison of Chemical Bonding between the Al-Containing Compounds

Table 1 list the theoretical band-gap energy, the Al effective charge, and the sum of BOPs for Al–X (X = Al, H, O, F) to characterize the chemical bonding of the calculated Al compounds. The other hydride phases, which are not shown in Table 1, had similar values to those of the α and α' phases. The Al effective charge represents the number of valence electrons belonging to the Al orbitals. A value smaller than 3 indicates cationic Al. In all of the three rows, the values for the hydrides are intermediate between those of metallic Al and Al_2O_3. The ionic bonding or the covalent bond strengths of the hydrides are intermediate between those of Al and Al_2O_3.

Except for the smallest BOP of Al_2O_3, the listed data indicate stronger ionic bonding toward the right-hand side. The exception may be ascribed to the crystal structure, since the Al–Al distances are much shorter in Al_2O_3 than in the hydride and fluoride (see Figure 3). The significant overlap between the Al orbitals and their antibonding interactions results in the lowest total BOP.

The threshold of the Al-$L_{2,3}$ edges in Figure 1 is shifted in a manner that is consistent with the overall trend for the chemical bonding in the compounds. In the theoretical spectrum of metallic Al, we can see a number of small peaks located continuously. In contrast, the number of fine peaks in the spectra decreases in the order AlH_3, Al_2O_3, and AlF_3 with increasing peak intensities and peak distances. Thus, the consistency between the spectra and theoretical data in Table 1 indicates that both the chemical shift and the fine peak structure of the Al-$L_{2,3}$ spectra may be experimental indicators for the degree of covalent bonding in the Al-containing compounds. These indicators may help clarify the chemical bonding changes of AlH_3 in a localized area (for example, changes during the accelerated dehydrogenation).

Table 1: Comparison of calculated band gap, effective charge, and BOP of Al-containing compounds (including AlH_3).

	Al	AlH_3 (α, α')	Al_2O_3	AlF_3
LDA band gap (eV)	0.0	1.4, 2.1	5.9	7.2
Al effective charge	3.0	2.8, 2.8	2.1	1.8
Sum of BOP Al-X (1/Al atom)	3.6	2.9, 2.7	−0.2	1.0

EXPERIMENTAL AND THEORETICAL SECTION

EELS Measurements

The observed hydride sample was prepared by the chemical reaction between $LiAlH_4$ and $AlCl_3$ in an ether (99.5% purity Et_2O) solution. Prior to preparing TEM specimens, we confirmed that the powder x-ray diffraction pattern of the sample was that of α-AlH_3 [3,12]. Further details about the sample preparation and characterization are given in references [12,13].

Electron-irradiation-induced dehydriding of AlH_3 has been observed [13]. A single crystal of AlH_3 was decomposed into metallic Al nanoparticles, while its external shape remained almost unchanged. We carefully measured EELS from the hydride by reducing the electron dose to be sufficiently low to avoid its instantaneous decomposition and also to identify the phase by electron diffraction and EELS. Al-$L_{2,3}$ ELNES was then isolated after subtracting the pre-edge background by a power law. Since the Al-$L_{2,3}$ ELNES of the hydride has a very low signal to noise ratio (<5) due to the low electron dose, the spectrum were processed by the Pixon method [18] to remove statistical noise.

Theoretical Calculations

The theoretical ELNES was calculated by first-principles calculations with a local approximation to the density functional theory so as to investigate the chemical bonding between Al and its surrounding atoms. Prior to the spectral calculations, the crystal parameters of the structures were fully optimized by another first-principles procedure, the projected augmented-wave method [19,20] to reduce the computational cost. For the theoretical ELNES calculation, we adopted the orthogonalized linear combination of atomic orbital (OLCAO) band method [21] because the atomic orbital basis of the method straightforwardly provides the chemical bonding of the structures, as described above. To reasonably account for the core-hole effects [22], an Al $2p$ hole was introduced to a supercell consisting of approximately 100 atoms. The size of the supercell was sufficiently large to neglect the unrealistic interactions between the core holes [23]. The transition probabilities from the Al $2p$ to the unoccupied states were calculated within the electric dipole approximation. These were integrated over the whole Brillouin zone of the supercell using a $2 \times 2 \times 2$ k-point mesh. The transition energy of the theoretical spectrum was obtained from the difference of the total electronic energy of the supercell at the final core hole induced state and that of the ground state. The final spectrum was broadened by convoluting it with a Gaussian function with a full width at half maximum (FWHM) of 0.8 eV. The small energy splitting of 0.07 eV between the Al $2p_{1/2}$ and $2p_{3/2}$ levels was neglected for simplicity.

The chemical bonding differences probed by the spectral difference between AlH_3 and other Al containing compounds were analyzed based on Mulliken population analysis of atomic orbitals [24]. The PDOS, BOP, and effective charge were calculated by using the overlap integral between atomic orbitals and their coefficients for wavefunctions [25]. This population analysis of the OLCAO method aids intuitive understanding because it uses atomic orbital basis sets, which is one of the main benefits of the OLCAO method. PDOS was constructed by convoluting the orbital population with a Gaussian function (FWHM: 0.4 eV). We chose a smaller FWHM to clearly evaluate the atomic orbital contribution near the band gap.

CONCLUSIONS

We have investigated the chemical bonding of AlH_3 by means of Al-$L_{2,3}$ ELNES and its first-principles band calculation. The results obtained are summarized as follows:

(1). Experimental Al-$L_{2,3}$ ELNES exhibited clear chemical shifts between metallic Al, AlH_3, and Al_2O_3. The threshold of the hydride spectrum was intermediate between those of Al and Al_2O_3. The first-principles calculation reasonably reproduced the shape and chemical shift of the experimental spectra.

(2). Theoretical chemical bonding data obtained by Mulliken population analysis of the calculated electronic structures revealed that the ionic bonding and covalent bonding strengths of the hydride were intermediate between those of Al and Al_2O_3.

(3). The theoretical Al-$L_{2,3}$ spectra and chemical bonding between the four Al compounds (Al, AlH_3, Al_2O_3, and AlF_3) exhibited systematic trends between the chemical bonding and their spectral features (fine peak structure and chemical shift). These trends are expected to be useful for further TEM-EELS nanoscale analysis of the chemical bonding changes that occur during the dehydrogenation of AlH_3 hydrides.

ACKNOWLEDGMENTS

This work was supported in part by Grants-in-Aid for Scientific Research (KAKENHI) in Priority Areas (#474) "Atomic Scale Modification" from the Ministry of Education, Culture, Sports, Science and Technology (MEXT), Japan. We thank W.Y. Ching for allowing us to use the OLCAO code.

REFERENCES

1. Sandrock, G.; Reilly, J.; Graetz, J.; Zhou, W.M.; Johnson, J.; Wegrzyn, J. Accelerated thermal decomposition of AlH_3 for hydrogen-fueled vehicles. *J. Appl. Phys. A* **2005**, *80*, 687–690.

2. Graetz, J.; Reilly, J.J. Decomposition Kinetics of the AlH3 Polymorphs. *J. Phys. Chem. B* **2005**, *109*, 22181–22185.

3. Orimo, S.; Nakamori, Y.; Kato, T.; Brown, C.; Jensen, C.M. Intrinsic and mechanically modified thermal stabilities of α-, β- and α-aluminum trihydrides AlH_3. *Appl. Phys. A Mater.* **2006**, *83*, 5–8.

4. Ke, X.; Kuwabara, A.; Tanaka, I. Cubic and orthorhombic structures of aluminum hydride AlH3 predicted by a first-principles study. *Phys. Rev. B* **2005**, *71*, 184107:1–184107:7.

5. Pickard, C.J.; Needs, R.J. Metallization of aluminum hydride at high pressures: A first-principles study. *Phys. Rev. B* **2007**, *76*, 144114:1–144114:5.

6. Graetz, J.; Chaudhuri, S.; Lee, Y.; Vogt, T.; Muckerman, J.T.; Reilly, J.J. Pressure-induced structural and electronic changes in α-AlH_3. *Phys. Rev. B* **2006**, *74*, 214114:1–214114:7.

7. Graetz, J.; Reilly, J.J. Thermodynamics of the α, β and γ polymorphs of AlH_3. *J. Alloys Comp.* **2006**, *424*, 262–265.

8. Brinks, H.W.; Langley, W.; Jensen, C.M; Graetz, J.; Reilly, J.J.; Hauback, B.C. Synthesis and crystal structure of β-AlD_3. *J. Alloys Compounds* **2006**, *424*, 262–265.

9. Yartys, V.A.; Denys, R.V.; Maehlen, J.P.; Frommen, C.; Fichtner, M.; Bulychev, B.M.; Emerich, H. Double-bridge bonding of aluminium and hydrogen in the crystal structure of γ-AlH_3. *Inorg. Chem.* **2007**, *46*, 1051–1055.

10. Vajeeston, P.; Ravindran, P.; Fjellvåg, H. Novel high pressure phases of β-alh_3: A density-functional study. *Chem. Mater.* **2008**, *20*, 5997–6002.

11. Takeda, Y.; Saitoh, Y.; Saitoh, H.; Machida, A.; Aoki, K.; Yamagami, H.; Muro, T.; Kato, Y.; Kinoshita, T. Electronic structure of aluminium trihydride studied using soft X-ray emission and absorption spectroscopy. *Phys. Rev. B* **2011**, *84*, 153102:1–153102:4.

12. Ikeda, K.; Muto, S.; Tatsumi, K.; Menjo, M.; Kato, S.; Bielman, M.; Züttel, A.; Jensen, C.M.; Orimo, S. Dehydriding reaction of AlH_3: *In situ* microscopic observations combined with thermal and surface analyses. *Nanotechnology* **2009**, *20*, 204004:1–204004:4.

13. Muto, S.; Tatsumi, K.; Ikeda, K.; Orimo, S. Dehydriding process of α-AlH$_3$ observed by transmission electron microscopy and electron energy-loss spectroscopy. *J. Appl. Phys.* **2009**, *105*, 123514:1–123514:4.

14. Keast, V.J.; Scott, A.J.; Kappers, M.J.; Foxon, C.T.; Humphyreys, C.J. Electronic structure of GaN and In$_x$Ga$_{1-x}$N measured with electron energy-loss spectroscopy. *Phys. Rev. B* **2002**, *66*, 125319:1–125319:7.

15. Mauchamp, V.; Boucher, F.; Ouvrard, G.; Moreau, P. *Ab initio* simulation of the electron energy-loss near-edge structures at the Li K edge in Li, Li$_2$O, and LiMn$_2$O$_4$. *Phys. Rev. B* **2006**, *74*, 115106:1–115106:4.

16. Olovsson, W.; Tanaka, I.; Mizoguchi, T.; Radtke, G.; Puschnig, P.; Ambrosch-Draxl, C. Al $L_{2,3}$ edge X-ray absorption spectra in III–V semiconductors: Many-body perturbation theory in comparison with experiment. *Phys. Rev. B* **2011**,*83*, 195206:1–195206:8.

17. Filippetti, K.; Spaldin, N.A. Self-interaction-corrected pseudopotential scheme for magnetic and strongly-correlated systems. *Phys. Rev. B* **2003**, *67*, 125109:1–125109:15.

18. Muto, S.; Puetter, R.C.; Tatsumi, K. Spectral restoration and energy resolution improvement of electron energy-loss spectra by Pixon reconstruction: I. Principle and test examples. *J. Electron Microsc.* **2006**, *55*, 215–223. Kresse, G.; Furthmüller, J. Efficient iterative schemes for *ab initio* total-energy calculations using a plane-wave basis set. *Phys. Rev. B* **1996**, *54*, 11169–11181.

19. Kresse, G.; Joubert, D. From ultrasoft pseudopotentials to the projector augmented-wave method. *Phys. Rev. B* **1999**,*59*, 1758–1775.

20. Ching, W.Y. Theoretical studies of the electronic properties of ceramic materials. *J. Am. Ceram. Soc.* **1990**, *73*, 3135–3160.

21. Mo, S.D.; Ching, W.Y. *Ab initio* calculation of the core-hole effect in the electron energy-loss near-edge structure. *Phys. Rev. B* **2000**, *62*, 7901–7907.

22. Mogi, M.; Yamamoto, T.; Mizoguchi, T.; Tatasumi, K.; Yoshioka, S.; Kameyama, S.; Tanaka, I.; Adachi, H. Theoretical investigation of Al K-edge X-ray absorption spectra of Al, AlN and Al$_2$O$_3$. *Mater. Trans.* **2004**, *45*, 2031–2034.

23. Mulliken, R.S. Electronic population analysis on LCAO-MO molecular wave functions. *J. Chem. Phys.* **1955**, *23*, 1833–1840.

24. Mizoguchi, T.; Tatsumi, K.; Tanaka, I. Peak assignments of ELNES and XANES using overlap population diagrams.*Ultramicroscopy* **2006**, *106*, 1120–1128.

Chapter 4

CHEMICAL REACTIVITY DYNAMICS AND QUANTUM CHAOS IN HIGHLY EXCITED HYDROGEN ATOMS IN AN EXTERNAL FIELD: A QUANTUM POTENTIAL APPROACH

P. K. Chattaraj and B. Maiti

Department of Chemistry, Indian Institute of Technology, Kharagpur, 721302, India

ABSTRACT

Dynamical behavior of chemical reactivity indices like electronegativity, hardness, polarizability, electrophilicity and nucleophilicity indices is studied within a quantum fluid density functional framework for the interactions of a hydrogen atom in its ground electronic state (n = 1) and an excited electronic state (n = 20) with monochromatic and bichromatic laser pulses. Time dependent analogues of various electronic structure principles like the principles of electronegativity equalization, maximum hardness, minimum polarizability and maximum entropy have been found to be operative. Insights into the variation of intensities of the generated higher order harmonics on the color of the external laser field are obtained. The quantum signature of chaos in hydrogen atom has been studied using a quantum theory of motion and quantum fluid dynamics. A hydrogen atom in the electronic ground state (n = 1) and in an excited electronic state (n = 20) behaves differently when placed in external oscillating monochromatic and bichromatic electric fields. Temporal evolutions of Shannon entropy, quantum Lyapunov exponent and Kolmogorov – Sinai entropy defined in terms of the distance between two initially close Bohmian trajectories for these two cases show marked differences. It appears that a larger uncertainty product and a smaller hardness value signal a chaotic behavior.

INTRODUCTION

The chaotic ionization of hydrogen atoms [1,2,3] in highly excited states by microwave fields has become an important area of research for both experimentalists [1,2,3,4,5,6,7] and theoreticians [4]. In 1974 Bayfield and Koch [8] first studied the chaotic ionization of hydrogen atoms which has been considered to be very important in atomic theory [1,2,4,5,9,10,11,12, 13,14,15,16,17,18,19,20,21,22,23,24,25,26,27,28]. Sanders and Jensen [4] have studied the chaotic ionization of hydrogen and helium using classical mechanics [4]. When the hydrogen atom is promoted to a highly excited state it gets ionized in case the field intensity is above some threshold value and the ionization probability depends on the field intensity [4,6,7]. Standard diagnostics used for the present study include electronegativity (χ), hardness (η), polarizability (α), phase – volume (V_{ps}), electrophilicity index (W), nucleophilicity index (1/W), Shannon entropy (S), quantum Lyapunov exponent (Λ) and Kolmogorov – Sinai entropy (H) defined in terms of the distance between two initially close Bohmian trajectories. In this paper we have generated the higher – order harmonics [3,29,30]. The response of the atom when it interacts with the external field vis–á–vis the variation of its reactivity is an important area of research. Electornegativity (χ)[31] and hardness (η) [32] are two cardinal indices of chemical reactivity. Pauling [33] introduced the concept of electronegativity as the power of an atom in a molecule to attract electrons to itself. The concept of hardness was given by Pearson [34] in his hard – soft acid – base (HSAB) principle which states that, "hard likes hard and soft likes soft". These popular qualitative chemical reactivity concepts have been quantified in density functional theory (DFT) [35]. Another important hardness – related principle is the maximum hardness principle (MHP) [36,37], which states that, " there seems to be a rule of nature that molecules arrange themselves so as to be as hard as possible". The quantitative definitions for electronegativity [38] and hardness [39] for an N – electron system with total energy E can respectively be given as

$$\chi = -\mu = -\left(\frac{\partial E}{\partial N}\right)_{v(\vec{r})}$$

(1)

and

$$\eta = \frac{1}{2}\left(\frac{\partial^2 E}{\partial N^2}\right)_{v(\vec{r})} = \frac{1}{2}\left(\frac{\partial \mu}{\partial N}\right)_{v(\vec{r})}.$$

(2)

In eqs. (1) and (2) μ and $v(\vec{r})$ are chemical potential (Lagrange multiplier associated with the normalization constraint of DFT [34,36]) and external potential respectively. An equivalent expression [40,41] for hardness is

$$\eta = \frac{1}{N} \iint \eta(\vec{r}, \vec{r}') f(\vec{r}') \rho(\vec{r}) d\vec{r} d\vec{r}'$$

(3)

where $f(\vec{r})$ is the Fukui function [40] and $\eta(\vec{r}, \vec{r}')$ is the hardness kernel given by [40]

$$\eta(\vec{r}, \vec{r}') = \frac{1}{2} \frac{\delta^2 F[\rho]}{\delta\rho(\vec{r})\delta\rho(\vec{r}')}$$

(4)

where $F(\rho)$ is the Hohenberg - Kohn universal functional of DFT [35].

The complete characterization of an N – particle system acted on by an external potential $v(\vec{r})$ requires only N and $v(\vec{r})$. The response of the system subjected to a change in N at fixed $v(\vec{r})$ is given by χ and η while the linear response function [34] measures the response of the system when $v(\vec{r})$ is varied at constant N. If the system is kept under the influence of the weak electric field, polarizability (α) takes care of the corresponding response. During molecule formation the electronegativities of the pertinent atoms get equalized [42,43]. A stable configuration or a favorable process is generally associated with maximum hardness [36,37], minimum polarizability [44,45,46,47] and maximum entropy [48] values. The conditions for maximum hardness and entropy and minimum polarizability complement the usual minimum energy criterion for stability.

Recently Parr et. al. [49] have defined the electrophilicity index (W) as

$$W = \frac{\mu^2}{2\eta}$$

(5)

We also study the behavior of (1/W), a valid candidate for the nucleophilicity index. Note that the quantity (1-W) will also serve the purpose of a nucleophilicity index. It has also been shown recently [50] that the uncertainty product or the phase space volume (V_{ps}) is a measure of quantum fluctuations and hence has bearing in the studies of quantum domain behavior of classically chaotic systems.

It has been already demonstrated [51] that in case we focus our attention to a specific atom / molecule taking part in a chemical reaction the whole procedure can be simulated by the interaction of an atom / molecule with an external field of the strength of the order of the "chemical reaction field". A molecular reaction dynamics can be envisaged [44] by monitoring the time evolution of the electronegativity of a specific atom from its isolated atom value to the equalized molecular electronegativity value as well as by studying the dynamic profiles of hardness and entropy and how they get maximized and that of the minimization of polarizability during the course of the chemical reaction. In the present work we study the interaction of a hydrogen atom in its ground electronic state and an excited electronic state with laser fields of different colors. The effect of the frequency of the external laser field on the overall reactivity of the atom in its various electronic states vis–á–vis the validity of the associated electronic structure principles in a dynamical context as well as the intensities of the generated higher order harmonics [52] would be understood in this study.

Dynamics of these reactivity parameters (η and α) have been studied [44,46,53] in the contexts of various time dependent processes. Whether η and α can provide some insight into the quantum domain behavior of a classically chaotic system is yet to be analyzed. Hydrogen atoms and molecules in an oscillating electric field have been considered to be "veritable gold mines for exploring the quantum aspects of chaos" [54]. Depending on the frequency and the field intensity, hydrogen [54,55] atoms in the presence of an external field have been shown to exhibit regular / chaotic dynamics. Both quantum fluid dynamics (QFD) [56,57] and quantum theory of motion (QTM) [58,59] have provided quantum signatures of chaos in hydrogen atoms. In QFD [56] the overall motion of the system under consideration is mapped onto that of a "probability fluid" having density $\rho(\bar{r},t)$ and current density $j(\bar{r},t)$ under the influence of the external classical potential augmented by a quantum potential [55,56,57,58,59] and $\rho(\bar{r},t)$ and $\chi(\bar{r},t)$ $(j = \rho\nabla\chi)$ are respectively obtained [55,56,57,58,59] from the amplitude and the phase of the wave function. In QTM [58], the wave motion is governed by the solution to the time dependent Schrödinger equation (TDSE) and the particle motion is followed by solving the pertinent Newton's equation of motion with forces originating from both classical and quantum potentials. Important insight into the chaotic dynamics has been obtained [57] through ρ vs $-\chi$ plots which can be considered to be "canonically conjugate". In QTM it is obtained [59] in terms of the distance between two initially close Bohmian trajectories and the associated Kolmogorov – Sinai entropy.

In the present paper we monitor the possible regular / chaotic dynamics through the time evolution of various reactivity indices of a hydrogen atom in the ground and highly excited electronic states in the presence of one – color and two – color laser pulses. The theoretical background of the present work is provided in section II. Section III presents the numerical details, and the results and discussions are given in section IV. Finally, section V contains some concluding remarks.

THEORETICAL BACKGROUND

Classical interpretation of quantum mechanics is as old as the quantum mechanics itself. In the Madelung representation [55] the time – dependent Schrödinger equation for a single particle moving under potential $V(\vec{r})$ (in au), viz.

$$\left[-\frac{1}{2}\nabla^2 + V(\vec{r})\right]\psi(\vec{r},t) = i\frac{\partial\psi(\vec{r},t)}{\partial t}, \; i = \sqrt{-1}$$

(6)

is transformed into two fluid dynamical equations. Substituting the following polar form of the wave function

$$\psi(\vec{r},t) = \rho^{1/2}(\vec{r},t)\exp(i\chi(\vec{r},t))$$

(7)

in eq. (6) and separating the real and the imaginary parts, one obtains an equation of continuity

$$\frac{\partial\rho}{\partial t} + \nabla.j = 0$$

(8a)

and an Euler – type equation of motion

$$\frac{\partial\upsilon}{\partial t} + (\upsilon.\nabla)\upsilon = -\nabla(V + V_{qu}).$$

(8b)

In eqs (8) the charge density, $\rho(\vec{r},t)$ and current density, $j(\vec{r},t)$ is

$$j(\vec{r},t) = \rho(\vec{r},t)\upsilon(\vec{r},t)$$

(9a)

where the velocity $\upsilon(\vec{r},t)$ can be defined in terms of the phase of the wave function as

$$\upsilon(\vec{r},t) = \dot{\vec{r}} = \nabla\chi(\vec{r},t)$$

(9b)

The quantity V_{qu} appearing in eq. (8b) is called the quantum potential or Bohm potential of hidden variable theory [60] and defined as

$$V_{qu} = -\frac{1}{2}\frac{\nabla^2\rho^{1/2}}{\rho^{1/2}}$$

(9c)

Therefore, in this quantum fluid dynamics [55] the overall motion of the system under consideration can be thought of as a motion of a "probability fluid" having density $\rho(\vec{r},t)$ and velocity $\upsilon(\vec{r},t)$ under the influence of the external classical potential augmented by a quantum potential, V_{qu}.

For the ground state of a many – particle system, $\rho(\vec{r},t)$ contains all information [35]. In a time – dependent situation also the time – dependent density functional theory [52] asserts that any physical observable can be expressed as a functional of $\rho(\vec{r},t)$ and $j(\vec{r},t)$ and thus allows us to formulate the dynamics in terms of "classical – like" 3D quantities. Although Madelung transformation in terms of $\rho(\vec{r},t)$ and $j(\vec{r},t)$ is not straightforward in a many particle situation, we can make use of the time dependent density functional theory in constructing two fluid – dynamical equations in 3D – space. The formalism is termed as quantum fluid density functional theory [61] which has been applied in understanding ion – atom collisions [61,62,63], atom – field interactions [64,65] and electronegativity [51,66], hardness [66,67,68] and entropy dynamics [68] in a chemical reaction.

Quantum potential plays a crucial role in the quantum theory of motion [58] as well. In this representation of quantum mechanics developed by de Broglie [69] and Bohm [70], the overall motion of the system is understood in terms of the motion of a particle experiencing forces originating from the classical and quantum potentials. The Newton's equation of motion for this particle guided by a wave (represented by $\psi(\vec{r},t)$ a solution to eq. (6)) can be written as

$$\left(\frac{\partial}{\partial t} + \dot{\vec{r}}.\nabla\right)\left(\dot{\vec{r}}\right) = -\nabla\left(V + V_{qu}\right)\big|_{\vec{r}=\vec{r}(t)}$$

(9d)

At a particular instant the solution to the time dependent Schrödinger equation (6) fixes the velocity of the particle (cf. eq. 9b) and, hence, for a given initial position the particle motion can be studied through the solution $\vec{r}(t)$ to the eq. (9b).

Theories based on quantum potential idea have been applied in solving various physico – chemical problems [58,71,72,73,74,75,76,77,78,79,80, 81,82,83]. Because of the presence of nonlinearity and also the "classical language", these theories have been found [57,58,59,81,82,83,84,85] to be helpful in understanding the quantum domain behavior of classically chaotic systems which is described as quantum chaology by Berry [86]. The quantum theory of motion, however, allows one to study the quantum chaos in a system without any resort to its classical domain dynamics [58].

The time – dependent Schrödinger equation (in a u.) for the present problem is

$$\left[-\frac{1}{2}\nabla^2 + V(\vec{r})\right]\psi(\vec{r},t) = i\frac{\partial\psi(\vec{r},t)}{\partial t}$$

(10a)

where the potential $V(\vec{r},t)$ is given by

$$V(\vec{r},t) = -\frac{1}{r} + v_{ext}(\vec{r},t)$$

(10b)

In eq. (10b) the external potential for the monochromatic and bichromatic laser pulses may be written as

$$v_{ext}(\vec{r},t) = \varepsilon 1 z \text{ , for monochromatic pulse}$$

(10c)

$$= \varepsilon 2 z \text{ , for bichromatic pulse}$$

(10d)

Where

$$\varepsilon 1 = \varepsilon \cos(\omega_0 t)$$

(10e)

and

$$\varepsilon 2 = 0.5\varepsilon[\cos(\omega_0 t) + \cos(\omega_1 t)]$$

(10f)

To have slow oscillations during and after the source being switched on, ε is written in terms of the maximum amplitude $\varepsilon 0$ and the switch – on time t' as

$$\varepsilon = \varepsilon_0 t / t' \quad \text{for} \quad 0 \le t \le t'$$

(10g)

$$= \varepsilon_0 \quad \text{otherwise.}$$

(10h)

It may be noted that for a many – electron problem one may either solve the associated TDSE or the corresponding generalized nonlinear Schrödinger equation within a quantum fluid density functional framework [46,53,55,63,64,65,66,69,89], the latter being three dimensional even in the case of a many – electron system. To construct the hardness kernel (eq 4), we need the Hohenberg – Kohn universal functional F[ρ]. For a many – electron system F[ρ] may be taken as [53]

$$F[\rho] = \frac{1}{2}\int \rho(\vec{r},t)|\nabla\chi(\vec{r},t)|^2 \, d\vec{r} + T[\rho] + \frac{1}{2}\iint \frac{\rho(\vec{r},t)\rho(\vec{r}',t)}{|\vec{r}-\vec{r}'|} d\vec{r} \, d\vec{r}' + E_{xc}[\rho]$$

(11a)

where the first term is the macroscopic kinetic energy, the last term is the exchange – correlation energy, and T[ρ] is the intrinsic kinetic energy given by [53]

$$T[\rho] = T_0[\rho] + T_w[\rho] - a(N)\lambda \int \frac{\rho^{4/3}/\bar{r}}{1+\frac{\bar{r}\rho^{1/3}}{0.043}}\, d\bar{r}$$

(11b)

where $T_0[\rho]$ is the Thomas – Fermi functional [88], $T_w[\rho]$ is the Weizsäcker functional [88], λ is a constant [53], a(N) is an N – dependent parameter [53].

For obtaining the global hardness η (eq. 3) we also require the Fukui function $f(\bar{r})$. We employ the following local formula for $f(\bar{r})$

$$f(\bar{r}) = \frac{s(\bar{r})}{\int s(\bar{r})d\bar{r}}$$

(11c)

Where the local softness $s(\bar{r})$ is given as follows as prescribed by Fuentealba [89]

$$s(\bar{r}) = \frac{\delta(\bar{r} - \bar{r}')}{2\eta(\bar{r}, \bar{r}')}.$$

(11d)

For calculating $\eta(\bar{r}, \bar{r}')$ of the above equation the following local form for F[ρ] is used [53]:

$$F[\rho] = T^{local}[\rho] + V_{ee}^{\ local}[\rho]$$

(11e)

where the local kinetic energy [90] and the electron – electron repulsion energy [91] may be taken as [53]

$$T^{local}[\rho] = T_0[\rho] + \frac{3}{4\pi}(3\pi^2)^{1/2} \int \frac{\rho^{4/3}/\bar{r}}{1+\frac{\bar{r}\rho^{1/3}}{0.043}}\, d\bar{r}$$

(11f)

and

$$V_{ee}^{\ local}[\rho] = 0.7937(N-1)^{2/3}\int \rho^{4/3} d\bar{r}.$$

(11g)

Note that the above treatment is applicable to many – electron systems and all electron – electron interaction terms would be absent in the case of a hydrogen atom.

To follow the polarizability dynamics the dynamic polarizability is defined as [44,53]

$$\alpha(t) = |D_{ind}^{\ z}(t)|/|\Im_z(t)|$$

(12a)

where $D_{ind}{}^z(t)$ is the electronic part of the induced dipole moment given as

$$D_{ind}{}^z(t) = \int z\rho(\vec{r},t)d\vec{r}$$

(12b)

and $\Im_z(t)$ is the z - component of the external field.

The phase space volume or the uncertainty product V_{ps} has been shown [92] to be an important diagnostic of the quantum signature of classical chaos [92] as related to the compactness of the electron cloud [93]. For the present problem it may be defined as

$$V_{ps} = \left\{ <(p_{\tilde{\rho}}-<p_{\tilde{\rho}}>)^2><(p_z-<p_z>)^2><(\tilde{\rho}-<\tilde{\rho}>)^2><(z-<z>)^2> \right\}^{1/2}.$$

(13)

A sharp increase in $V_{ps}(t)$ implies a chaotic motion [92] since it is a measure of the associated quantum fluctuations [92].

To generate the harmonic spectrum the induced dipole moment, $D_{ind}{}^z(t)$ is Fourier transformed to obtained d(ω). It has been shown [94] that the absolute square of the Fourier transform, |d(ω)|2 is roughly proportional to the experimental harmonic distribution.

The Shannon entropy is given by

$$S == k\int \rho \ln(\rho)d\vec{r},$$

(14)

where k is the Boltzmann constant.

We can generate the "quantum trajectory" of a particle for a given initial position from equation (9b). Now, we are in a position to analyze the sensitive dependence on initial condition, a characteristic of a chaotic system. Equation (9b) is solved with two different initial positions of the particle, $(\tilde{\rho},z)$ and $(\tilde{\rho}+d\tilde{\rho},z+dz)$, $d\tilde{\rho} = 0$ $dz = 0.01$. Initial momentum of the particle is taken as zero in all cases. We study the time evolution of phase space distance (D) for the corresponding quantum trajectories defined as [56,59,82,83]

$$D(t) = \left[(\tilde{\rho}_1(t)-\tilde{\rho}_2(t))^2 + (z_1(t)-z_2(t))^2 + (p_{\tilde{\rho}_1}(t)-p_{\tilde{\rho}_2}(t))^2 + (p_{z_1}(t)-p_{z_2}(t))^2 \right]^{1/2},$$

15a)

where $(\tilde{\rho},p_{\tilde{\rho}},z,p_z)$ refers to a point in phase space.

We also calculate the associated Kolmogorov – Sinai entropy as defined [82,83] below

$$H = \sum_{\Lambda_+ > 0} \Lambda_+ \,,$$

(15b)

where the Lyapunov exponent is given by [82,83]

$$\Lambda = \lim_{\substack{D(0) \to 0 \\ t \to \alpha}} \frac{1}{t} \ln[D(t)/D(0)]$$

(15c)

According to the Hamilton – Jacobi formulation of quantum mechanics, a positive KS entropy is associated with a chaotic quantum dynamics [59,87].

NUMERICAL SOLUTION

The TDSE (eq. 10a) is solved numerically in cylindrical polar coordinates $(\tilde{\rho}, \tilde{\phi}, z)$, , as an initial boundary value problem using an alternating direction implicit method [95]. The solution procedure begins with the ψ_{1s} and ψ_{20s} analytical wave functions of the hydrogen atom. Since the electron density varies rapidly near the nucleus and relatively slowly elsewhere, we transform the variables as follows

$$y = \tilde{\rho}\phi$$

(16a)

and

$$\tilde{\rho} = x^2.$$

(16b)

Eq. (10a) takes the following form in the transformed variables once an analytical integration is carried out over $0 \le \tilde{\phi} \le 2\pi$,

$$\left\{ \left(\frac{3}{4x^3} \right) \frac{\partial y}{\partial x} - \left(\frac{1}{4x^2} \right) \frac{\partial^2 y}{\partial x^2} - \frac{\partial^2 y}{\partial z^2} \right\} - \left(\frac{1}{x^4} - 2\upsilon_{eff} \right) y = 2i \frac{\partial y}{\partial t}.$$

(17)

The resulting tridiagonal matrix equation is solved using a Thomas algorithm. The mesh sizes adopted here are $\Delta x = \Delta z = 0.4$ au and $\Delta t = 0.01$ au, ensuring the stability of the forward – time – central – space type numerical scheme adopted here.

The initial and boundary conditions associated with this problem are

y(x,z) is known for ∀ x,z at t = 0   (18a)

y(0,z) = 0 = y(∞,z) ∀ z,t   (18b)

y(x,±∞) = 0 ∀ x,t.   (18c)

The numerical scheme is stable [96] due to the presence of $i = \sqrt{-1}$. As a further check of the numerical accuracy, we have verified the conservation of norm and energy (in zero field cases). The wave function is moved forward to the end of the simulation and then taken back to its initial position by reversing the time direction, where the original profile is reproduced well within the tolerance limit of the present calculation. We have also solved eq. (9b) using a second order Runge – Kutta method to generate the "quantum trajectories" of a given initial position. The field parameters are in atomic units unless otherwise specified.

RESULTS AND DISCUSSIONS

The time evolution of different reactivity parameters are depicted in Figure 1, Figure 2, Figure 3, Figure 4, Figure 5, Figure 6, Figure 7, Figure 8, Figure 9, Figure 10, Figure 11 and Figure 12. All quantities are in atomic units. Unless otherwise specified, in all figures a and b refer to the ground state (n=1) and excited state (n=20) of the hydrogen atom, respectively, and a red colored solid line and blue colored solid line respectively signify monochromatic and bichromatic pulses.

Figure 1 presents the time dependence of the external field with different frequencies and the same amplitude.

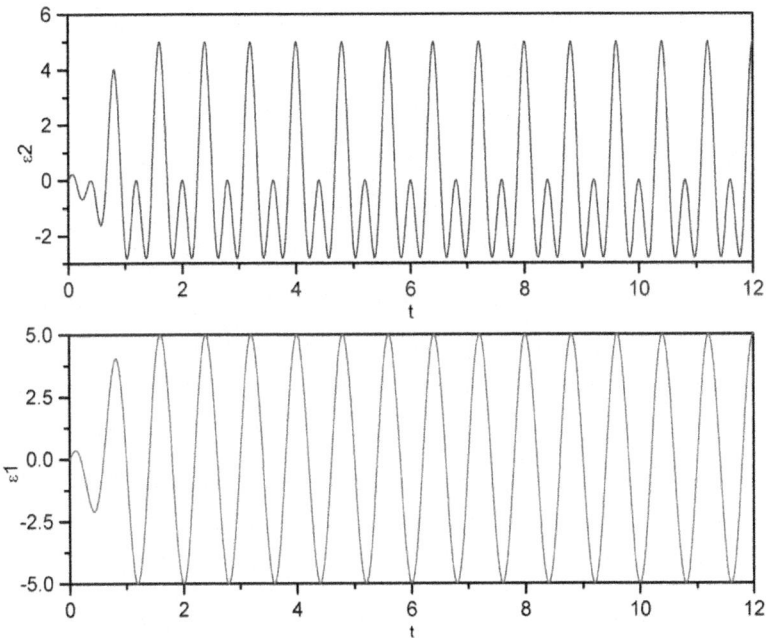

Figure 1: Time evolution of the external electric field: ε1 (▬) monochromatic pulse, ε2 (▬) bichromatic pulse. Field parameters: $\varepsilon_0 = 5.0$; $\omega_0 = 2.5\pi$, $\omega_1 = 2\omega_0$.

Temporal evolution of the chemical potential is depicted in Figure 2. It exhibits characteristic oscillations. The oscillations in μ is not in phase with the external field. It is important to note that the amplitude of μ- oscillations becomes very large for both the electronic states and both monochromatic and bichromatic pulses.

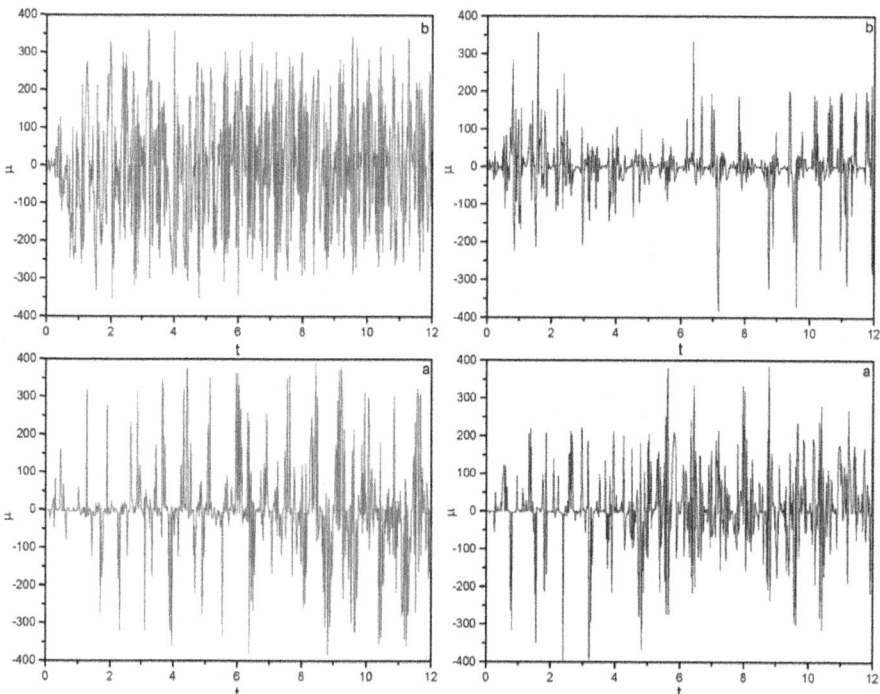

Figure 2: Time evolution of chemical potential (μ) when a hydrogen atom is subjected to external electric fields: a – Ground state; b – Excited state. (———) Monochromatic pulse, (———) bichromatic pulse. Field parameters: $\varepsilon_0 = 5.0$; $\omega_0 = 2.5\pi$, $\omega_1 = 2\omega_0$.

Chemical hardness (η) is presented in Figure 3. For both one and two – color cases η is much larger for n=1 state than that of the n=20 state for the whole time range. This may be considered to be a dynamical variant of the MHP. Hardness oscillates in time in all the cases. However, the oscillation is neither in phase nor out of phase with respect to the oscillations in the external one – and two – color fields. It is expected because of the fact that as soon as the laser is switched on, there starts a tug – of – war between the atomic nucleus and the external field to govern the electron – density distribution. The nucleus tries to make the density distribution spherically symmetric owing to the central nature of the nuclear coulomb field while the cylindrical symmetry of the applied electric field tries to create an oscillating dipole that emits radiation including higher harmonics. Overall density oscillation becomes nonlinear due to the interplay of two different types of effects. Hardness for the n=1 state decreases (for both one – and two - color situations) and attains a more or less steady value at the end of the simulation, which is still large in comparison to the corresponding value for the n=20 state. For both one– and two–color situations, η values relative to the corresponding values in absence of the field (not shown) are much larger for the n=1 state. It appears that a relatively smaller η value signals a possible chaotic dynamics.

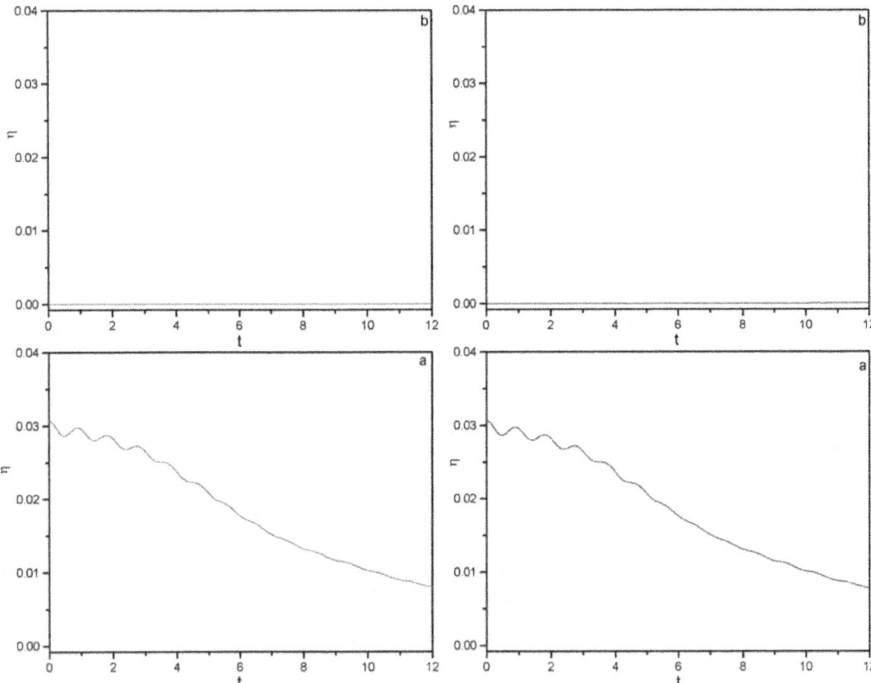

Figure 3: Time evolution of hardness (η) when a hydrogen atom is subjected to external electric fields: a – Ground state; b – Excited state. (▬▬) Monochromatic pulse, (▬) bichromatic pulse. Field parameters: $\varepsilon_0 = 5.0$; $\omega_0 = 2.5\pi$, $\omega_1 = 2\omega_0$.

Polarizability values as they evolve in the course of time are presented in Figure 4. It oscillates with a frequency that is double that of the external field. The extrema in the external field corresponding to the minima in α and the latter blows up when the field is zero. Here also if we compare the respective minimum α values (α_{min}) for the two electronic states, α_{min} for the ground state is smaller than that of the excited state which is conspicuous for the bichromatic pulse. This is in conformity with minimum polarizability principle (MPP). The MPP reveals itself in a time–dependent situation.

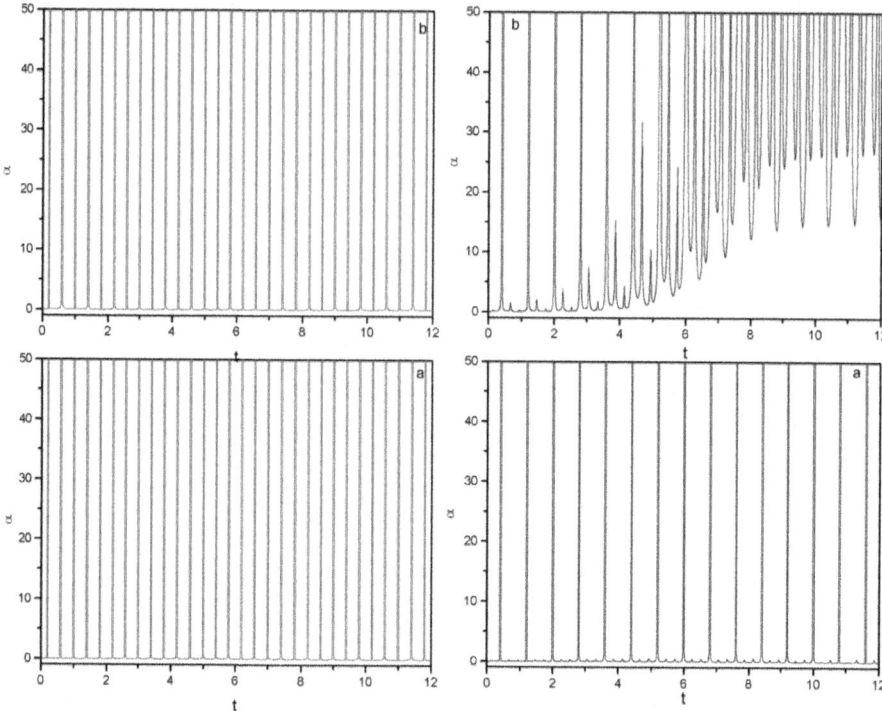

Figure 4: Time evolution of polarizability (α) when a hydrogen atom is subjected to external electric fields: a – Ground state; b – Excited state. (▬) Monochromatic pulse, (—) bichromatic pulse. Field parameters: $\varepsilon_0 = 5.0$; $\omega_0 = 2.5\pi$, $\omega_1 = 2\omega_0$.

Figure 5 depicts the dynamics of the uncertainty product (phase volume). As in the cases of μ and η, V_{ps} also oscillates neither in phase nor out of phase with the external field. The magnitude of V_{ps} retains its initial (t=0) small value for the n=1 state whereas for the n=20 state it increases quickly to a very large value. Since V_{ps} measures the quantum fluctuations, a chaotic trajectory is generally associated with large V_{ps} values [92]. "......
large increases in V_{ps} can be expected to accompany a chaotic trajectory. Conversely, small to moderate increases in V_{ps} can be evidence that given quantum mechanical trajectory should be regarded a nonchaotic [92a]". In general, the electrons are "tightly bound" and hence the distribution is "less diffuse" for the n=1 state and "loosely bound" for the n=20 state and the system is expected to be harder and less polarizable for the ground state

[32,34,46,53,88,93]. Again, the electron density being more compact in the ground state, the corresponding uncertainty product is expected [93] to be small. Once the external field is switched on, the ground state density would be distributed over a larger volume and consequently there would be a decrease in η and increase in α and V_{ps} of the system. Since a smaller η value is accompanied with a large V_{ps} value and vice versa and V_{ps} is known [92] to bear the signature of the classical chaos in the corresponding quantum domain behavior, hardness can as well be considered to be a diagnostic of the chaotic dynamics in a quantum system.

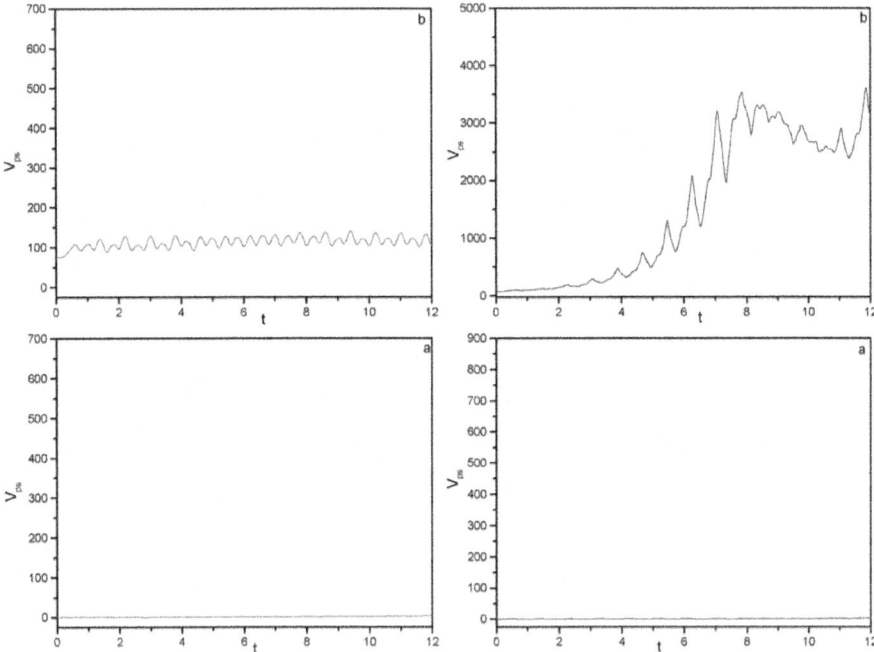

Figure 5: Time evolution of phase volume (V_{ps}) when a hydrogen atom is subjected to external electric fields: a – Ground state; b – Excited state. (▬) Monochromatic pulse, (▬) bichromatic pulse. Field parameters: $\varepsilon_0 = 5.0$; $\omega_0 = 2.5\pi$, $\omega_1 = 2\omega_0$.

Figure 6 and Figure 7 depict respectively the dynamical profiles of electrophilicity and nucleophilicity indices respectively. Both W and 1/W show oscillations characteristic of the resultant field of two competing ones for both the electronic states and for the one– and two–color pulses.

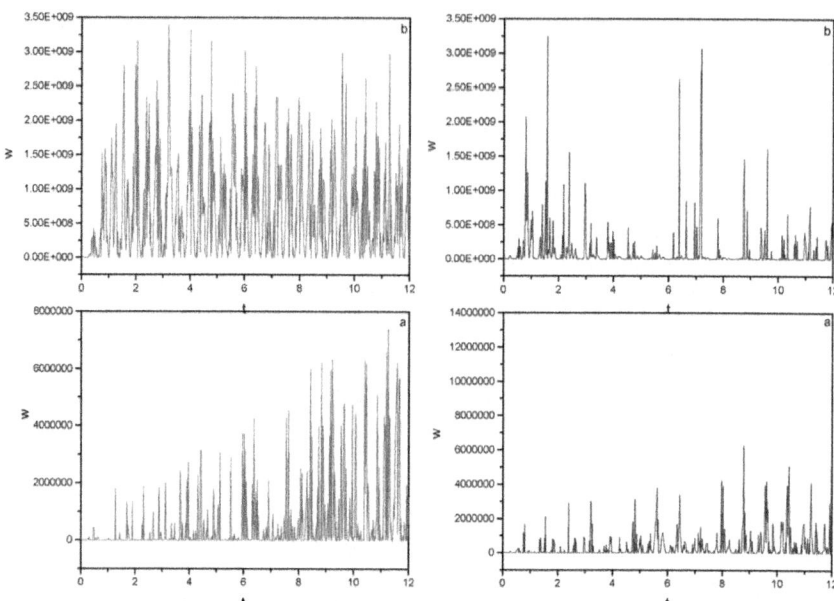

Figure 6: Time evolution of electrophilicity index (W) when a hydrogen atom is subjected to external electric fields: a – Ground state; b – Excited state. (▬) Monochromatic pulse, (▬) bichromatic pulse. Field parameters: $\varepsilon_0 = 5.0$; $\omega_0 = 2.5\pi$, $\omega_1 = 2\omega_0$.

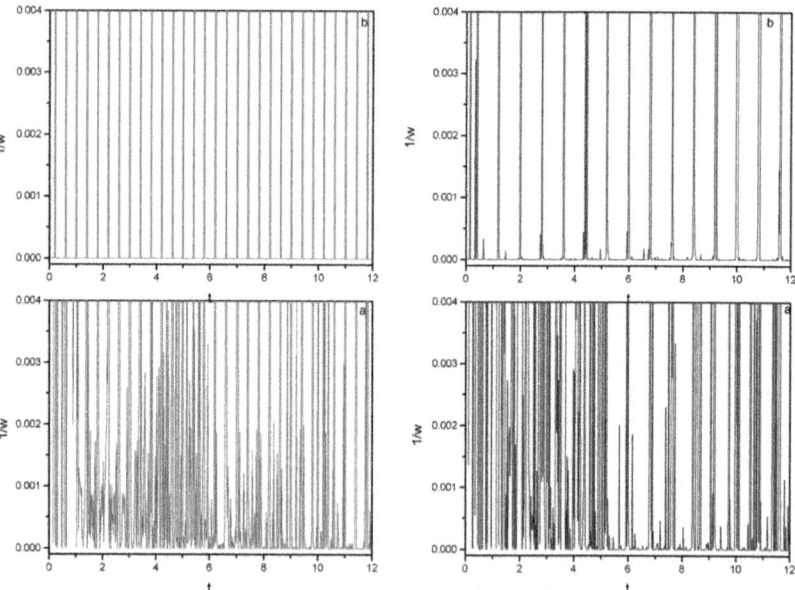

Figure 7: Time evolution of nucleophilicity index (1/W) when a hydrogen atom is subjected to external electric fields: a – Ground state; b – Excited state. (▬) Mono-

chromatic pulse, (▬) bichromatic pulse. Field parameters: $\varepsilon_0 = 5.0$; $\omega_0 = 2.5\pi$, $\omega_1 = 2\omega_0$.

The harmonic spectra are presented in Figure 8. The overall domain of the spectra and their envelopes look like those reported by Erhard and Gross [52]. We found that the harmonics generated by the monochromatic and bichromatic pulses look similar and those generated from the former is less intense than those resulted from the latter [52].

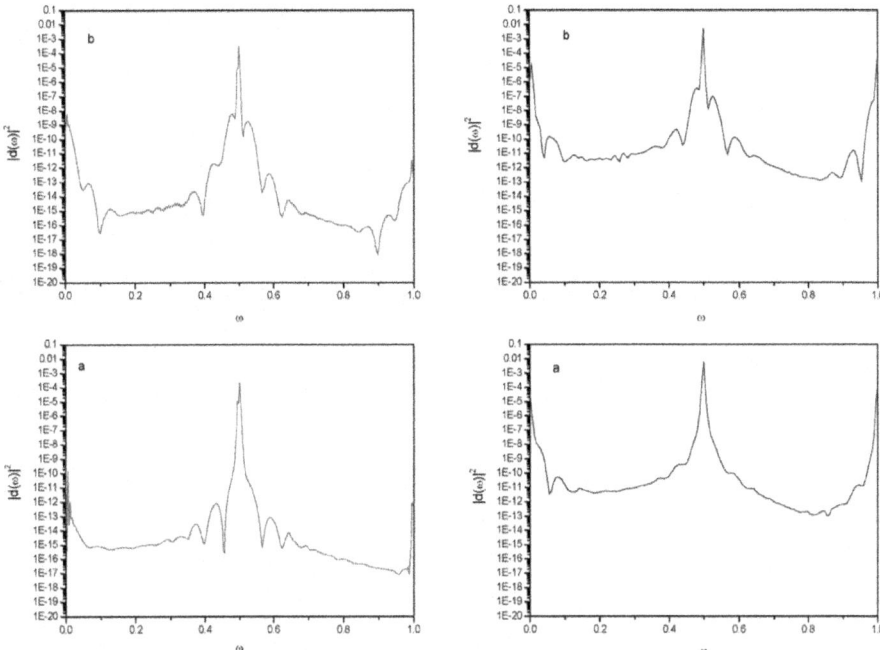

Figure 8: $|d(\omega)|^2$ vs ω plot when a hydrogen atom is subjected to external electric fields: a – Ground state; b – Excited state. (▬) Monochromatic pulse, (▬) bichromatic pulse. Field parameters: $\varepsilon_0 = 5.0$; $\omega_0 = 2.5\pi$, $\omega_1 = 2\omega_0$.

Figure 9 depicts the phase $(p_{\tilde{\rho}}$ vs $\tilde{\rho}$ and p_z vs z$)$ of cases a and b for monochromatic laser pulse, The fraction of the total phase space visited by the Bohmian trajectories is much more for the excited state. These plots reflect that the case a is for regular motion whereas the case b is for chaotic motion.

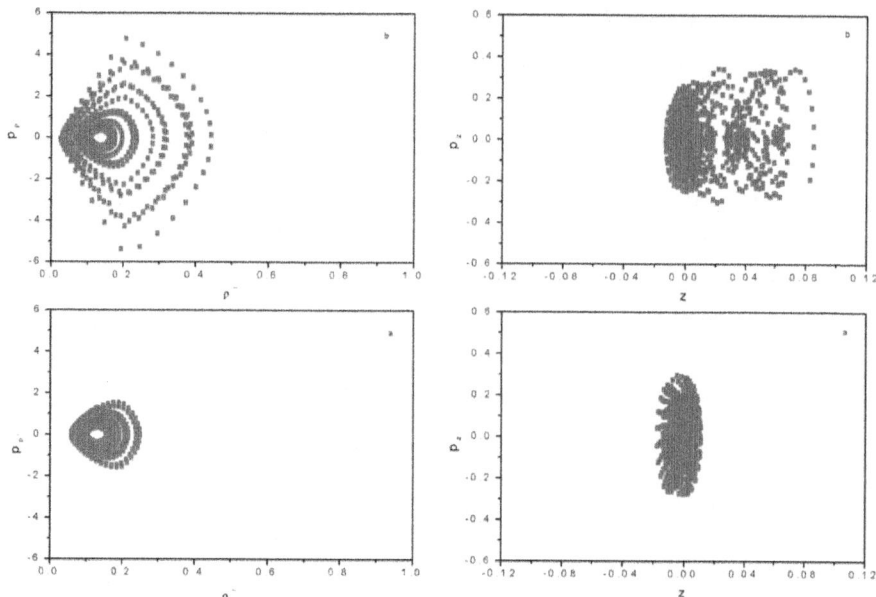

Figure 9: Phase space trajectories when a hydrogen atom is subjected to external electric field: a – Ground state; b – Excited state. (• • • • • • •) Monochromatic pulse. Field parameters: $\varepsilon_0 = 5.0$; $\omega_0 = 2.5\pi$, $\omega_1 = 2\omega_0$.

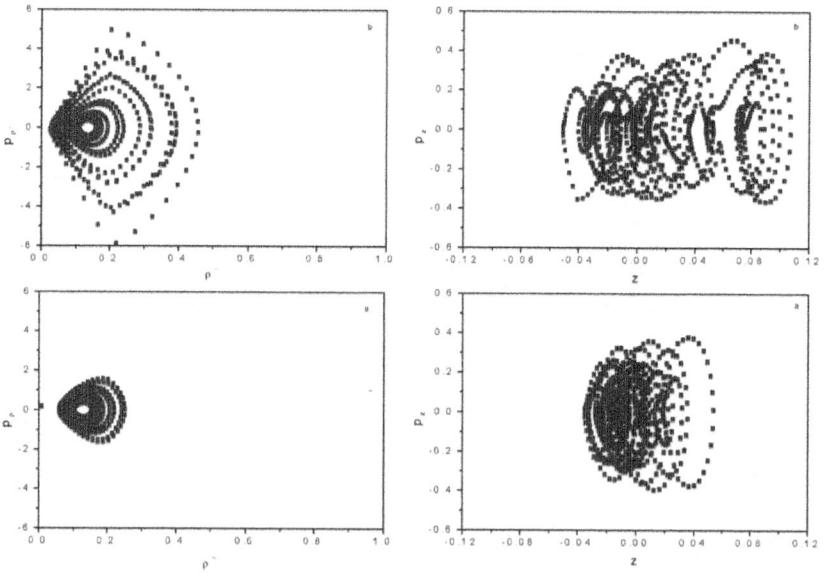

Figure 10: Phase space trajectories when a hydrogen atom is subjected to external electric field: a – Ground state; b – Excited state. (• • • • • • • • •) Bichromatic pulse. Field parameters: $\varepsilon_0 = 5.0$; $\omega_0 = 2.5\pi$, $\omega_1 = 2\omega_0$.

Figure 10 depicts the phase $(p_{\tilde{\rho}}$ vs $\tilde{\rho}$ and p_z vs $z)$ plots case a and b for bichromatic laser pulse. These plots also reflect that the case a is for regular motion whereas the case b is for chaotic motion.

Figure 11 depicts the Kolmogorov – Sinai (KS) entropy for both ground (n=1) and excited (n=20) states for monochromatic and bichromatic laser pulses. For both monochromatic and bichromatic laser pulses the KS entropy (H) retains its initial very small value for n=1. For n=20 case H remains small initially and then increases rapidly to a high positive value. The small H value in the former case vis – á – vis the very large H value in the latter provides unmistakable signature of chaos in the highly excited state of the hydrogen atom in presence of an external electric field.

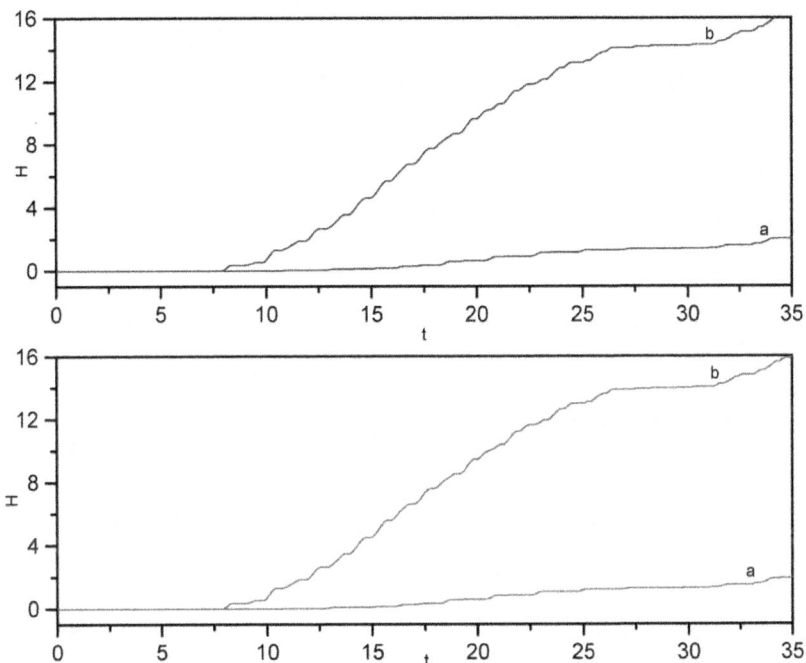

Figure 11: Time evolution of KS entropy (H) when a hydrogen atom is subjected to external electric fields: a – Ground state; b – Excited state. (——) Monochromatic pulse, (——) bichromatic pulse. Field parameters: $\varepsilon_0 = 5.0$; $\omega_0 = 2.5\pi$, $\omega_1 = 2\omega_0$.

Shannon entropy has been shown in Figure 12. In the figure a and b refer to the ground and n=20 states of the hydrogen atom respectively. It increases in the ground state and decreases in the excited state for both the laser pulses, a possible signature of the maximum entropy principle vis – a – vis chaotic ionization from the highly excited state.

It is important to note that the calculations have been carried out up to 3500 au with no change in the qualitative trends. Plots are truncated at a much smaller time steps for easy visualization.

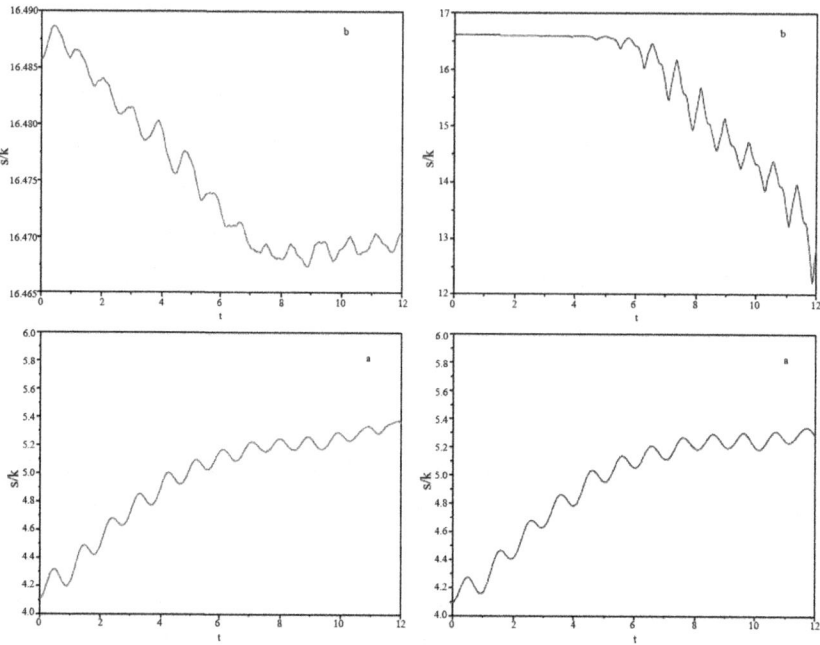

Figure 12: Time evolution of S/k, where S is the Shannon entropy and k is the Boltzmann constant when a hydrogen atom is subjected to external electric fields: a – Ground state; b – Excited state. (—) Monochromatic pulse, (—) bichromatic pulse. Field parameters: $\varepsilon_0 = 5.0$; $\omega_0 = 2.5\pi$, $\omega_1 = 2\omega_0$.

CONCLUDING REMARKS

Quantum potential based theories are adopted to study the reactivity dynamics and chaos of a hydrogen atom in its ground and excited electronic states interacting with z – polarized laser pulses of different colors. Dynamical variants of the principles of electronegativity equalization, maximum hardness, minimum polarizability and maximum entropy manifest themselves. A tug – of – war between the spherically symmetric nuclear coulomb field and cylindrically symmetric external electric field to govern the electron density distribution is delineated through the dynamical profiles of various reactivity indices like electronegativity, hardness, polarzability, electrophilicity, nucleophilicity and phase volume for the external field and in different electronic states. Harmonic spectra of the higher order harmonics included in the radiation emitted by the resulting oscillating dipole have been analyzed. Temporal evolution of Bohmian trajectory, KS entropy and Shannon entropy has easily differentiated the regular and chaotic behavior of hydrogen atom respectively in ground and excited states in presence of an oscillating electric field. For both the laser pulses the increase in the uncertainty product for the excited state is very large, which implies a possible chaotic dynamics. A large hardness value, on the other hand, is expected to characterize a regular behavior.

ACKNOWLEDGMENT

We thank CSIR, New Delhi for financial assistance and Mr. U. Sarkar for help in computation.

REFERENCES AND NOTES

1. Jensen, R. V.; Susskind, S. M.; Sanders, M. M. *Phys. Rep.* 1991, *201*, 1–56.

2. Koch, P. M.; van Leeuwen, K. A. H. *Phys. Rep.* 1995, *255*, 289–403.

3. *Atoms in Intense Laser Fields*; Gavrila, M., Ed.; Academic Press: Boston, 1992.

4. Sanders, M. M.; Jensen, R. V. *Am. J. Phys.*

5. Mariani, D. R. *The Ionisation of Hydrogen and Helium Atoms by Static and Microwave ionization of highly excited hydrogen atoms: Experimental and theory, in the physics of phase space*; Kim, Y. S., Zachary, W. W., Eds.; Springer: New York, 1987; pp. 106–113.

6. Mariani, D. R.; van de Water, W.; Koch, P. M.; Bergeman, T. *Phys. Rev. Lett.* 1983, *50*, 1261.

7. van de Water, W.; Yoakum, S.; van Leeuwen, K. A. H.; Sauer, B. E.; Moorman, L.; Galvez, E. J.; Mariani, D. R.; Koch, P. M. *Phys. Rev. A.* 1990, *42*, 573.

8. Bayfield, J. E.; Koch, P. M. *Phys. Rev. Lett.* 1974, *33*, 258–261.

9. Koch, P. M.; van Leeuwen, K. A. H.; Rath, O.; Richards, D.; Jensen, R. V. *Microwave ionization of highly excited hydrogen atoms: Experiment and theory, in the physics of phase space*; Kim, Y. S., Zachary, W. W., Eds.; Springer: New York, 1987; pp. 106–113.

10. van Leeuwen, K. A. H.; Oppen, G. V.; Renwick, S.; Bowlin, J. B.; Koch, P. M.; Jensen, R. V.; Rath, O.; Richards, D.; Leopold, J. G. *Phys. Rev. Lett.* 1985, *55*, 2231–2234.

11. Galvez, E. J.; Sauer, B. B.; Moorman, L.; Koch, P. M.; Richards, D. *Phys. Rev. Lett.* 1988, *61*, 2011–2014.

12. Leopold, J. G.; Richards, D. J. *Phys. B At. Mol. Phys.* 1989, *24*, 1209–1240.

13. Sanders, M. M. Chaotic Ionisation of One and Two Electron Atom. Ph. D. thesis, Yale University, New Haven, CT, 1991.

14. Bayfield, J. E. *Am. Sci.*; 1983; Volume 71, pp. 375–383.

15. Leopold, J. G.; Percival, I. C. *J. Phys. B At. Mol. Phys.* 1979, *12*, 709–721.

16. Rath, O.; Richards, D. *J. Phys. B.* (submitted).

17. Born, M. The Mechanics of the Atom. Frederick Ungar: New York, 1960.

18. Leopold, J. G.; Richards, D. *J. Phys. B. At. Mol. Phys.* 1986, *19*, 1125.

19. Jensen, R. V. *Phys. Rev. Lett.* 1982, *49*, 1365–1368.

20. Jensen, R. V. *Phys. Rev. A.* 1982, *30*, 386–397.

21. Schwieters, C. D.; Delos, J. B. *Phys. Rev. A* 1995, *51*, 1030–1041.

22. Shepelyansky, D. L. Chaotic Behavior in Quantum System: Theory and Application. Plenum: New York, 1985; pp. 187–197.

23. Lichtenberg, A. J.; Liebergman, M. A. *Regular and Stochastic Motion*; Springer: New York, 1983.

24. Casati, G.; Chirikov, B. V.; Shepelyansky, D. L. *Phys. Rev. Lett.* 1984, *53*, 2525–2528.

25. Casati, G.; Guarneri, L.; Shepelyansky, D. L. *IEEE J. Quantum Electron.* 1988, *QE – 24*, 1420–1444.

26. Jensen, R. V. *Phys. Scr.* 1987, *35*, 668–673.

27. Landau, L. D.; Lifshitz, E. M. *Mechanics, Course of Theoretical Physics* Pergamon: New York, 1976; Vol. 1, 3rd ed.

28. Goldstein, H. *Classical Mechanics* Addison – Wesley: Reading, MA, 1980, 2nd ed.

29. Mcphersion, A.; Gibson, G.; Jara, H.; Johann, U.; Luk, T. S.; McIntyre, I. A.; Boyer, K.; Rhodes, C. K. *J. Opt. Soc. Am.* 1987, *B4*, 595.

30. *Super – Intense Laser – Atom Physics*; Piraux, B., L'Huillier, A., Rzazewski, K., Eds.; NATO ASI Series B316; Plenum Press: New York, 1993.

31. Electronegativity: Struct. Bonding. Sen, K. D., Jorgenson, C. K., Eds.; Springer–Verlag: Berlin, 1987; Vol. 66.

32. *Chemical Hardness: Struct. Bonding*; Sen, K. D, Mingos, D. M. P., Eds.; Springer–Verlag: Berlin, 1987; Vol. 66.

33.   Pauling, L. *The Nature of the Chemical Bond*, 3rd ed.; Cornell University Press: Ithaca, NY, 1960.

34.   Pearson, R. G. *Coord. Chem. Rev.* 1990, *100*, 403.*Hard and Soft Acids and Bases*; Dowden, Hutchinson and Ross: Stroudsberg, PA, 1973.

35.   Hohenberg, P.; Kohn, W. *Phys. Rev. B* 1964, *136*, 864.Kohn, W.; Sham, L. J. *Phys. Rev. A* 1965, *140*, 1133.Parr, R. G.; Yang, W. *Density Functional Theory of Atoms and Molecules*; Oxford University Press: Oxford, 1989. *Annu. Rev. Phys. Chem.* 1995, *46*, 701.Chattaraj, P. K. *J. Indian. Chem. Soc.* 1992, *69*, 173.Kohn, W.; Becke, A. D.; Parr, R. G. *J. Phys. Chem.* 1996, *100*, 12974.

36.   Pearson, R. G. *J. Chem. Educ.* 1987, *64*, 561.

37.   Parr, R. G.; Chattaraj, P. K. *J. Am. Chem. Soc.* 1991, *113*, 1854.Chattaraj, P. K.; Liu, G. H.; Parr, R. G. *Chem. Phys. Lett.*1995, *237*, 171.Pearson, R. G. *Chemtracts Inorg. Chem.* 1991, *3*, 317.Liu, S.; Parr, R. G. *J. Chem. Phys.* 1997, *106*, 5578.Chattaraj, P. K. *Proc. Indian Natl. Sci. Acad. Part A* 1996, *62*, 513.Pearson, R. G. *Chemical Hardness: Application from Molecules to Solid*; Wiley–VCH Verlag GMBH: Weinheim, 1997. Ayers, P. W.; Parr, R. G. *J. Am. Chem. Soc.* 2000, *122*, 2010.

38.   Parr, R. G.; Donnelly, D. A.; Levy, M.; Palke, W. E. *J. Chem. Phys.* 1978, *68*, 3801.

39.   Parr, R. G.; Pearson, R. G. *J. Am. Chem. Soc.* 1983, *105*, 7512. Berkowitz, M.; Ghosh, S. K.; Parr, R. G. *J. Am. Chem. Soc.* 1985, *107*, 6811.Ghosh, S. K.; Berkowitz, M. *J. Chem. Phys.*1985, *83*, 2976.

40.   Parr, R. G.; Yang, W. *J. Am. Chem. Soc.* 1984, *106*, 4049.

41.   Sanderson, R. T. *Science* 1951, *114*, 670.*Science* 1955, *121*, 207.*J. Chem. Educ.* 1954, *31*, 238.Chattaraj, P. K.; Maiti, B. *J. Chem. Educ.* 2001, *78*, 811–813.

42.   Politzer, P.; Weinstein, H. *J. Chem. Phys.* 1979, *70*, 3680.Parr, R. G.; Bartolotti, L. J. *J. Am. Chem. Soc.* 1982, *104*, 3081.Nalewajski, R. F. *J. Phys. Chem.* 1985, *89*, 2831.Mortier, W. J.; Ghosh, S. K.; Shankar, S. *J. Am. Chem. Soc.* 1986, *108*, 4315.

43.   Chattaraj, P. K.; Sengupta, S. *J. Phys. Chem.* 1996, *100*, 16126.

44.   Ghanty, T. K.; Ghosh, S. K. *J. Phys. Chem.* 1996, *100*, 12296.

45.   Chattaraj, P. K.; Sengupta, S. *J. Phys. Chem. A* 1997, *101*, 7893.

46.   Chattaraj, P. K.; Poddar, A. *J. Phys. Chem. A*.Chattaraj, P. K.; Fuentealba, P; Gomez, B.; Contreas, R. *J. Am. Chem. Soc.*2000, *122*, 348.Fuentealba, P.; Simon – Manso, Y.; Chattaraj, P. K. *J. Phys. Chem. A* 2000, *104*, 3185. Chattaraj, P. K.; Fuentealba, P.; Jaque, P.; Toro – Labbe, A. *J. Phys. Chem.*

A 1999, *103*, 9307.Chattaraj, P. K.; Maiti, B. *J. Phys. Chem. A*2001, *105*, 169–183.

47.   Jaynes, E. T. *Statistical Physics*; Ford, K. W., Ed.; Brandeis Lectures, Vol – 3; Benjamin: New York, 1963. Levine, R. D.; Bernstein, R. B. *Dynamics of Molecular Collisions*; Miller, W. H., Ed.; Plenum Press: New York, 1976. Gadre, S. R.; Bendale, R. D. *Curr. Sci.* 1985, *54*, 970.

48.   Parr, R. G.; Szentpaly, L. v.; Liu, S. *J. Am. Chem. Soc.* 1999, *121*, 1922.

49.   Chattaraj, P. K.; Sengupta, S. *J. Phys. Chem. A* 1999, *103*, 6122.

50.   Chattaraj, P. K.; Nath, S. *Int. J. Quantum Chem.* 1994, *49*, 705.

51.   Runge, E.; Gross, E. K. U. *Phys. Rev. Lett.*; 1984; Volume 52, p. 997. Dhara, A. K.; Ghosh, S. K. *Phys. Rev. A*; 1987; Volume 35, p. 442. Erhard, S.; Gross, E. K. U. *In. Multiphoton Processes 1996*; Lambropoulos, P., Walther, H., Eds.; IOP Publishing: London, 1997; pp. 37–45, and reference therein.

52.   Chattaraj, P. K.; Sengupta, S.; Poddar, A. *Int. J. Quantum Chem.* 1998, *69*, 279.*Nonlinear Dynamics and Computational Physics*; Sheorey, V. B., Ed.; Narosa: New Delhi, 1999; pp. 45–53.

53.   Lakshmanan, M.; Ganeshan, K. *Curr. Sci.* 1995, *68*, 38.

54.   Casati, G.; Chirikov, B. V.; Guarneri, I; Shepelyansky, D. L. *Phys. Rep.* 1987, *154*, 77.Hasegawa, H.; Robnik, M.; Wunner, G. *Prog. Theo. Phys. Suppl.* 1989, *98*, 198.Ganeshan, K.; Lakshmanan, M. *Phys. Rev. A* 1990, *42*, 3940.Howard, J. E.; Farelly, D. *Phys. Lett. A* 1993, *178*, 62.Delande, D.; Gay, J. C. *Phys. Rev. Lett.* 1987, *59*, 1809.Delande, D. *Chaos and Quantum Physics*; Elsevier: Amsterdam, 1991. Friedrich, H.; Wintgen, D. *Phys. Rep.* 1989, *183*, 37.Holle, A.; Weibusch, G.; Main, J.; Hager, B.; Rottke, H.; Welge, K. H. *Phys. Rev. Lett.* 1986, *56*, 2594.Holle, A; Marini Rottke, H.; Welge, K. H. *Phys. Rev. Lett.* 1986, *56*, 2594.Holle, A.; Marini, J.; Weibusch, G.; Rottke, H.; Welge, K. H.*Phys. Rev. Lett.* 1988, *61*, 161.

55.   Madelung, E. *Z. Phys.* 1926, *40*, 322.Ghosh, S. K.; Deb, B. M. *Phys. Rep.* 1982, *92*, 1–44.

56.   Chattaraj, P. K.; Sengupta, S. *Phys. Lett. A* 1993, *181*, 225.*Ind. J. Pure Appl. Phys.* 1996, *34*, 518.Chattaraj, P. K. *Ind. J. Pure Appl. Phys.* 1994, *32*, 101.

57.   Holland, P. R. *The Quantum Theory of Motion*; Cambridge University Press: Cambridge, U. K., 1993.

58.   Sengupta, S.; Chattaraj, P. K. *Phys. Lett. A* 1996, *215*, 119.Chattaraj, P. K.; Sengupta, S. *Curr. Sci.* 1996, *71*, 134.Chattaraj, P. K.; Sengupta, S. *J.*

Phys. Chem. A 1999, *103*, 6122–6126.

59. Belinfante, F. J. *A Survey of Hidden Variable Theories*; Pergamon Press: New York, 1973.

60. Deb, B. M.; Chattaraj, P. K. *Phys. Rev. A* 1989, *39*, 1696–1713.

61. Deb, B. M.; Chattaraj, P. K. *Chem. Phys. Lett.* 1988, *148*, 550–556.

62. Deb, B. M.; Chattaraj, P. K.; Mishra, S. *Phys. Rev. A* 1991, *43*, 1248–1257.

63. Chattaraj, P. K. *Int. J. Quant. Chem.* 1992, *41*, 845–859.

64. Deb, B. M.; Dey, B. K. *Int. J. Quant. Chem.* 1995, *56*.

65. Nath, S.; Chattaraj, P. K. *Pramana* 1995, *45*, 65–73.

66. Chattaraj, P. K.; Nath, S. *Chem. Phys. Lett.* 1994, *217*, 342–348.

67. Chattaraj, P. K.; Nath, S. *Proc. Indian. Acad. Sci. (Ch. Sci.)* 1994, *106*, 229–249.

68. de Broglie, L. *Nonlinear Wave Mechanics: A Causal Interpretation*; Elsevier: Amsterdam, 1993.

69. Bohm, D. *Phys. Rev.* 1952, *85*, 166 – 179, 180 – 193.

70. Gutzwiller, M. C. Chaos in Classical and Quantum Mechanics. Springer: Berlin, 1990.

71. Eckhardt, B. *Phys. Rep.* 1988, *103*, 205–297.

72. Jensen, R. V. *Nature*.

73. Dewdney, C.; Hiley, B. *J. Found. Phys.* 1982, *12*, 27–48.

74. Takabayasi, T. Prog. *Theor. Phys.*

75. Kan, K. K.; Griffin, J. *Phys. Rev. C* 1977, *15*, 1126–1157.

76. Weiner, J. H.; Partom, Y. *Phys. Rev.* 1969, *187*, 1134–1147.Weiner, J. H.; Askar, A. *J. Chem. Phys.* 1972, *54*, 3534–3541.

77. McCullough, E. A.; Wyatt, R. E. *J. Chem. Phys.* 1971, *54*, 3534–3541.

78. Hirschfelder, J. O.; Christoph, A. C.; Palke, W. E. *J. Chem. Phys.* 1974, *61*, 5435–5455.Hirschfelder, J. O.; Tang, K. T. *J. Chem. Phys.*

79. Skodie, R. T.; Rohrs, H. W.; VanBuskirk, J. *Phys. Rev. A* 1989, *40*, 2894–2916.

80. Parmenter, R. H.; Valentine, R. W. *Phys. Lett. A* 1995, *201*, 1–8.

81. Schwengelbeck, U.; Faisal, F. H. M. *Phys. Lett. A* 1995, *199*, 281–286.

82. Faisal, F. H. M.; Schwengelbeck, U. Phys. Lett. A, 1995, 207, 31–36.

83. Misner, C. W.; Thorne, K. S.; Wheeler, J. A. *Gravitation*; W. H. Freeman and Company: San Francisco, 1973.

84. McDonald, S. W.; Kaufman, A. N. *Phys. Rev. Lett.* 1979, *42*, 1189–1191.

85. Berry, M. V. *Proc. R. Soc. A, London* 1987, *413*, 183–198.

86. Chattaraj, P. K. *Symmetries and Singularity Structures: Intregability and Chaos in Nonlinear Dynamical Systems*; Lakshmanan, M., Daniel, M., Eds.; Springer–Verlag: Berlin, 1990; pp. 172–182.

87. Parr, R. G.; Yang, W. *Density Functional Theory of Atoms and Molecules*; Oxford University Press: Oxford, U. K., 1989.

88. Fuentealba, P. *J. Chem. Phys.* 1995, *103*, 6571.

89. Ghosh, S. K.; Deb, B. M. *J. Phys. B* 1994, *27*, 381.

90. Parr, R. G. *J. Phys. Chem.* 1988, *92*, 3060.

91. Feit, M. D.; Fleck, J. A., Jr. *J. Chem. Phys.* 1984, *80*, 2578.Choudhury, S.; Gangopadhayay, G.; Ray, D. S. *Ind. J. Phys. B*1995, *69*, 507.Graham, R.; Hohnerbach, M. *Phys. Rev. A* 1991, *43*, 3966.*Idem. Phys. Rev. Lett.* 1990, *64*, 637.

92. Pearson, R. G. *Chemical Hardness: Applications from Molecules to Solids*; Wiley – VCH Verlag GMBH: Weinheim, 1997; pp. 116–119.

93. L'Huillier, A.; Lompre, L. A.; Mainfray, G.; Manus, C. *Atoms in Intense Laser Fields*; Gavrila, M., Ed.; Academic Press: Boston, 1992; p. 139.

94. Ames, W. F. Numerical Methods for Partial Differential Equations. Academic: New York, 1977; p. 252.

95. Chattaraj, P. K.; Rao, K. S.; Deb, B. M. *J. Comput. Phys.* 1987, *72*, 504.

Chapter 5

CHEMICAL REACTIVITY AS DESCRIBED BY QUANTUM CHEMICAL METHODS

P. Geerlings and F. De Proft

Eenheid Algemene Chemie, Free University of Brussels (VUB), Pleinlaan 2, 1050 Brussels, Belgium Tel: 32. 2.629 33 14, Fax: 32.2. 629 33 17,

ABSTRACT

Density Functional Theory is situated within the evolution of Quantum Chemistry as a facilitator of computations and a provider of new, chemical insights. The importance of the latter branch of DFT, conceptual DFT is highlighted following Parr's dictum "to calculate a molecule is not to understand it". An overview is given of the most important reactivity descriptors and the principles they are couched in.Examples are given on the evolution of the structure-property-wave function triangle which can be considered as the central paradigm of molecular quantum chemistry to (for many purposes) a structure-property-density triangle. Both kinetic as well as thermodynamic aspects can be included when further linking reactivity to the property vertex. In the field of organic chemistry, the ab initio calculation of functional group properties and their use in studies on acidity and basicity is discussed together with the use of DFT descriptors to study the kinetics of S_N2 reactions and the regioselectivity in Diels Alder reactions. Similarity in reactivity is illustrated via a study on peptide isosteres. In the field of inorganic chemistry non empirical studies of adsorption of small molecules in zeolite cages are discussed providing Henry constants and separation constants, the latter in remarkable good agreement with experiments. Possible refinements in a conceptual DFT context are presented. Finally an example from biochemistry is discussed: the influence of point mutations on the catalytic activity of subtilisin.

QUANTUM MECHANICS, QUANTUM CHEMISTRY, COMPUTATIONAL CHEMISTRY, DENSITY FUNCTIONAL THEORY: WHO IS WHO

From Quantum Mechanics to Quantum Chemistry and Computational Chemistry

The failure of classical physics (mechanics and electromagnetism) at the end of the 19th century led to the introduction of the Quantum Concept by Planck, Einstein, Bohr,... culminating in the birth of "modern" quantum mechanics around 1925 due to the work of Schrödinger, Heisenberg, Born, ... Schrödinger's equation occupied a central position in this new theory and, although later on complemented by its relativistic analogue by Dirac, stood the test of time and has been for now 75 years the central equation for the description both of the internal structure of atoms and molecules and their interactions. In his famous quote Dirac already in 1929 went so far to state [1] "The underlying physical laws necessary for the mathematical theory of a large part of physics and the whole of chemistry are thus completely known, and the difficulty is only that the exact application of these laws leads to equations too complicate to be soluble."

The step from Quantum Mechanics to Quantum chemistry can in principle be situated in the pioneering work by Heitler and London [2] on the hydrogen molecule in 1928 providing insight into, to quote Pauling, the Nature of the Chemical Bond [3]. However Quantum Chemistry is, at least in our opinion, more than the mere application of quantum mechanical principles to molecules and their interaction. In the years between 1930 and 1950 Pauling [3], Huckel [4], Coulson [5] indeed used quantum mechanical principles but combined them with their chemical intuition thereby gradually creating a new discipline, nowadays called Quantum Chemistry. The Valence Bond approach (after Heitler and London) was prominent in those days, to the detriment of Hund's and Mulliken's MO method [6]. A revolution was provoked by Roothaan's matrix formulation of the MO method in 1951 [7]. Its elegance together with the increasing computer power paved the way for the large scale introduction of the MO-LCAO method within the framework of the Hartree Fock Self Consistent Field (SCF) approach [8a] as excellently summarized in Pople's comprehensive treatise [8b].

The late seventies, eighties and nineties saw the development and/or the adaptation for systematic use of beyond SCF methods including (part of) electron correlation: MØller Plesset Perturbation Theory [9], the method of Configuration Interaction [10] and various types of Coupled Cluster Theory

[11]. The introduction of initially freely distributed, later on commercially available, computer programs which became more and more user friendly (cf Pople's GAUSSIAN series) [12] definitely promoted Quantum Chemistry, from a branch of Theoretical Chemistry almost exclusively reserved for "pure-sang" theoreticians and concentrating on diatomic and small polyatomic molecules, to a field also creating tools for non-specialists, in many other subfields of Chemistry (Inorganic, Organic, Biochemistry). The new subfield "Computational Chemistry" with P. Schleyer as a prominent figure [13] particularly stresses the "applied" aspects of Quantum Chemistry.

The combination of conceptual and methodological improvements, and the growing performance of soft- and hardware led to an ever increasing accuracy in the treatment of problems of a given complexity. Orders of magnitude in the complexity of problems that could be treated at a given level were gained. These evolutions are beautifully illustrated in Pople's two dimensional chart of Quantum Chemistry, given below in Figure 1 in a slightly adapted version [14].

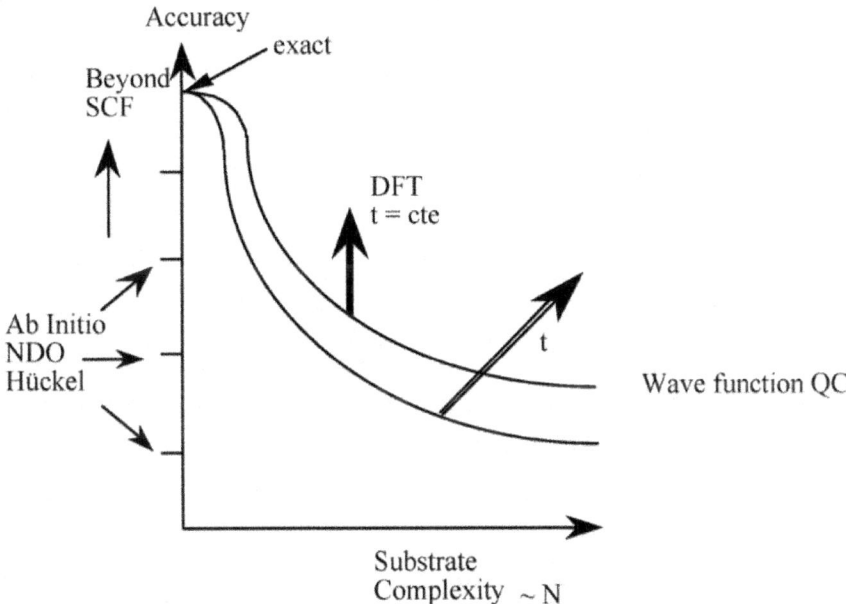

Figure 1: Accuracy versus complexity chart of Quantum Chemistry (After J.A.Pople)

A central paradigm in Quantum Chemistry is the Structure-Properties-Wave function triangle(Fig.2), Structure and Properties further determining reactivity. The third corner of the triangle is the ground state wave-function Ψ_0, or more generally also all excited state wavefunctions determining all properties of the system.

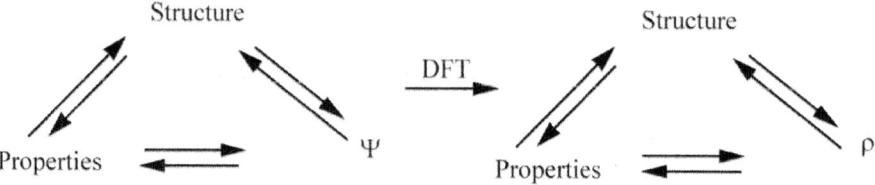

Figure 2: The central paradigm in Quantum Chemistry and its evolution upon the introduction of DFT

Density Functional Theory revolutionarized Quantum Chemistry from a computational point of view

A step of immense importance has been taken by Kohn and Hohenberg [15] in 1964. They proved that the information content in the electron density function $\rho(r)$, depending on only 3 variables, determines all ground state properties, thus replacing the crucial position of the complex wavefunction, Ψ, function of 4N variables (where N is the number of electrons).

Equation (1) clearly shows how much information in the wave function of a N-electron system is integrated out when passing to the electron density (x stands for a four vector containing three spatial coordinates r and one spin coordinates of an electron) [16]

$$\rho(r) = N \int \Psi^{*}(x,\ x_2,x_3\ ...,\ x_N)\,\Psi(x,\ x_2,\ x_3,\ ...,\ x_N)\,ds\,dx_2\,dx_3\,...\,dx_N$$

$$(1)$$

The problem of searching an optimal ρ instead of the much more complex optimal Ψ is most conveniently done within the Kohn-Sham formalism [17]) introducing orbitals φ_i, whose squares sum up to the electron density.

$$\rho = \sum_i |\varphi_i|^2$$

$$(2)$$

A variational procedure yields a pseudo-one electron equation, the analogue of the Hartree-Fock equations, which is written as

$$\left(-\frac{1}{2}\nabla_i^2 + v(r) + \int \frac{\rho(r)}{|r-r'|}\,dr + v_{xc}(r)\right)\varphi_i = \varepsilon_i\varphi_i$$

$$(3)$$

Here, besides the electronic kinetic energy term $(-\frac{1}{2}\nabla_i^2)$, the nuclear attraction term $v(r)$ and the classical electronic repulsion term, the exchange correlation term $\upsilon_{xc}(r)$ appears whose form is actually unknown. One of the key features of present day DFT is the search for the best performing exchange correlation functionals [18]. Although this task is hampered by the lack of a unifying principle as present in wave-function theory (see e.g. Pople's Model Chemistry chart) [8b] impressive progress has been made in recent years among others via the so called hybrid functionals [19] which gained widespread use. Extensive testing of their capability in reproducing molecular properties has been performed [20,21]. The whole of these efforts led to a methodology which affords the calculation of molecular ground state properties of high quality (in fact often way beyond SCF) at a much lower computational cost. Parr and Yang termed this branch of DFT "Computational DFT" [18]. A "témoignage par excellence" of the ever increasing importance of DFT is (the title of) Koch's book "A Chemist's Guide to DFT" [22] published in 2000 offering an overview of the performance of DFT for various properties to the practicing organic or inorganic chemist. As a result of this evolution the triangle is Fig. 2 can be adapted at one of its vertices, putting ρ (and its obtention via computational DFT) at equal footing with the wave-function Ψ.

DFT as a Provider of New Insights: Conceptual DFT

From Computational Chemistry to Chemical Insight

Both wave function Quantum Chemistry and Density Functional Theory, when being used to compute atomic and molecular properties, yield results which often and for most chemists are not always directly exploitable. The numbers they produce should in many cases be translated, or casted into a language or formalism pointing out their chemical relevance. As simply stated by Parr [23] "Accurate calculation is not synonymous with useful interpretation. To calculate a molecule is not to understand it". Quite often this translation involves terms going back to the early days of theoretical chemistry but still in use as a guideline for chemists in the interpretation of experimental data: hybridization, electronegativity, aromaticity,

A beautiful example in wave function quantum chemistry, dating from the sixties and seventies is the transformation of the Molecular Orbitals resulting from the Hartree-Fock equations, which are, usually delocalized over the entire molecule, to a set of localized Molecular Orbitals using a localization criterion [24]. The resulting MO picture is much closer to the Lewis picture of great use in organic chemistry (Figure 3), e.g. in the study of the electronic structure of bonds and its relation to spectroscopic properties. The relation between NMR coupling constants and the percentage of s character of the carbon atom hybrid involved in a CH bond [27], is a classical example, partly addressed in our own work [28]. Of importance in both branches (wave function and Density Functional Theory) is the visualization of the results (e.g. electron density or density difference plots (as in Figure 3), its extensive use also going hand in hand with hard- and software developments.

Also in this area of obtaining a better chemical insight via Quantum Chemical calculations, a prominent role was played in recent years by DFT.

DFT as a source of Chemical Concepts

As already mentioned before, computational DFT is founded on a variational principle, more precisely for the energy functional

$$E = E[\rho]$$

(4)

Looking for an optimal ρ, i.e. the one which minimizes E, is thereby subjected to the constraint that ρ should at all times integrate to N, the number of electrons.

$$\int \rho(\underline{r}) d\underline{r} = N$$

(5)

Within a variational calculation this constraint is introduced via the method of Lagrangian multipliers, yielding the variational condition

$$\delta[E - \mu\rho] = 0$$

(6)

where μ is the Lagrangian multiplier, a constant gaining its physical significance in the differential equation (the Euler equation) resulting from (6)

$$v(\underline{r}) + \frac{\delta F_{HK}}{\delta\rho} = \mu$$

(7)

Here $v(\underline{r})$ is the external potential (i.e. due to the nuclei) and FHK is the Hohenberg Kohn functional containing the electronic kinetic energy and the electron-electron interaction operators [29]. It has been Parr›s impressive contribution to identify this abstract Lagrange multiplier as [30]

$$\mu = \left(\frac{\partial E}{\partial N}\right)_v = -\chi$$

(8)

i.e. the derivative of the energy of the atom or molecule with respect to its number of electrons at constant external potential (i.e. identical nuclear charges and positions) (Figure 4). In this seminal paper, cited already more than 500 times, Parr thereby regained Iczkowski and Margrave's definition of electronegativity ($\chi=-\partial E/\partial N$) [31], Mulliken›s 1934 definition

$$\chi = \frac{1}{2}(IE + EA)$$

(9)

can be considered as an approximation to it [32]. The Mulliken values, the arithmetic average of ionization energy (IE) and electron affinity (EA), were already shown before to correlate with the Pauling values [33] and received more and more importance in recent years on the basis of its simpler foundation. It can easily be seen that they correspond to the average slope of the E=E(N) curve at the N value considered.

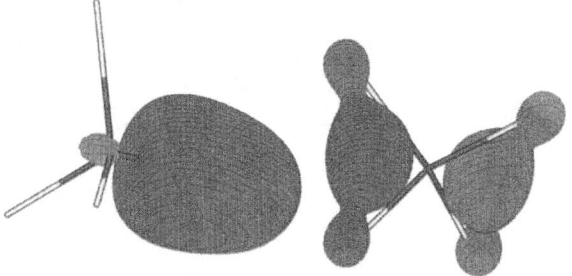

Figure 3: Delocalized Hartree Fock versus Localized Orbitals : one of the triply degenerate HOMO orbitals of methane versus a CH bond orbital. Electron density plot (Hartree Fock STO-3G/Boys localization procedure [25]) obtained with the software package [26]

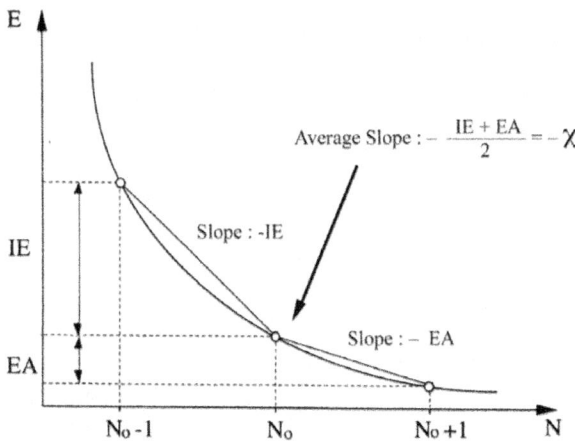

Figure 4: Atomic or molecular energy (E) versus number of electrons (N) at constant external potential : the modern definition of electronegativity

In analogy with the thermodynamic potential

$$\mu_{Therm} = \left(\frac{\partial G}{\partial n}\right)_{P,T}$$

(10)

where G represents the Gibbs Free Energy function and n the number of moles, μ was termed electronic chemical potential which turns out to be the negative of the electronegativity.

Within a few years after Parr›s contribution various other quantities representing the response of a system›s energy to perturbation in its number of electrons and/or its external potential (cf the index v in 8), which both lie at the heart of chemistry, were published. They can nicely be ordered according to Nalewajski›s charge sensitivity analysis [34] (Figure 5).

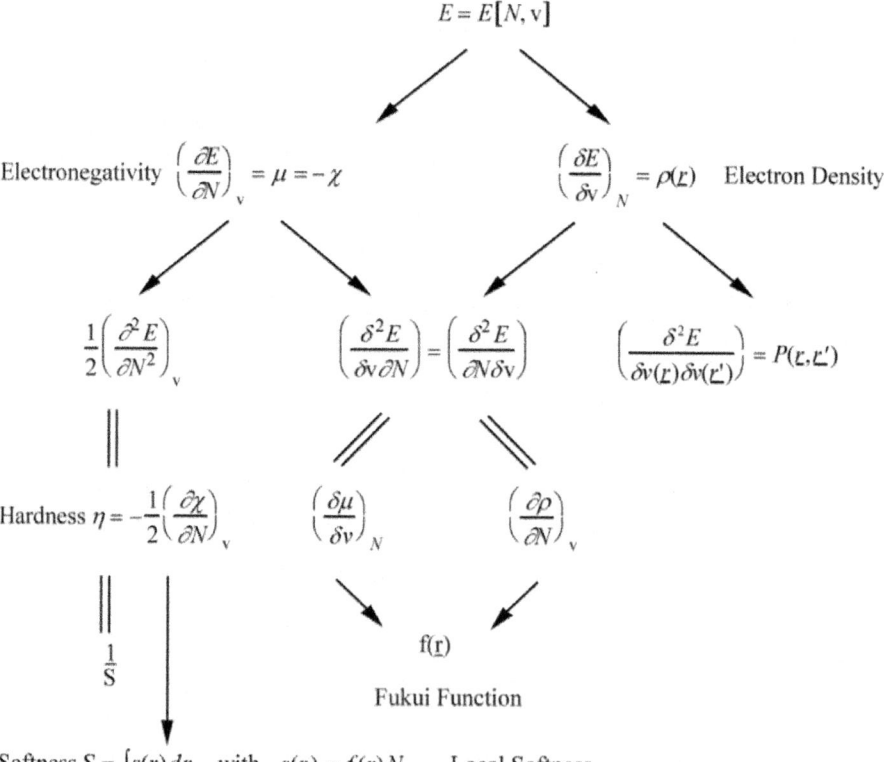

Figure 5: Nalewajski's Sensitivity Analysis: atomic and molecular properties as energy derivatives with respect to N and v.

Appearing in a natural way are the chemical hardness η, an identification proposed by Parr and Pearson [35] for the second derivative which respect to N, $\left(\frac{\partial^2 E}{\partial N^2}\right)_v$ and representing the resistance of a system to changes in its number of electrons. The chemical softness S is naturally defined as the inverse of η

$$S = 1/2\, \eta \qquad (11)$$

The analogue of equation (9) turned out to be

$$\eta = \frac{1}{2}(IE - EA) \qquad (9')$$

the Fukui function f(r) [36], representing the change in electron density ρ at a given point r when the total number of electrons is changed, a generalization of Fukui's frontier orbital concept [37]

a local version of S, s(r), obtained by multiplying S and f(r), the latter function distributing the local softness over various domains in space [38]

(Other concepts introduced in this framework are reviewed in [39,40])

In this way it is shown that DFT gave the possibility to sharply define concepts known for a long time in chemistry, but to which inadequate precision could be given to use them with confidence in quantitative studies.

The last 15 years showed growing importance of this branch of DFT, conceptual DFT, where these concepts were used as such or within the context of three important principles,

Sanderson's electronegativity equalization principle [41,42] stating that upon molecule formation, atoms (or more general arbitrary portions of space of the reactants) with initially different electronegat $\chi_i^0 (i = 1,...., M)$ combine in such a way that their "atoms-in-molecule" electronegativities are equal. The corresponding value is termed the molecular electronegativity χM. Symbolically:

$$\chi_1^0,\ \chi_2^0,\ ....\ \chi_M^0 \qquad \longrightarrow \qquad \chi_1 = \chi_2 = ... = \chi_M$$

isolated atoms   molecule formation

Electron transfer thereby takes place from atoms with lower electronegativity to those with higher electronegativity, the latter reducing their χ value, the former increasing it (Fig.6)

Pearson's Hard and Soft Acids and Bases Principle (HSAB) [43,44] stating that Hard (Soft) Acids (electron pair acceptors), preferentially interact with hard (soft) Bases (electron pair donors).

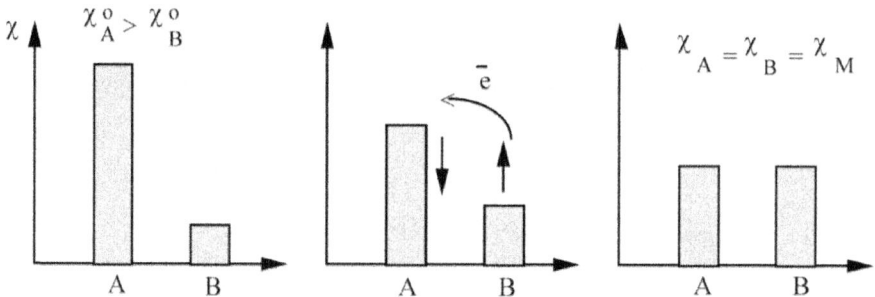

Figure 6: Sanderson's Electronegativity Equalization Principle

Both principles were proven [30,45] as was also the third one, the Maximum Hardness Principle, stating "that molecules try to arrange themselves to be as hard as possible" [44,46].

In recent years our group was active in the development and/or use of DFT based concepts as such or within the context of the afore mentioned and other principles. Also performance testing was one of our objectives : setting standards for computational DFT in order that it can be used for a given type of problem with the same level of confidence as the combination of level and basis set in the case of Pople's model Chemistry for wave function theories. Studies were undertaken on IR frequencies and intensities, dipole and quadrupole moments, ionization energies and electron affinities and Molecular Electrostatic Potentials [20,47,48,49].

THE STRUCTURE -PROPERTY (REACTIVITY)- ELECTRON DENSITY TRIANGLE: SOME EXAMPLES

Introduction

In this second part of the contribution examples are given on the role DFT studies, both conceptual and computational, can play in exploiting the structure, property -density triangle in Figure 2 where we concentrate on properties directly related to reactivity (both seen from a thermodynamic and kinetic point of view). Examples will be taken essentially from our own work with reference to work of other groups if relevant to the discussion. As such this part is not aiming at completeness at all. The reader should consult other sources to have a complete overview of applications of conceptual DFT [39,40]. Illustrations will be taken from organic, inorganic and biochemistry.

Organic Chemistry

Group Properties and Their Use in Acidity and Basicity Studies

Functional groups are playing a fundamental role in rationalizing structure and reactivity, thus dictating transformations in synthetic chemistry [50], both in organic and inorganic chemistry. An insight in the properties of these molecular building blocks is of utmost importance in the design of a rational chemistry.

Whereas group electronegativity has already a longstanding history [33], the field of group softness and/or hardness is much less developed. Moreover a non-empirical uniform computational scheme obeying the working equations (9) and (9') was lacking in the period we started this work.

We therefore presented a non-empirical computational scheme for group electronegativity, hardness and softness [51] for more than 30 functional groups

CH_3 ; CH_2CH_3; $CH=CH_2$; $C\equiv CH$; CHO ; $COCH_3$; $COOH$; $COCl$; $COOCH_3$; $CONH_2$; $C\equiv N$; $NH2$; CH_2-NH_2 ; NO_2 ; OH ; CH_2OH ; OCH_3 ; F ; CH_2F ; CHF_2 ; CF_3 ; SiH_3 ; PH_2 ; SH ; CH_2SH ; SCH_3 ; Cl ; CH_2Cl ; $CHCl_2$; CCl_3.

Starting from the geometry the group usually adopts when being embedded in the molecule we calculated its η, S and χ values via (9) and (9›), considered as a radical, both at the Hartree Fock and CISD level using Pople›s 6-31++G** basis [8].Figure 7 shows the correlation of the CISD calculated group electronegativities (showing a correlation coefficient r of 0.943 with the HF values for the same basis) with what is recently [52] considered as the most appropriate «experimental scale», the ^{13}C ^1JCC (ipso ortho) coupling constants in monosubstituted benzenes [53]. As can be seen the correlation fails for OCH_3 and SiH_3 for reasons that could not be detected. It is less convincing for groups containing triple bonds clearly to the higher demands for correctly describing electron correlation effects. Dropping these values a correlation coefficient r of 0.941 is obtained for the remaining groups.

Typical trends to be observed are

-the central atom effect

$$\chi_{CH3} < \chi_{NH} < \chi_{OH} \quad cf. \; \chi_C < \chi_N < \chi_O$$

indicating that upon saturation of two different atoms with hydrogens the electronegativity of the resulting groups parallels that of the naked atoms.

-the second row effect

$$\chi_{CH3} > \chi_{SiH3}$$

$$\chi_{NH2} > \chi_{PH2}$$

$$\chi_{OH} > \chi_{SH}$$

showing increasing electronegativity of a group the higher the central atom is positioned in a given column of the periodic table.

-the hybridisation effect

$$\chi_{CH_2}CH_3 < \chi_{CH=CH2} < \chi_{CH+CH}$$

$$\chi_{CH_2}NH_2 < \chi_{CH=NH2} < \chi_{C+N}$$

indicating increasing electronegativity upon increasing s-character of the central atom of the group.

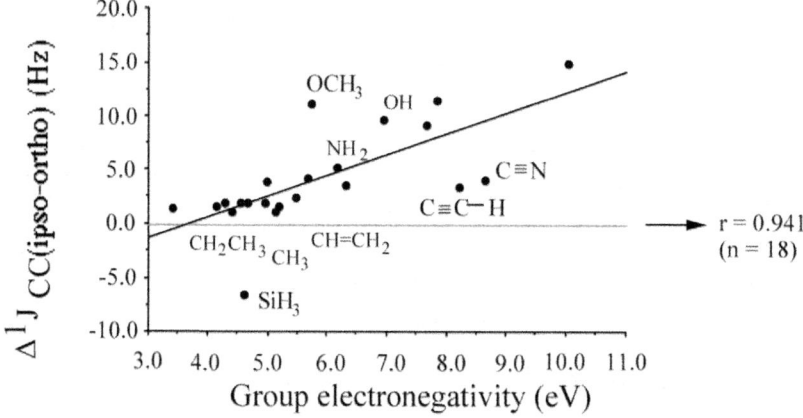

Figure 7: Calculated vs. Experimental Group Electronegativity values.

For alkylgroups the intuitively expected decrease of χ upon increasing chain length or branching is found : Me : 5.12 ; Et : 4.42 ; n-Pr : 4.39 ; i-Pr : 3.86 (CISD values).

Turning now to the hardness values, experimental scales are scarce, the best candidate being the corresponding radical hardness [54], although possible differences in geometry (e.g. for the CH3 radical) indicate that this correlation should be looked upon with much reserve. Figure 8 shows a correlation, upon withdrawing the extremely hard CF3 group as an outlier, of 0.926 for the 14 remaining cases for which experimental values are available. Typical trends which can be discussed are again (vide supra)

-the central atom effect

$$\chi_{CH_3} < \chi_{NH_2} > \eta_{OH} < \eta_F \quad cf. \quad \eta_C < \eta_N > \eta_O < \eta_F$$

-the second row effect (extremely important in forthcoming discussions but here already illustrating the opposite behaviour of χ and S)

$$\eta_{OH} \; > \; \eta_{SH}$$

$$\eta_{CCl_3} \; < \; \eta_{CH_3} \; < \; \eta_{CF_3}$$

$$\eta_{CH_3} \; > \; \eta_{SiH_3}$$

-the "volume" effect in alkyl groups, showing decreasing hardness (increasing softness) upon increasing chain length or branching. CISD η values illustrating this trend are

6.60 (H) ; 5.34 (Me) ; 4.96 (Et) ; 4.70 (n-Pr) ; 4.62 (i-Pr)

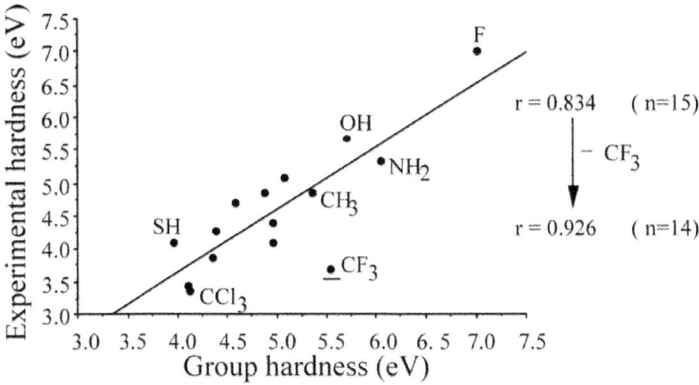

Figure 8: Calculated vs. Experimental Group Hardness values (see text).

As an example of the use of group properties we consider [55] the experimental acidity sequence of alkylalcohols showing an opposite behaviour in aqueous solution and gas phase [56] [57]. Whereas in aqueous solution the acidity decreases upon increasing carbon chain length and branching, the opposite behaviour is encountered in the gas phase. The former trend is usually traced back to the electron donating character of alkylgroups (the +I effect as well known in electrophilic aromatic substitutions on benzene [58]), whereas the latter tendency should imply an at first sight unexpected electron withdrawing character of alkyl groups. We correlated the (gas phase) acidity quantified by the ΔG°_{acid} value with the calculated alkylgroup properties χ and η (all calculations were done at a uniform 6-31G* MP4 level). The sequences parallel those of the 6-31++G** CISD values in [51].

The values shown in Table 1 indicate the expected tendencies of increasing softness upon increasing chain length and branching accompanied by decreasing electronegativity, reflecting the traditional idea of an alkylgroup as electron donor (+I effect). It was tempting to have a more detailed look at the correlation between charge distribution, both in the acidic form ROH and the alkoxide anion RO-, and the acidity variation upon varying R. We therefore made use of χ and η values given in Table 1 with an electronegativity equalization scheme.

Working at functional group resolution and neglecting external perturbation higher order terms one obtains that upon embedding a functional group A with intrinsic χ_A^0 and η_A^0 into a molecule its electronegativity changes to χ_A with

$$\chi_A = \chi_A^0 - 2\eta_A^0 \Delta N_A \tag{11}$$

Table 1: 6-31G* MP4 alkylgroup properties: electronegativity χ (eV) and hardness η (eV)

	χ	η
Me	4.35	5.98
Et	3.76	5.45
n-Pr	3.68	5.17
i-Pr	3.44	5.04
n-Bu	3.57	5.05
t-Bu	3.30	4.19

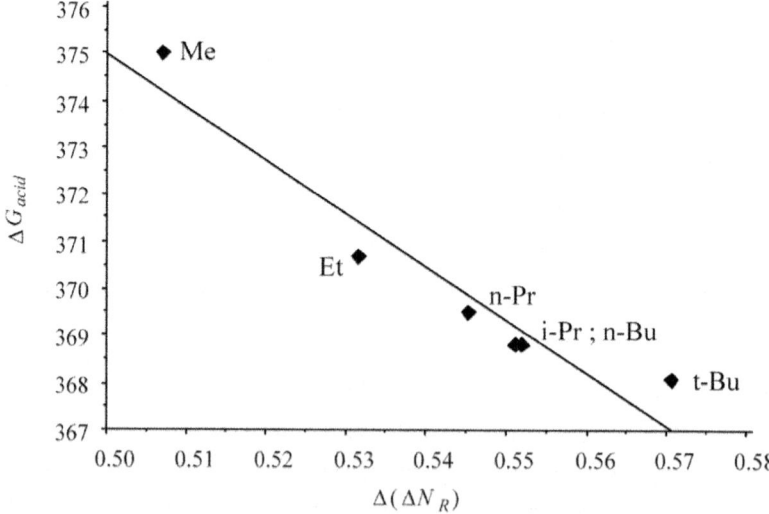

Figure 9: Experimental gas phase acidity of alkylalcohols (in kcal mol-1) vs. $\Delta(\Delta N_R)$ (see text) (reprinted with permission by Pergamon/Elsevier-Reference [55]).

This equation is easily derived using a Taylor series expansion of the E = E[N,v] functional (cf. Fig.5) around a reference number of electrons and at constant external potential. Applying this relationship to both the acidic form of an alcohol ROH both for the R and OH groups and equalizing the electronegativities of both molecular building blocks, one gets for ΔN_{R_I} (the number of electrons transferred to the alkylgroup in the acidic form)

R:OH $$\Delta N_{R_I} = \frac{\chi_R^0 - \chi_{OH}^0}{2\left(\eta_R^0 + \eta_{OH}^0\right)} < 0 \qquad (12)$$

In the case of the conjugate base RO⁻ one obtains in a completely analogous way (taking into account again charge conservation, but now not at charge 0 but -1)

R:O⁻ $$\Delta N_{R_{II}} = \frac{\chi_R^0 - \chi_O^0}{2\left(\eta_R^0 + \eta_O^0\right)} + \frac{\eta_O^0}{\eta_R^0 + \eta_O^0}$$

$\qquad\qquad\qquad\qquad$ (a) $\qquad\qquad$ (b) $\qquad\qquad\qquad\qquad$ (13)

In (12) ΔN_{R_I} is negative as χ_{OH} is higher than χ_R^0 in the case of an alkylgroup. The intramolecular charge transfer is electronegativity dominated, as far as its direction is concerned, hardness playing a role in the amount of charge transferred. In the conjugate base form eqn.(13) indicates that a second, completely hardness-determined term, shows up which turns out to be larger in absolute value than the first term in all cases considered. This hardness dominated term consequently accounts for an electron transfer towards the alkyl group, the value being more important when the softness of the alkylgroup increases. Equating $\eta 0$ and $\chi 0$ both for OH and O, the difference between ΔNZ in the case of an alkylgroup. The intramolecular charge transfer is electronegativity dominated, as far as its direction is concerned, hardness playing a role in the amount of charge transferred. In the conjugate base form eqn. (13) indicates that a second, completely hardness-determined term, shows up which turns out to be larger in absolute value than the first term in all cases considered. This hardness dominated term consequently accounts for an electron transfer towards the alkyl group, the value being more important when the softness of the alkylgroup increases.

Equating η^0 and χ^0 both for OH and O, the difference between $\Delta N_{R_{II}}$ and ΔN_{R_I}, $\Delta(\Delta N_R)$, can be approximately written as

$$\Delta\left(\Delta N_R\right)= \Delta N_{R_{II}} - \Delta N_{R_I} \approx \frac{\eta_O^0}{\eta_R^0 + \eta_O^0}$$

(14)

Indicating that the difference in electron transfer from R to the oxygen part of the alcohol upon deprotonation becomes more important when the alkylgroup is softer. Otherwise stated: $\Delta(\Delta N_R)$ is R-softness dominated. It may be therefore concluded that in order to correctly describe acid-base properties of molecules, the electronic properties of the charged form of the acid base equilibrium are of utmost importance. Figure 9 shows the excellent correlation between the experimental gas phase acidity and $\Delta(\Delta N_R)$ for the simplest alkylalcohols. In the present case alkylgroups act as electron acceptors; these electron withdrawing properties of alkylgroups were previously encountered in the literature in cases where alkylgroups are placed on negative ely charged carbon atoms or reaction centers such as in α-alkylbenzyl carbanions [59].

We finally turned to the effect of solvent. We therefore recalculated group electronegativity and softness values, using a Self Consistent Reaction Field Model [60], introducing the solvent dielectric constant ε [61]. In Figure 10 we plot the $\Delta(\Delta N)$ quantity as a function of the Kirkwood function ε−12ε+1 related to the free energy of solvatation.

It is seen that an approximate linear relationship between $\Delta(\Delta N)$ and the Kirkwood function is found for the simplest alkylalcohols, with a crossing of the curves near the H_2O case. The gas phase acidity sequence is thereby inverted when passing to aqueous solution as found experimentally [62].

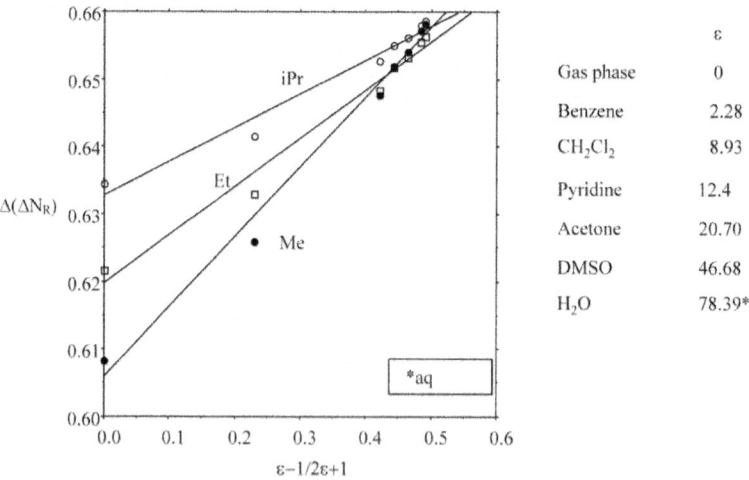

	ε
Gas phase	0
Benzene	2.28
CH_2Cl_2	8.93
Pyridine	12.4
Acetone	20.70
DMSO	46.68
H_2O	78.39*

Figure 10: Plot of $\Delta(\Delta N_R)$ versus the Kirkwood function $\frac{\varepsilon-1}{2\varepsilon+1}$ (Reprinted with permission by the American Chemical Society - Reference [61]).

Kinetics of S_N2 reactions

The nucleophilic substitution reaction is one of the basic transformations in organic chemistry. It has been extensively studied both theoretically and experimentally, both in the gas phase and in solution [63].

In the case of gas phase reactions e.g. $X^- + CH_3Y \rightarrow Y^- + CH_3X$ a double well profile is present in the Reaction Profile (Fig. 11)

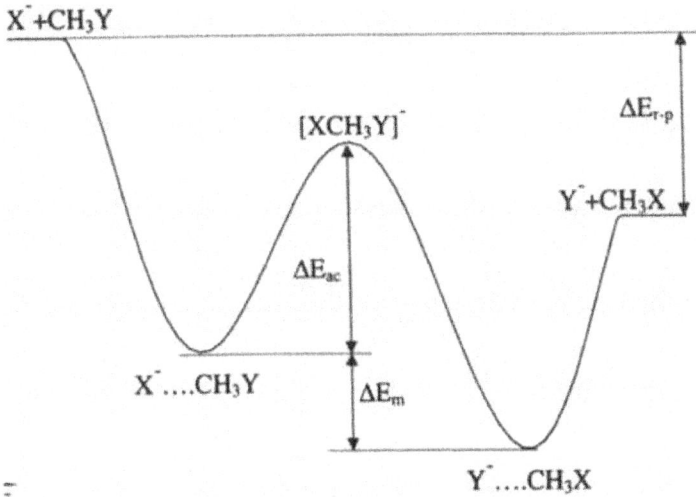

Figure 11: Reaction Profile for a S_N2 reaction in the gas phase (Reprinted with permission by the American Chemical Society - Reference [64]).

We tried to interpret the reaction pathway via DFT based concepts (η, S ...) and principles (HSAB, MHP), using a variety of X, Y combinations [64]. 6-31+G* optimized structures were generated for reactants, products, transition states (TS) and ion-molecule complexes. A remarkable correlation is found between ΔE_m both with the calculated ΔE_{p-r} ($r^2=0.97$) and the experimental heat of reaction ($r^2=0.99$) indicating the relevance of ΔE_m. This quantity shows, e.g. in the case of the $X^- + CH_3F \rightarrow CH_3X + F^-$ reaction, a remarkable correlation with the difference in hardness between F and the group X (Figure 12). The hard F atom in CH_3F can be expected also to harden its neighborhood, *in casu* the carbon atom thereby favoring an interaction with a harder X- according to the HSAB principle. One thereby regains Pearson's statement about S_N2 reactions that "when nucleophile and leaving group have similar hardnesses, reaction rates are relatively high".

Gazquez [65] developed a formalism to relate the reaction energy and the activation energy to differences in hardness between reagents, products and TS. He showed that ΔE_{p-r} should be proportional to $\frac{1}{\Sigma S_r} - \frac{1}{\Sigma S_p}$ where ΣSr and ΣSp denote the softness sum for reactants and products, respectively

$$\Delta E_{p-r} \sim \frac{1}{\Sigma S_r} - \frac{1}{\Sigma S_p}$$

(15)

It is clearly seen that an exothermic reaction yields

$$\Sigma S_r > \Sigma S_p$$

(16)

i.e. products which are harder than reagents, thus recovering the MHP. It should however be remarked that the stringent conditions to be satisfied in the proof of the MHP (constancy of both the external and chemical potential) put severe restrictions on situations in which the principle is applied. For a recent detailed discussion we refer to Chandra and Uchimaru [67]. When applying (15) to the ion-reagent and product complexes one expects a relationship

$$\Delta E_m \approx \frac{1}{S_{\substack{prod \\ complex}}} - \frac{1}{S_{\substack{reag \\ complex}}} \sim \frac{1}{\alpha_{\substack{prod \\ complex}}} - \frac{1}{\alpha_{\substack{reag \\ complex}}} \approx \Delta E_{r,\alpha}$$

(17)

where the proportionality between softness and polarizability (α) [66] has been exploited. In Fig. 11 we observe that $\Delta E_{r,\alpha}$ is always negative (evolution from a complex with lower hardness to one with higher hardness) and that ΔE_m correlates with $\Delta E_{r,\alpha}$ (separate correlations for hard and soft R groups have been drawn).

Turning now to kinetic aspects the central barrier can be regarded as providing the activation energy ΔE_{ac}. Adopting a similar ansatz as before in eq. (17) we looked for two correlations between ΔE_{ac} and the difference $\frac{1}{\alpha} - \frac{1}{\alpha_{TS}}$. Again two correlations can be drawn, one for soft groups, one for hard groups as seen in Fig. 14.

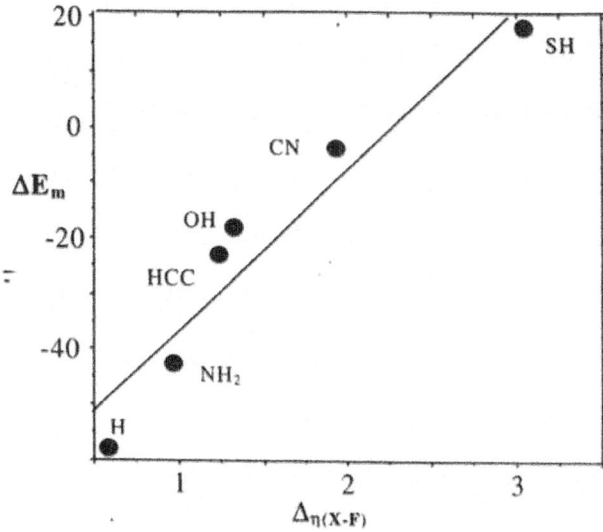

Figure 12: Calculated reaction energy ΔE_m (in kcal mol^{-1}) as a function of the group hardness difference between the X group and fluorine, $\Delta\eta_{X-F}$ (Reprinted by permission by the American Chemical Society - Reference [64]).

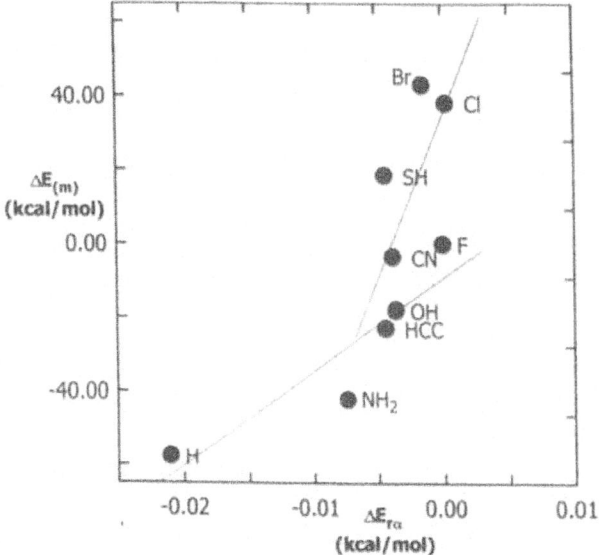

Figure 13: Calculated reaction energy ΔE_m (in kcal mol^{-1}) as a function of the reaction energy $\Delta E_{r,\alpha}$ (in a.u.) obtained via the Gazquez approach. Separate linear correlations for hard (X = H, NH$_2$, OH, F) and soft (X = HCC, CN, SH, Cl) nucleophiles are shown (Reprinted by permission by the American Chemical Society - Reference [64]).

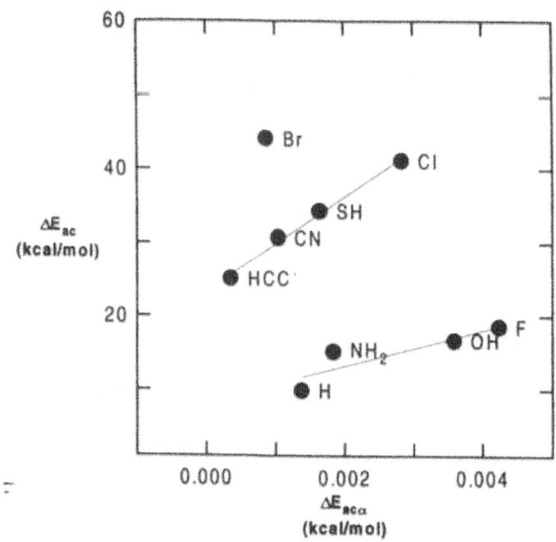

Figure 14: Calculated values of the central barrier energies ΔE_{ac} (kcal mol^{-1}) as a function of the central barrier $\Delta E_{ac,\alpha}$ value (in a.u.) obtained via the Gazquez approach. Separate linear correlations for hard and soft nucleophiles are shown as inFigure 12. (Reprinted by permission by the American Chemical Society - Reference [64]).

These recent results show that the use of DFT based concepts such as softness might shed further light on the reactivity within the context of both the HSAB and MHP principles.

In the next paragraph the HSAB principle is further exploited in the study of regioselectivity in organic reactions.

Regioselectivity in organic reactions: the HSAB principle

As stated above the HSAB principle has been given theoretical support by Parr and Pearson, later on by Gázquez. Based on the global properties of the reacting systems (acid (A) and base (B)) with global softness values S_A and S_B it was shown, based among others on the idea of equalization of electronic chemical potentials (cf. §1.3.2), that the largest interaction energy for constant S_A occurs when S_B equals S_A thereby regaining the HSAB principle [68] . If the interaction sites in A and B are specified (say k in A and l in B) it was shown by Gazquez and the present author [68] that the demand

$$S_A = S_B$$
(18a)

is converted into an analogous equation at local level , i.e.

$$s_{A,k} = s_{B,l}$$
(18b)

where $s_{A,k}$ and $s_{B,l}$ are the local softness values at sites k and l.

Obviously this expression is ready for use in the study of regioselectivity problems. As an example (for further applications see [69]) we briefly discuss the regioselectivity of Diels Alder reactions using the softness matching (18b) approach in at local level [68b].

The predominance of orthoregioisomers in the cycloaddition of 1-substituted dienes **1** and asymmetrical dienophiles **2** cannot be explained by electronic effects, as replacement of an electron donating substituent by an electron-attracting one does not alter the regioselectivity as discussed by Anh and coworkers [70].

In the double local-local approach of the HSAB principle we considered the local softness resemblance of the termini combinations 1 and 1', 4 and 2', as compared to the 1-2', 4-1' ones. The latter combination yields the meta product, the former the paracycloadduct, considered (cf sterical hindrance) as the contrathermodynamic reaction product.

In the cases R=-Me, -OMe, -COOH, -CN, -NH$_3$, -NMe$_2$, -NEt$_2$, and -OEt and R' = -COOH, -COOMe, -CN, -CHO, -NO$_2$ and -COMe (ensuring in most cases a Normal Electronic Demand reaction type, the dienophile being the electrophilic partner and the diene the nucleophilic one), we calculated local softness values s_1^-, s_1^+, s_2^+, and s_4^- from 3-21G optimized structures at the same level*.

In order to look for a simultaneous fulfillment of the local HSAB principle at both termini the following local softness similarly indicators were evaluated.

$$S_{ortho} = \left(s_1^- - s_{1'}^+\right)^2 + \left(s_4^- - s_{2'}^+\right)^2$$

$$S_{meta} = \left(s_1^- - s_{2'}^+\right)^2 + \left(s_4^- - s_{1'}^+\right)^2$$

$$(19)$$

Upon analysis of the Frontier Molecular Orbital-energies, the diene could always be identified as the electron donating system and the dienophile acting as electron acceptor.

It is seen that in the 8×6=48 cases studied, corresponding to all R and R' combinations, S_{ortho} is always smaller than S_{meta} except when a CN substituent is present either in the diene or the dienophile. However the DFT related reactivity parameter for the CN group was shown to be highly sensitive to correlation effects. Moreover the idea when writing down equations (19) is based on the hypothesis of a reaction with both couples of termini reacting at the same rate (synchronicity). Concertedness however is not a synonym to synchronicity prompting us to look for the smallest of the four quadratic forms in (19) which in almost all cases turns out to be the $\left(s_4^- - s_2^+\right)^2$ term.

This result points into the direction of the C_4-C_2' bond forming faster than the C_1-C_1' bond, in obvious agreement with the demand of equal softness of interacting termini as there are most remote from R and R'. The asynchronicity in the mechanism suggested on this basis is confirmed by Houk›s transition state calculations [71] where in all cases with R and R›≠H asymmetric transition states were found with the C_4-C_2' distance being invariably shorter than C_1-C_1'. This HSAB study performed at the local-local level provides an answer to Anh's long pending hypothesis [70] stating that "it is likely that the first bond would link the softest centers". Important work broadening the scope of this type of approach has been delivered in recent years among others by Nguyen and coworkers, partly in collaboration with the authors, including free radical additions to olefins [72], hydration of cumulenes [73], 2+1 cycloadditions between isocyanide and heteronuclear dipolarophiles [74], cycloaddition reactions between a 1,3-dipole and a dipolarophile [75] and excited state [2+2] photocycloaddition reactions of carbonyls [76]. On the other hand Ponti [77] presented a generalization of the ansatz via eqn (19).

Similarity in Reactivity

Similarity in a fundamental concept in chemistry and pharmacology. In the design of drugs for example one supposes that molecules with similar structures will also exhibit similar biological or physiological activities [78]. The rapid development of computational techniques in recent years has enabled a systematic investigation of the similarity concept suited for quantitative studies of molecular activity [79]. Several similarity indices have been proposed among which the Carbo Quantum Molecular Similarity Index Z_{AB} has played a prominent role [80]. In simplest form it can be written as

$$Z^\rho_{AB} = \frac{\int \rho_A(r)\rho_B(r)dr}{\left[\int \rho_A^2(r)dr \cdot \int \rho_B^2(r)dr\right]^{1/2}} \tag{20}$$

or introducing the shape factor $\sigma(r)$ [81]

$$\rho(r) = \sigma(r)N$$

$$Z^\rho_{AB} = \frac{\int \sigma_A(r)\sigma_B(r)dr}{\left[\int \sigma_A^2(r)dr \cdot \int \sigma_B^2(r)dr\right]^{1/2}} \tag{21}$$

indicating that the similarity index only depends on the shape of the density distribution and not on its extent. To obtain a more reactivity related similarity index we proposed to replace the electron density in eq. (21) by the local softness $s(r)$ (cf. § 1.3.2.) [82] Yielding the following expression

$$Z^s_{AB} = \frac{\int f_A(r)f_B(r)dr}{\left[\int f_A^2(r)dr \int f_B^2(r)dr\right]^{1/2}} \tag{22}$$

Where the local softness has been filtered out.

The index (22) has been tested via similarity analysis of peptide isosteres first using the Carbo and Hodgkin Richards indices [82a]. In the field of peptidomimetics analogues of the peptide bond [83] are looked for, showing however resistance to nucleophilic attack. Functional groups proposed in the vast literature in this domain are $CH=CH$, $CF-CH$, CH_2-CH_2, CH_2-S, $COCH_2$, CH_2-NH etc. We tested the similarity of a model system $CH_3-CO-NH-CH_3$ with its analogues on the basis of the similarity indices involving ρ (20) and s (22). DFT has been used here also as a computational tool using the B3PW91 exchange correlation potential [19a] [84] combined with a 6-311G* basis set [8b]. Table 2 shows that the sequences for Z^ρ and Z^s are not identical. The highest density similarity is observed for trans-butene, which is not unexpected in view of the trans structure of the peptide bond and the resulting matching of the terminal chains. However, turning to the softness similarity, the Z fluorinated alkene structure shows higher resemblance with the amide bond due to the similarity in polarity with the carboxyl group. The potential use of C=C-F as a peptidomimetric is in accordance with the results of [85].

Notice that the softness similarity has been calculated on the basis of an average of s+(r) and s-(r). The high similarity index thus represents a similarity in an average interaction of the model system, including terminating groups, with the surroundings, whereas (not shown) the similarity for a nucleophilic attack is very low (0.016 for the Z_{AB}^{s+} index) [82]. Summarizing, a combined search for optimal Z_{AB}^{ρ} (identical shape) - if necessary with a different type of index, such as Hodgkin's one [82] [86] including extent - and optimal Z_{AB}^{s} might be a valuable approach not only in this case but in rational drug design as a whole. In a more recent study [82b] one of the remaining problems in this approach, the orientational and translational dependence of Z, was avoided by introducing the concept of an autocorrelation function of the property considered.

Table 2: Similarity indices Z_{AB}^{ρ} and Z_{AB}^{s} for peptide isosteres of CH_3-CO-NH-CH_3 (B3PW91/ 6-311G* calculation). The centers of the central bond were brought to coincidence.

- CO-NH -	Z_{AB}^{ρ}	Z_{AB}^{s}
- CH=CH - (trans)	0.582	0.576
- CH=CH - (cis)	0.506	0.501
- CF=CH - (Z)	0.418	0.635
- CH_2-CH_2 -	0.466	0.495
- CH_2-S -	0.034	0.337
- CO-CH_2 -	0.531	0.108
- CH_2-NH -	0.412	0.559

Inorganic Chemistry: Zeolites

Zeolites are alumino-silicates consisting of SiO_4 and AlO_4 tetrahedra linked to each other by their corner oxygens. The alumino silicate structure is negatively charged due to the isomorphic subsitution of silicon by aluminum [87]. This negative charge is balanced by exchangeable cations.

When protons are introduced as counterions, the zeolite becomes a Brønsted acid, the protons being positioned on an oxygen atom connecting an aluminum and a silicon atom. These bridging hydroxyls [88] are at the origin of the acid catalysis application of zeolites [89]. Another aspect of the structure of zeolites is the occurrence of large vacant interconnected spaces forming long, wide channels of varying size depending on the type considered allowing the crystal to act as a molecular sieve [90].

For many years [91] the concepts of electronegativity, softness, hardness .... were exploited by us in the study of the acidity of bridging hydroxyl in zeolites, of utmost importance in their catalytic properties (for a review see [92]).

Due to space limitations and because some fundamental aspects of this kind of study are already present in § 2.2.1. on the acidity of alcohols, we concentrate in this paragraph on a recently aborded topic especially highlighting the possibilities of Computational Chemistry, switching at the end however again to Conceptual DFT : adsorption in zeolites.

The problem we recently addressed [93,94,95] was the selectivity of adsorption of gases [96] for which up to now no parameter free ab initio quantum chemical studies were performed yet [90]. Below we summarize the results for the interaction of small molecules (such as N_2, O_2, CO), with a NaY faujasite type zeolite, more specifically with the α cage.

At sufficiently low pressure adsorption is governed by Henry's law [96]

$$q=Kp \qquad (23)$$

where q is the amount of substance per unit volume in the adsorbed phase and p the pressure. Henry's constant K can be written as [97]

$$K = \frac{BI}{aRT} \qquad (24)$$

where B is the number of cavities in which adsorption can take place per unit mass of zeolite. The quantity "a" equals 1 for a monoatomic gas, 4π for a linear molecule and $8\pi^2$ for a non linear molecule. The configuration integral I is defined as

$$I = \int e^{-E(\underline{r},\underline{\Phi})/RT} d\underline{r}d\underline{\Phi} \qquad (25)$$

where E represents the interaction energy of the adsorbing molecule with the zeolite cage at position r and with orientation $\underline{\Phi}$. The integration is performed over the volume V of the supercage.

The cluster representing the supercage and its nearest environment was chosen to be "as large as possible" (within computational limits) and with an Si/Al ratio of 3. The 232 atoms cluster, involving OH groups as terminators is shown in Figure 15, where also four cationic adsorption sites of type II, the ones "active" in the conditions that we consider are present, one of them being indicated.

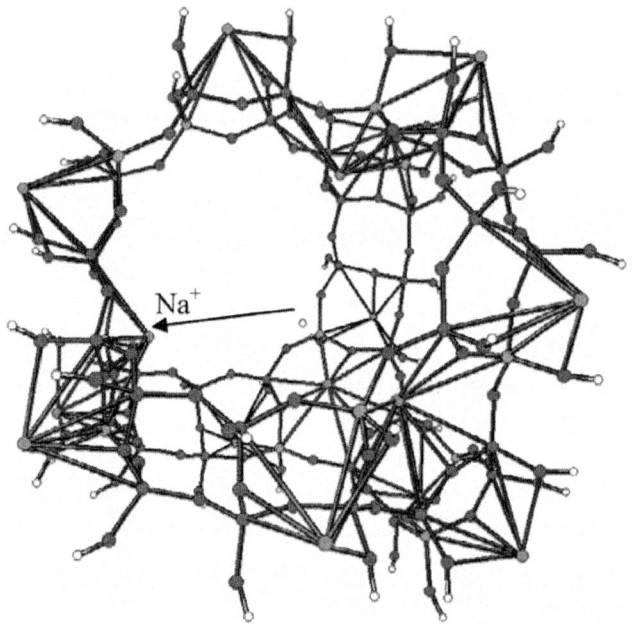

Figure 15: Cluster used to model the large cage of faujasite Y (Reprinted with permission by J. Wiley - Reference [94])

An embedded cluster method was then adopted, considering the molecule in the field generated by the cage atom charges, the latter being obtained in a single run at Hartree Fock STO-3G or 3-21G level. The points at which the interaction energy is evaluated are selected by constructing a cubic grid with a grid distance of 0.50 Å in each direction and considering the center of each cube generated in this way. The expression (25) is thereby approximated as

$$I \approx \frac{4\pi}{3}\sum_i\left(e^{-E_{x,i}/RT} + e^{-E_{y,i}/RT} + e^{-E_{z,i}/RT}\right)\Delta V_i$$

(26)

where the three terms account for three mutually perpendicular orientations of the molecule, each one given an equal weight to simplify the orientational integration $d\Phi$. ΔV_i equals 0.125 Å^3.

A supplementary condition was introduced imposing a minimal distance between an atom of the adsorbing molecule and a cage atom (cf. the treatment of the repulsive term in the potential, for details see [94]). At these small distances E becomes positive and its contribution to I, via the exponential function, very small. This procedure leads to the consideration of not less than ± 3000 cage-points necessitating an in-depth search for an optimal quality/cost ratio in the procedure for the interaction energy evaluation. The optimal procedure, from a quality/cost ratio point of view, turned out the following one [94,95,98]. The starting point was the energy expansion of a molecule in a non-uniform electric field [99].

$$E(F_\alpha, F_{\alpha\beta}, F_{\alpha\beta\gamma}, \dots) = E - \mu_\alpha F_\alpha - \frac{1}{3}\Theta_{\alpha\beta}F_{\alpha\beta} - \frac{1}{15}\Omega_{\alpha\beta\gamma}F_{\alpha\beta\gamma}$$
$$- \frac{1}{105}\Phi_{\alpha\beta\gamma\delta}F_{\alpha\beta\gamma\delta} - \frac{1}{2}\alpha_\alpha F_\alpha F_\beta - \frac{1}{6}\beta_{\alpha\beta\gamma}F_\alpha F_\beta F_\gamma \tag{27}$$

where F_α, $F_{\alpha\beta}$, ...represent the field, field gradient, ... components whereas μ_α, $\Theta_{\alpha\beta}$, $\Omega_{\alpha\beta\gamma}$, ... stand for the dipole, quadrupole, octadecapole, ... components and α_α, $\beta_{\alpha\beta}$, ... are the dipole polarizability and the first hyperpolarizability. In the case of a neutral molecule of the $C_{\infty,v}$ type (CO) the expression simplifies to (the molecular axis is taken to be the z-axis)

$$E(q, R, \theta) = E^0 + \sum_i \mu_z q_i R_i^{-2}\cos\theta_i + \sum_i \Theta_{zz}q_i R_i^{-3}(3\cos^2\theta_i - 1)/2$$
$$+ \sum_i \Omega_{zzz}q_i R_i^{-4}(5\cos^3\theta_i - 3\cos\theta_i)/2$$
$$+ \sum_i \Phi_{zzzz}q_i R_i^{-5}(35\cos^4\theta_i - 30\cos^2\theta_i + 3)/8$$
$$- \sum_i E_i^2(\alpha_{zz}\cos^2\theta_i + \alpha_{xx}\sin^2\theta_i)/2 \tag{28}$$

F_i represents the electric field at the origin of the molecule due to the surrounding point charges. In $D_{\infty,h}$ cases (N_2, O_2, CO_2, ...) terms in μ and Ω vanish.

The quality of this approach in dependent on the number of terms retained in the expressions (27) and (28) and the quality of the multi-pole moments, polarizabilities ... These were calculated with DFT methodology (B3LYP functional [19]) combined with Dunning's extremely large basis sets (the augmented -correlation consistent and polarized - valence - quadruple or quintuple basis sets (AVQZ or AV5Z) [100][101]).

This methodology was proved by us to yield multipole moments which are in excellent agreement with experiment [98] [102].

In Table 3 the main results are summarized for N_2, O_2, CO, CO_2, C_2H_2. Besides Henry constants, also separation constants α (ratio of two Henry constants) and the isosteric heats of adsorption are tabulated. The latter values were obtained via the Van't Hoff equation

$$\frac{\partial \ln K}{\partial \left(\frac{1}{T}\right)} = -\frac{\Delta H^\circ}{R}$$

(29)

for which K was calculated in the temperature interval between 260 and 340K with an increment of 10K and using a linear regression.

Table 3: Henry constants, separation constants and isosteric heats of adsorption on NaY : comparison between theory and experiment [94][98].

	N_2	O_2	CO	CO_2	C_2H_2
K	3.320	1.830	9.951	63.33	196.8
K_{exp}	31.4	15.4	85		
ΔH°	-12.9	-7.9	-18	-27	-29
ΔH_{exp}	-14	-9.4	-20	-37	

	N_2/O_2	CO/N_2	CO_2/CO
α	1.81	2.99	6.36
α_{exp}	2.04	2.70	

As a whole these results yield K values which are systematically one order of magnitude too small but showing correct sequences. The order of magnitude should be considered within the correct context : a uniform underestimation of the interaction energy of only 1.4 kcal mol[-1] already yields an order of magnitude underestimation of K. This extremely high sensitivity is of course due to the sensitivity of the potential in the repulsion part, plugged into an exponential in the configuration integral. The ratio K_{exp}/K is almost constant with values of 9.45, 8.41, 8.54 respectively. As a consequence the separation constants are in excellent agreement with experiment. The isosteric heats of adsorption also show very good agreement in absolute value and reproduce the experimental sequence. In our opinion these investigations pave the way for future studies involving more complex molecules where the lines drawn here, together with increasing computer hard and software and DFT based methodology, may finally yield to calculations which will be of great use in the design of zeolites for well defined purposes (e.g. gas separation).

An alternative for the interaction energy evaluation including electrostatic and polarization effects was proposed in a conceptual DFT context. Using a DFT perturbational approach we obtained the following expression for the interaction energy

$$\Delta E = \sum_i q_i V(\underline{R}_i) + S \sum_i \sum_j q_i q_j \left(\int \frac{f(\underline{r})}{|\underline{r} - \underline{R}_i|} d\underline{r} \int \frac{f(\underline{r}')}{|\underline{r}' - \underline{R}_j|} d\underline{r}' \right) - \int \frac{f(\underline{r})}{|\underline{r} - \underline{R}_i||\underline{r} - \underline{R}_j|} d\underline{r}$$

(30)

The first term corresponds to the electrostatic interaction V (R$_i$) as given by the molecular electrostatic potential at position R$_i$where charge q$_i$ is located, multiplied by this charge. The second term is the polarization term [103] in which both the total softness, S, and the fukui function, f(r), appear. Further work to evaluate in an efficient way the integral in the second term is in progress, the basic requirements for the evaluation of the interaction energy being the molecular electrostatic potential and fukui function (V(r) and f(r)) and the total softness nowadays ready available at a high precision level.

Conceptual DFT can thus be exploited in the study of adsorption behaviour of zeolites, as also especially witnessed in the work by Chatterjee and coworkers using the condensed Fukui function and local softness to estimate and rationalize the interaction energy of several small molecules with a zeolitic framework [104,105,106]. An attempt was made to explain selective permeation of these molecules [105] and the choice of the best template for a particular zeolite synthesis by estimating the reactivity of the templating molecule [106].

Biochemistry: Influence of Point Mutations on the Catalytic Activity of Subtilisin

As an example of our work in biochemistry we summarize very briefly recent studies in which the embedded cluster approach presented in § 2.3 was used to study the active sites in enzymes, more precisely the catalytic triad in subtilisin [107] and Ribonuclease T$_1$ [108].

The methodology followed in these studies shows similarities with part of the zeolite adsorption studies (high level quantum chemical calculation on that part of the system at which the active site is situated and a point charge environment for the residues farther away). The results of the subtilisin study are given as an example.

Subtilisin is a bacterial enzyme belonging to the class of the serine proteases characterized by a catalytical apparatus consisting of three amino acid residues, serine, histidine and aspartate : the catalytic triad (Asp32 - His64 - Ser221 for subtilisin) (Fig. 16). In the rate determining step of the hydrolysis reaction (see Fig. 17) the hydroxyl proton of Ser221 is transferred to the N$_{\varepsilon2}$ atom of His64. Simultaneously a nucleophilic attack by the hydroxyl oxygen of Ser221 at the carbon atom of the scissile peptide bond occurs. The role of the aspartate residue is to enhance the nucleophilicity of Ser221 due to the electric field of the charged aspartate side chain and to provide electrostatic stabilization of the tetrahedral intermediate.

The role of the catalytic triad amino acids was studied by Carter and Wells : both single and double alanine substitutions of these residues led to a lowering of the catalytic rate constant k_{cat} [109]. Russell and Fersht studied the effect on k_{cat} for mutations occurring outside the catalytic triad [110].

These effects were studied by placing the catalytic triad into an environment of ChelpG point charges representing all atoms of the amino acids within a 15 Å sphere around His64, and obtained from ab initio calculations on isolated amino acids; the structural data on the wild type enzyme were taken from X-ray diffraction studies [111], mutations were carried out *in computero*. The nucleophilicity of the Ser221 oxygen was investigated using local softness and (models for) the local hardness: the charge on the oxygen atom and the MEP.

Local softness turned out to be not successful when correlating s−o with k_{cat} values in line with Fersht›s statement that the nucleophilic attack of the serine on the substrate can be considered as an attack on a hard nucleophilic center [112]. We now further concentrate on local hardness descriptors. When comparing the wild type enzyme and the His64 Ala mutant with, respectively, the Asp32 Ala and Asp32 Ala : His64 Ala mutants, Table 4 shows a less negative charge and, much more pronounced, a less negative MEP value, indicating that the interaction of the catalytic serine with an electrophile is less advantageous in the aspartate mutants. These results are in agreement with experiments pointing out an enhancement of the serine nucleophlicity by the aspartate residue [113].

Table 4: Serine oxygen charge (q_0) (au) and MEP minimum around this oxygen atom ($V(R)_{min}$) (in kcalmol^{-1}) calculated for the wild-type and mutant enzymes (3-21G Hartree Fock with ChelpG charges).

	q_0	$V(\underline{R})_{Min}$
Wild Type	-0.7719	-121.72
Asp32 Ala	-0.7440	-75.64
His64 Ala	-0.7665	-113.13
Asp32 Ala : His64 Ala	-0.7487	-67.92

A more quantitative picture is obtained when studying the atomic charges and MEP vs. the experimental k_{cat} values [109] for the enzymes when the mutations were performed in the environment of the catalytic triad (Table 5). A correlation coefficient of 0.927 was obtained in the case of the charges (Figure 18).

Table 5. Experimental k_{cat} values (s^{-1}) and calculated charges q_O (au) and MEP minimum $V(R)_{min}$ (kcal mol^{-1}) for the serine oxygen for wild-type, aspartate 99 and glutamate 156 mutant enzymes.

	k_{cat}	q_O	$V(\underline{R})_{min}$
1. Wild Type	57	-0.7719	-121.72
2. Asp99 Ser	45	-0.7610	-121.33
3. Asp99 Lys	30	-0.7581	-121.30
4. Glu156 Ser	55	-0.7700	-122.96
5. Glu156 Lys	79	-0.7748	-124.57

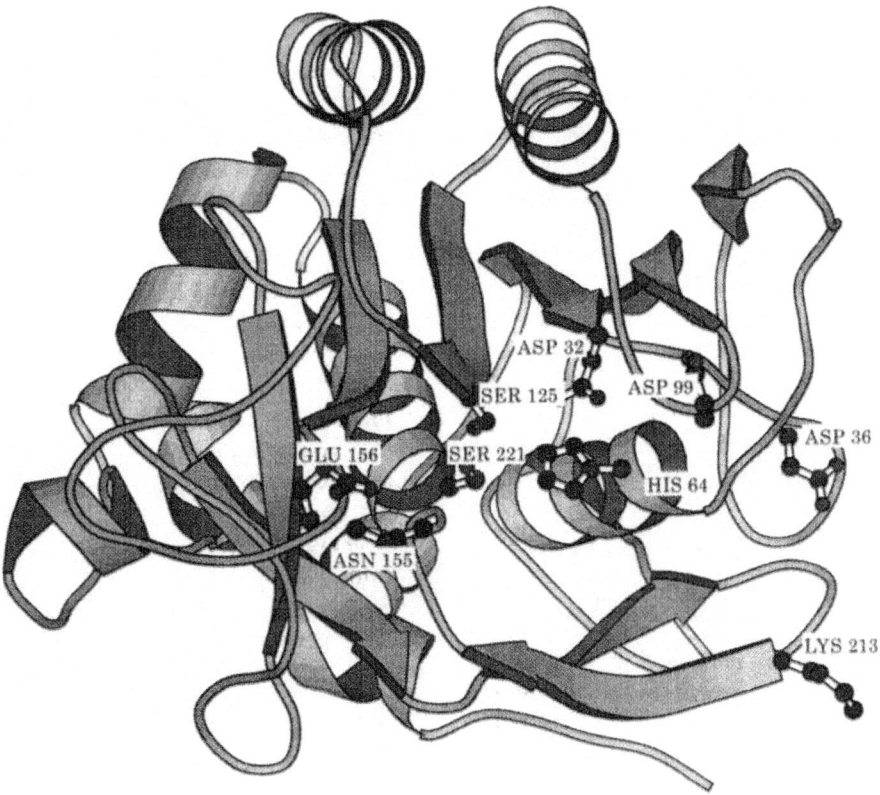

Figure 16: Schematic drawing of subtilisin. Side chains of the residues of importance in the discussion are shown explicitely (Reprinted with permission by Academic Press, Reference [107]).

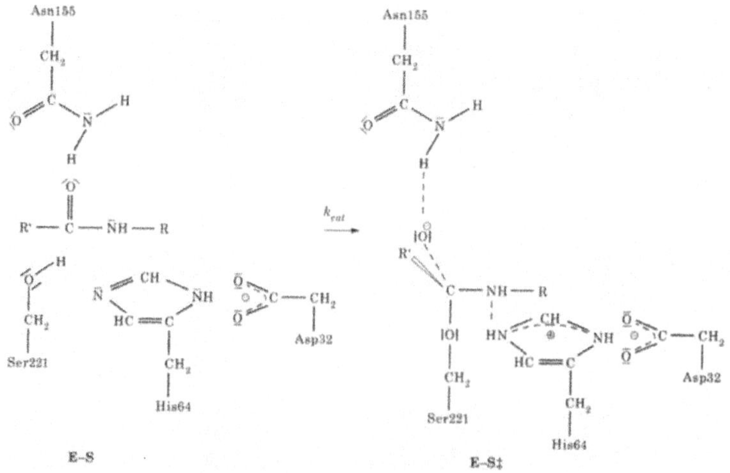

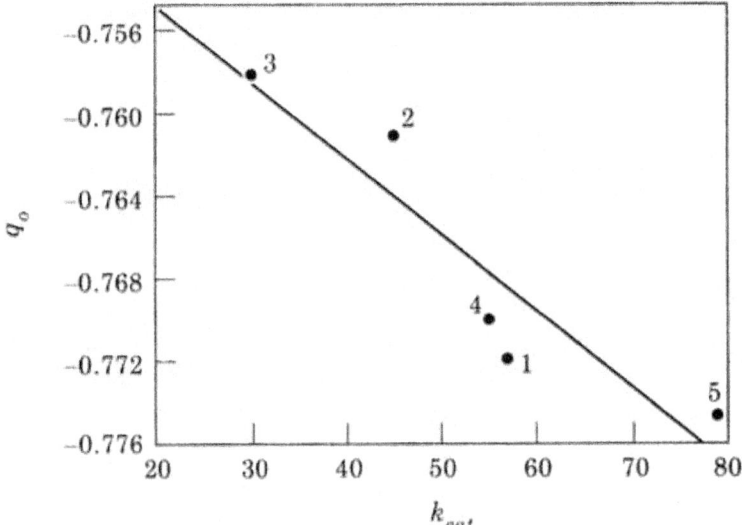

Figure 17: Schematic representation of the enzymatic reaction of serine proteases; E-S is the enzyme substrate complex, E-S$^{\neq}$ the tetrahedral reaction intermediate (Reprinted with permission by Academic Press - Reference [107])

Figure 18: Atomic charge on the serine oxygen atom q_0 vs. experimental k_{cat} (see Table 5).

CONCLUSIONS

In this contribution it is seen that Quantum Mechanics gave birth to new branches in chemistry, Quantum Chemistry and Computational Chemistry, which nowadays provide the theoretical/conceptual and computational framework for studying "real" chemical problems, i.e. systems whose size promotes them to valuable models for the actual reaction partners, their environment, and their interactions. Within this context Density Functional is a source of increasing importance both for concepts and computational techniques.

Application to reactivity problems (considered within a broad context including kinetic as well as thermodynamic aspects) in organic, inorganic and biochemistry shows that present day quantum chemistry becomes a priceless tool to compute and understand phenomena in various traditional subfields of chemistry. Especially the DFT ansatz provides an ideal situation to bridge the gap between chemistry and physics in the years to come [114].

ACKNOWLEDGEMENTS

Besides his coauthor PG wants to thank his past and present Ph.D. students and postdoctoral associates W. Langenaeker, A. Baeten, K. Choho, B. Safi, G. Boon, S. Damoun, F. Tielens, whose work was summarized in this contribution. Special thanks to Professor F. Mendez (Mexico) for the collaboration on organic reactivity during his stay in Brussels, to Professors W. Mortier (Exxon-Mobil-KU Leuven), R. Schoonheydt (KU Leuven) and G. Baron (VUB) for the longstanding collaboration on zeolites and Professors L. Wyns and J. Steyaert (VUB) for backing the quantum biochemical research.

REFERENCES

1. Dirac, P.A.M. Proc. Roy.Soc.(London) 1929, 123, 714.

2. Heitler, W.; London, F. Z. Phys. 1927, 44, 455.

3. Pauling, L. The Nature of the Chemical Bond, Third Edition ed; Cornell University Press: Ithaca, 1960.

4. Streitwieser, A. For a classic and authoritative account of Hückel's approach and its applications see A. In Molecular Orbital Theory for Organic Chemistry; John Wiley, 1961.

5. Mc Weeny, R. Coulson's Valence, Third Edition ed; Oxford, 1979.

6. Hund, F. Z. Phys. 1928, 51, 759.Mulliken, R. S. Phys Rev. 1928, 32, 186.

7. Roothaan, C. C. J. Rev. Mod. Phys. 1951, 23, 69.

8. For an overview see (a)Hartree, D.R. The Calculation of Atomic Structures; John Wiley: New York, 1957. b)Hehre, W. J.; Radom, L.; Schleyer, P. v. R; Pople, J. A. Ab Initio Molecular Orbital Theory; Wiley: New York, 1986.

9. MØller, C.; Plesset, M.S. Phys. Rev. 1934, 46, 618.

10. Shavitt, I. The Method of Configuration Interaction in Modern Theoretical Chemistry, Vol.3, Methods of Electronic Structure Theory; Schaefer, H.F., III, Ed.; Plenum Press: New York, 1977; p. 189.

11. Bartlett, R.J. J. Phys. Chem. 1989, 93, 1697.

12. Pople, J.A.; et al. GAUSSIAN 98; and previous releases (GAUSSIAN 94, GAUSSIAN 92, ..., GAUSSIAN 70); Gaussian Inc.: Pittsburgh A, 1998.

13. The Encyclopedia of Computational Chemistry; Schleyer, P. v. Rague, Allinger, N. L., Clark, T., Gasteiger, J., Kollman, P. A., Schaefer, H. F., Schreiner, P. R., Eds.; John Wiley & Sons Ltd.: Chichester, 1998.

14. Pople, J.A. J. Chem. Phys. 1966, 43, S229.

15. Hohenberg, P.; Kohn, W. Phys. Rev. B 1964, 136, 864.

16. Mc Weeny, R.; Sutcliffe, B.T. Methods of Molecular Quantum Mechanics; Academic Press: London, 1969; Chapter 4.

17. Kohn, W.; Sham, L. Phys. Rev. A 1965, 140, 1133.

18. Parr, R.G.; Yang, W. Ann. Rev. Phys. Chem. 1995, 46, 701.

19. Becke, A.D. J. Chem. Phys 1993, 98, 5648.Lee, C.; Yang, W.; Parr, R.G. Phys. Rev. B 1998, 37, 785.

20. Geerlings, P.; De Proft, F.; Langenaeker, W. Adv. Quantum Chem. 1999, 33, 303.

21. Kohn, W.; Becke, A.; Parr, R.G. J. Phys. Chem. 1996, 100, 12974.

22. Koch, W.; Holthausen, M.C. A Chemistry's Guide to Density Functional Theory; Wiley-VCH: Weinheim, 2000.

23. Parr, R.G. Density Functional Methods in Physics; Dreizler, R.M., da Providencia, J., Eds.; Plenum, 1985; p. 141.

24. England, W.; Salmon, L.S.; Rüdenberg, K. Fortschr. Chem. Forsch. 1971, 23, 3, and references therein.

25. Foster, M.; Boys, S.F. Rev. Mod. Phys. 1960, 32, 300.

26. Schaftenaar, G.; Noordik, J.H. Molden : a pre- and post-processing program for molecular and electronic structures. J. Comp.-Aided Mol. Design 2000, 14, 123.

27. Müller, N.; Pritchard, D.E. J. Chem. Phys. 1959, 31, 768.

28. Figeys, H. P.; Geerlings, P.; Raeymaekers, P.; Van Lommen, G.; Defay, N. Tetrahedron 1975, 31, 1731.

29. Parr, R.G.; Yang, W. Density Functional Theory of Atoms and Molecules; Oxford, 1989; Chapter.

30. Parr, R.G.; Donelly, R.A.; Levy, M.; Palke, W.E. J. Chem. Phys. 1978, 68, 3801.

31. Iczkowski, R.P.; Margrave, J.L. J. Am. Chem. Soc. 1961, 83, 3547.

32. Mulliken, R.S. J. Chem. Phys. 1934, 2, 782.

33. For a series of papers covering various aspects of electronegativity see : Structure and Bonding; Sen, K.D., JØrgensen, C.K., Eds.; Springer Verlag: Berlin, 1987; Vol.66.

34. Nalewajski, R.F.; Parr, R.G. J. Chem. Phys. 1982, 77, 399.

35. Parr, R.G.; Pearson, R.G. J. Am. Chem. Soc. 1983, 105, 7512.

36. Parr, R.G.; Yang, W. J. Am. Chem. Soc. 1984, 106, 4049.

37. Fukui, K.; Yonezawa, T.; Shinghu, H. J. Chem. Phys. 1952, 20, 722.

38. Yang, W.; Parr, R.G. Proc. Natl. Acad. Sci. 1985, 82, 6723.

39. Chermette, H. J. Comp. Chem. 1999, 20, 129.

40. De Proft, F.; Geerlings, P. Chem. Rev. 2001, 101, 1451.Geerlings, P.; De Proft, F.; Langenaeker, W. Chem. Rev. submitted.

41. Sanderson, R.T. Science 1955, 121, 207.

42. Sanderson, R.T. Polar Covalence; Academic Press: New York, 1983.

43. Pearson, R.G. J. Am. Chem. Soc. 1963, 85, 3533.

44. Pearson, R.G. Chemical Hardness; J. Wiley: New York, 1997.

45. Chattaraj, P.K.; Lee, P.K.; Parr, R.G. J. Am. Chem. Soc. 1991, 113, 1855.

46. Parr, R.G.; Chattaraj, P.K. J. Am. Chem. Soc. 1991, 113, 1854.

47. De Proft, F.; Martin, J.M.L.; Geerlings, P. Chem. Phys. Lett. 1996, 250, 393.

48. De Proft, F.; Geerlings, P. J. Chem. Phys. 1997, 106, 3270.

49. De Oliveira, G.; Martin, J.M.L.; De Proft, F.; Geerlings, P. Phys. Rev. 1999, A60, 1034.

50. See for example the impressive series The Chemistry of Functional Groups; Patai, S., Ed.; Interscience Publishers: London.

51. De Proft, F.; Langenaeker, W.; Geerlings, P. J. Phys. Chem. 1993, 97, 1826.

52. Datta, D.; Nabakishwar, S.S. J. Phys. Chem. 1990, 94, 2184.

53. Wray, V.; Ernst, L.; Luna, T.; Jakobsen, H.-J. J. Magn. Res. 1980, 40, 55.

54. Pearson, R.G. J. Am. Chem. Soc. 1998, 110, 7684.

55. De Proft, F.; Langenaeker, W.; Geerlings, P. Tetrahedron 1995, 55, 4021.

56. Lias, S.G.; Bartmess, J.-E.; Liebman, J.F.; Holmes, J.L.; Levin, R.D.;

Mallard, W.G. J. Phys. Chem. Ref. Data 1988, 17(Suppl 1).

57. Bartmess, J.E.; Scott, J.A.; Mc Iver Jr., R.T. J. Am. Chem. Soc. 1979, 101, 6056.

58. Taylor, R. Electrophilic Aromatic Substitution; J. Wiley: New York, 1990.

59. Vanermen, G.; Toppet, S.; Van Beylen, M.; Geerlings, P. J. Chem. Soc., Perkin Transactions 2 1986, 699.

60. Wiberg, K.; Keith, T.A.; Frisch, M.J.; Muacko, M. J. Phys. Chem. 1995, 99, 9072.

61. Safi, B.; Choho, K.; De Proft, F.; Geerlings, P. J. Phys. Chem. 1998, 5253, 5253.

62. Isaacs, N. Physical Organic Chemistry; Longman Scientific and Technical: Singapore, 1995.

63. Shaik, S.S.; Schlegel, H.B.; Wolfe, S. Theoretical Aspect of Physical Organic Chemistry; John Wiley, 1992. [

64. Safi, B.; Choho, K.; Geerlings, P. J. Phys. Chem 2001, A105, 591.

65. Gazquez, J.L. J. Phys. Chem. 1997, A101, 8967.

66. Politzer, P. J. Chem. Phys. 1987, 86, 1072.

67. Chandra, A. K.; Uchimaru, T. J. Phys. Chem. A 2001, 105, 3578.

68. Gazquez, J.L.; Mendez, F. J. Phys. Chem. 1994, 98, 459.Damoun, S.; Van de Woude, G.; Mendez, F.; Geerlings, P. J. Phys. Chem. 1997, A101, 886.

69. Geerlings, P.; De Proft, F. Int. J. Quant. Chem. 2000, 80, 227.

70. Eisenstein, O.; Lefour, J.M.; Anh, N.T.; Hudson, R.F. Tetrahedron 1977, 33, 523.

71. K.N. Houk, K.N.; Li, Y.; Evanseck., J.T. Angew. Chem. Int. Ed. Engl. 1992, 31, 682.

72. Chandra, A. K.; Nguyen, M. T. J. Chem. Soc. Perkin Trans. 2 1997, 1415.

73. Raspoet, G.; Nguyen, M. T.; McGarraghy, M.; Hegarty, A. F. J. Org. Chem. 1998, 63, 6867.Nguyen, M. T.; Raspoet, G.Can. J. Chem. 1999, 77, 817.Nguyen, M. T.; Raspoet, G.; Vanquickenborne, L. G. J. Phys. Org. Chem. 2000, 13, 46.

74. Chandra, A. K.; Geerlings, P.; Nguyen, M. T. J. Org. Chem. 1997, 62, 6417.Nguyen, L. T.; Le, T. N.; De Proft, F.; Chandra, A. K.; Langenaeker, W.; Nguyen, M. T.; Geerlings, P. J. Am. Chem. Soc. 1999, 121, 5992.

75. Chandra, A. K.; Nguyen, M. T. J. Comput. Chem. 1998, 19, 195.Nguyen, L. T.; Le, T. N.; De Proft, F.; Chandra, A. K.; Uchimaru, T.; Nguyen, M. T.; Geerlings, P. J. Org. Chem. 2001, 66, 6096.Le, T. N.; Nguyen, L. T.; Chandra, A. K.; De Proft, F.; Uchimaru, T.; Geerlings, P.; Nguyen, M.

T. J. Chem. Soc. Perkin Trans. 2 1999, 1249.Chandra, A. K.; Uchimaru, T.; Nguyen, M. T. J. Chem. Soc. Perkin Trans. 2 1999, 2117.Chandra, A. K.; Nguyen, M. T. J. Phys. Chem. A 1998, 102, 6181.

76. Sengupta, D.; Chandra, A. K.; Nguyen, M. T. J. Org. Chem. 1997, 62, 6404.

77. Ponti, A. J. Phys. Chem. A 2000, 104, 8843.

78. Rouvray, D.H. Top. Curr. Chem. 1995, 173, 2.

79. Mishra, P.C.; Kumar, A. Theor. Comput. Chem. 1996, 3, 257.

80. Carbo, M.; Arnau, M.; Leyda, L. Int. J. Quant. Chem. 1980, 17, 1185.

81. Parr, R.G.; Bartolotti, L.J. J. Phys. Chem. 1983, 87, 2810.

82. Boon, G.; De Proft, F.; Langenaeker, W.; Geerlings, P. Chem. Phys. Lett. 1996, 295, 122.Boon, G.; De Proft, F.; Langenaeker, W.; De Proft, F.; De Winter, H.; Tollenaere, J.P.; Geerlings, P. J. Phys. Chem.A 2001, 105, 8805.

83. Spatola, A.F. Chemistry and Biochemistry of Amino Acids, Peptides and Proteins; Weinstein, B., Ed.; Marcel Dekker: New York, 1983; vol. 7, p. 267.

84. Perdew, J.P.; Wang, Y. Phys. Rev. B 1992, 45, 13244.

85. Allmendiger, T.; Felder, E.; Hungerbühler, A. Tetr. Lett. 1995, 7301. Bartlett, P.A.; Otake, A. J. Org. Chem. 1995, 60, 3107.

86. Hodgkin, E.E.; Richards, W. G. Int. J. Q. Chem., Quantum Biology Symp. 1987, 14, 1051.

87. Breck, D.W. Zeolite Molecular Sieves : Structure, Chemistry and Use; John Wiley: Canada, 1974.

88. Uytterhoeven, J.B.; Christner, L.G.; Hall, W.K. J. Phys. Chem. 1965, 69, 2117.

89. Van Bekkum, H.; Flanigen, E.M.; Jansen, J.C. Introduction to Zeolite Science and Practice; Elsevier: Amsterdam, 1991.

90. Sherman, J.D. Proc. Natl. Acad. Sci. USA 1999, 96, 3471.

91. Van Genechten, K.; Mortier, W.J.; Geerlings, P. J. Chem. Phys. 1987, 86, 5063.

92. Langenaeker, W.; De Proft, F.; Geerlings, P. Recent Developments in Physical Chemistry; Transworld Research Network: Triviandum, India, 1998; Vol. 2, p. 1219.

93. Peirs, J.C.; De Proft, F.; Baron, G.; Van Alsenoy, C.; Geerlings, P. Chem. Comm. 1997, 531.

94. Tielens, F.; Langenaeker, W.; Ocakoglu, A. R.; Geerlings, P. J. Comp. Chem. 2000, 21, 909.

95. Tielens, F.; Geerlings, P. J. Mol. Catal. 2001, A166, 175.

96. Ruthven, D. M. Principles of Adsorption and Adsorption Processes; John Wiley: Canada, 1984.

97. Kiselev, A. V. Pure Appl. Chem. 1980, 52, 2161.

98. Tielens, F.; Geerlings, P. Int. J. Quant. Chem. 2002, 84, 58.Tielens, F.; Geerlings, P. Chem. Phys. Lett. 2002, 354, 474.

99. Buckingham, A.D. Intermolecular Interactions; Pullman, B., Ed.; Wiley, 1988.

100. Dunning Jr., T.H. J. Chem. Phys. 1989, 90, 1007.

101. Kendall, R.A.; Dunning Jr., T.H.; Harrison, R.J. J. Chem. Phys. 1992, 96, 6796.

102. De Proft, F.; Tielens, F.; Geerlings, P. J. Mol. Struct. (Theochem) 2000, 506, 1.

103. Langenaeker, W.; De Proft, F.; Tielens, F.; Geerlings, P. Chem. Phys. Lett. 1998, 288, 628.

104. Chatterjee, A.; Iwasaki, T.; Ebina, T. J. Phys. Chem. A 1999, 103, 2489.

105. Chatterjee, A.; Iwasaki, T. J. Phys. Chem. A 1999, 103, 9857.

106. Chatterjee, A.; Iwasaki, T. J. Phys. Chem. A 2001, 105, 6187.

107. Baeten, A.; Maes, D.; Geerlings, P. J. Theoret. Biol. 1998, 195, 27.

108. Mignon, P.; Loverix, S.; Van Houtven, S.; Steyaert, J.; Geerlings, P. in preparation.

109. Carter, P.; Wells, J.A. Nature 1988, 332, 565.

110. Russell, A.J.; Fersht, A.R. J. Mol. Biol. 1987, 193, 803.

111. Bott, R.; Vetsch, M.; Kossiakoff, A.; Graycar, T.; Katz, B.; Power, S. J. Biol. Chem. 1988, 263, 7895.

112. Fersht, A. Enzyme Structure and Mechanism; W.H. Freeman and Co.: New York, 1985.

113. Wells, J.A.; Estell, D.A. TIBS 1988, 13, 291, (1988).

114. Geerlings, P., De Proft, F., Langenaeker, W., Eds.; Density Functional Theory: a Bridge between Chemistry and Physics, Proceedings of a Two Day Symposium at the VUB, VUB-Press, Brussels, May 14-15, 1998; 1999.

Chapter 6

CHEMICAL KINETICS, AN HISTORICAL INTRODUCTION

Stefano Zambelli

University of Padova, Italy

INTRODUCTION

This Chapter would provide a methodological analysis of the historical developments of chemical kinetics from the beginnings to the achievements of Transition state theory and Kramers-Christiansen approach. Chemical kinetics is often treated as a side issue of the most important disciplines of chemical science. Students in most of the cases gain knowledge of Kinetics as part of Physical Chemistry introductory courses and find it again applied in many other contests. Despite that, it would necessitate a fundamental and main teaching course as we will see in the course of this chapter. This didactical and academic approach could have many reasons. A general one may be the philosophical and psychological disposition to put our attention more on objects rather than concepts, matter over processes. In Science History there are many examples of this tendency: the transmission of heat and electromagnetic waves are good examples. Phlogiston and Luminiferous Aether represents a materialization of processes that processes themselves do not need to be studied, however our mind need this primitive objectivization to grasp the concept in a simpler way. This represents a fundamental issue of scientific method: to do Science we need to go beyond banality and perception. The development of Chemical Kinetics is deeply involved in the counterfactual approach that brought from Alchemy to Chemistry as for Physics form Aristotelic Natural Philosophy to Galilean Science.

ORIGINS OF CHEMICAL KINETICS: THE DECLINATIONS OF AFFINITY

The chemical affinity principle, developed during the seventeenth century, derives from the alchemical concept of chemical wedding: similar substances will interact so we can categorize them. The real innovation at the end of 17th and during the 18th centuries was the application of that concept not only as a taxonomic principle but also for the comprehension of chemical reactivity. The interaction of bodies is simpler when there is a similitude between them, this is the base idea of Chemical Affinities and come from ancient and medieval alchemy and naturalism doctrine. At the end of 17th century this intuitive principle become a theory, although qualitative, that justify and classify interactions between different substances. In the same period also the observation of time become important for the determination of the nature of chemical reactions. Time of decurrence was clearly contemplated for the preparation of substances with long reactions but it was seen as an ordinary technical factor. The Opera of Alchemy, for example in the transmutations of metals, was considered as a means for the acceleration of the millenary gestation of precious metals in the bowels of Earth Mother. The underestimate of real times in the alchemist's conceptions resulted so natural in an activity that already theoretically reduced geological times. The paradox was that time, a fundamental principle for alchemic theory, resulted of little importance in the alchemical praxis. Probably the first scholar that introduced a dynamical vision of the chemical phenomena was Wilhelm Homberg (1652-1715). Homberg, a German scholar, worked in Magdeburg with Von Guericke, in Italy and later in England with Boyle. He introduced the first principles of quantitative measurement for chemical action: the strength of an acid towards a series of alkali depends on the time of neutralization of the various alkali.

Tabulae affinitatum

The lists of strength of alkali and the concept of chemical affinity brought Etienne Francois Geoffroy (1672-1731), a French scholar initiated to chemistry by Homberg himself, to the compilation of the Tables of Affinity, (or Tables of Rapports) that could be considered as the first ancestor of the periodic table. The first one was done by Geffroy (Geoffroy, 1718). You can see the Encyclopédie version in the following figure.

In the first row you can see the primary substances then going down along the columns the similar substances in order of affinity with the first one. The development of Affinity tables was inevitably considered in the light of the main scientific discussion of the 18th century: the debate between plenistic Cartesian vision and the Newtonian distance action principle. Important chemisters of this period took parts in that debate: Boerhaave and later Buffon among Newtonian side identified affinities as a special form of gravitational attraction, Stahl on the other side negated the distance action invoking the medium of Phlogiston.

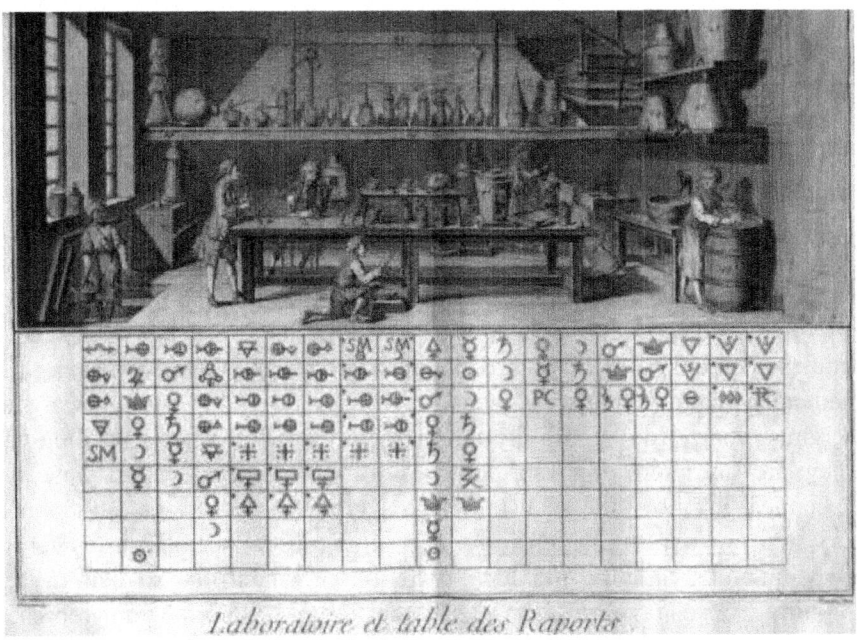

Figure: 1. Recueil de Planches, sur les Sciences, les Arts Libéraux, et les Arts Mécha-niques, 1772

Guyton de Morveau (1737-1816), a French scholar, sustained initially phlogiston theory, but leaved it after in favor of a distance action between the different elementary particles of substances bringing the chemical affinities to a microscopic level, a similar position was taken by Berthollet and Lavoisier. De Morveau classified the kind of affinities: simple or by aggregation, composed, decomposed, double, reciprocal, intermediate, dispositional. He listed also the laws of affinity:

- Molecules have to be in fluid state to respond to affinities influence.
- Affinities acts between the elementary particles of bodies.
- Affinities between two different substances may be different from that between their= composites.
- Affinity of substances acts only if it is bigger than the aggregation affinity of themselves.

- Two or more bodies united by affinity form a new body with different properties from precursors.
- Affinities action and velocity depends on temperature.

Basilar principles of Chemical Kinetics and Chemistry are going to take form. Of particular importance the last law: temperature and so ambient conditions have influence on chemical reactivity. The position of Torben Olof Bergman (1735-1784), a Swedish scholar, about the influence of temperature is particularly interesting. He assumed the affinity constant at constant temperature and suggested to compile different affinity tables depending on conditions: the affinities of dry phase is different from that in liquid phase. Bergman closed elegantly the debate on the nature of the affinities assuming a very wise position: it is not useful debating about the last nature of interacting forces between chemical particles because it will remain unknown until quantitative experiments will be done on affinities. Bergman so is the first scholar that made some hypothesis about a measure of the affinities, but their mathematical expressions and measures will be a duty for future researchers. Bergman compiled also affinity diagrams in his major opera, the Opuscula. They are an interesting representation of chemical reactions done with alchemical symbols: the ancestors of stoichiometric equations (although the very first one appeared even in 1615, but not systematically, in the famous Tyrocinium Chymicum, the first Textbook of Chemistry written by Jean Beguin). You can see an example in the figure 2. The diagram represents the reactions of sulfuric and hydrochloric acid with calcium carbonate and potassium hydroxide (Vitriolic and Marine for acids, Pure calcareous earth and Pure fixed vegetable alkali for the basis).

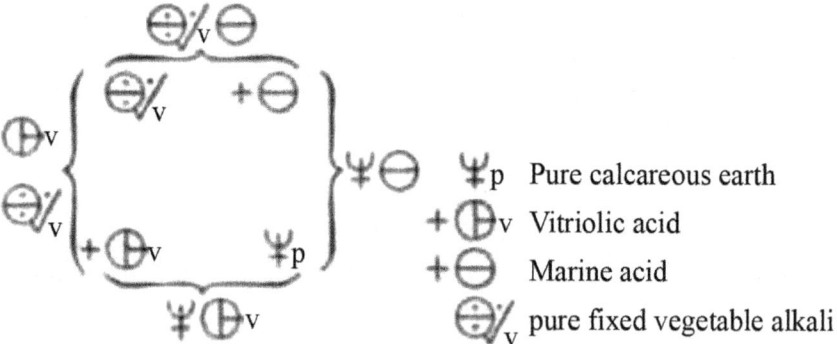

Figure: 2. This Affinity Diagram schematize two acid-base reactions

CHEMICAL EQUILIBRIUM CONCEPTION: THE LAW OF MASS ACTION

The end of 18th Century and the first half of 19th added other essential pieces to the puzzle of Chemical Kinetics and Chemistry in general. There is a surprising absent actor in the debate on Chemical Affinities, the father of modern Chemistry: Antoine-Laurent de Lavoisier (1743-1794). The Lavoisier Revolution brought quantitative measurements to Chemistry and so to Affinity Diagrams. We can see one of the first examples of stoichiometric equation from Lavoisier works in the following figure (Lavoisier 1782).

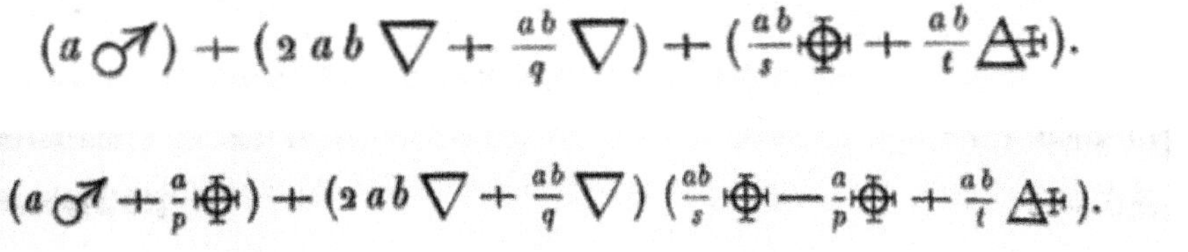

$$(a\,\sigma^{7}) + (2\,a\,b\,\nabla + \frac{ab}{q}\nabla) + (\frac{ab}{s}\oplus + \frac{ab}{t}\triangle^{\ddagger}).$$

$$(a\,\sigma^{7} + \frac{a}{p}\oplus) + (2\,a\,b\,\nabla + \frac{ab}{q}\nabla)\,(\frac{ab}{s}\oplus - \frac{a}{p}\oplus + \frac{ab}{t}\triangle^{\ddagger}).$$

Figure: 3. Stoichimetric Equations with Lavoisier's symbols

Those symbolic equations represent one of the passages of the oxidation of iron in nitric acid where Mars symbolize iron, the nabla water, the crossed circle oxygen, the triangle and cross nitrogen oxide. In this passage iron gains the same part of oxygen that nitric acid loses, an example of the Law of Mass Conservation. Why Lavoisier did not play a role in the debate about Affinities if he applied quantitative methods also for affinity diagrams? The causes may be many, for example the fact that he was outside main academic circles, (he was member of the French Academy of Sciences from the age of 25, but never gained an academic position). The reasons are explained by Lavoisier himself in the Traité élémentaire de chimie, and follow Bergan recommendations: In this writing I followed the principle of not arguing beyond experimentations, not taking over the silence of facts. So I cannot consider those parts of Chemistry that would probably become Exact Science before the others. Scholars as Bergman, Scheele, de Morveau and many others are conducting numerous studies about Chemical Affinities and Attractions, but basic, precise and general data are lacking at the moment. Affinities theory respect to ordinary Chemistry is as Transcendent Geometry respect to Elementary one and goes over the scope of this introductory book. Mr de Morveau is writing the voice Affinity in the Encyclopédie and I am worried to compete with him.

Characterising of Chemical Reactions

With the development of Lavoisier's methods in the second half of the 18th Century new definitions and properties are established. A concept that for today scientists results obvious was defined: the concentration of substances. The fist timid attempts to distinguish reactivity and equilibrium was made, for example sulphuric acid was considered the most powerful because it shifted other acids from their salts, the most strong because it absorbs most water, the least active because Oleum needs water or hydrated compounds to take effect. The researches about the reactions between acids and metals are of particular interest in this period. For example many scholars did not consider more metals as primary substances thinking they was compounds with an alkaline parts (it will need nearly a Century for the comprehension of redox reactions).

In the work of Carl Friedrich Wenzel (1740-1793), a dresden metallurgist, we can find the first link between reaction velocity and quantity of the reactants. He investigated the reactions between metals considering the time of dissolution of little metal cylinders inside dilute acid solutions. Using Buffon theories Wenzel considered the affinity of the acids inversely proportional to the time of dissolution but considered also the role of the solvent (water). The velocity of reaction results proportional to the affinity or the strength of the acid while inversely proportional to the resistance of the solvent. In modern terms reaction velocity is proportional to concentration. Wenzel made also interesting considerations about thermal conditions, imposing the same temperature for all the dissolutions to compare them correctly. Some scholars, Wilhelm Ostwald between the most famous, awarded Wenzel for the first qualitative definition of the Law of Mass Action, although the primacy is commonly given to Berthollet. Count Claude Louis Berthollet (1748-1822), member of Academy of France and founder of the Ecole Polytechnique, collaborated with Lavoisier but was more lucky than him. He had no problems during the revolution and got in the good books of Napoleonic government. He followed Bonaparte's expedition to Egypt. Visiting the Natron Lakes, Berthollet observed soda deposits on the surrounding limestone hills. He supposed a chemical reaction occurring between salt (sodium chloride) and the limestone (calcium carbonate) in the hills to produce soda (sodium carbonate) and an accompanying product, calcium chloride, which seeped away into the ground. The reaction was the reverse of the one that chemists knew under laboratory conditions, and this indicated to Berthollet that physical conditions, such as heat and pressure, and quantities of reactants could affect the course of a chemical reaction. From these and other considerations he exposed the first qualitative form of the Law of Mass Action during 1803 in two famous publications: "Essai de statique chimique" (fig. 4) e "Recherches sur les lois des affinités chimiques".

The progress of a chemical reaction depends on the quantity and conditions of reacting substances. Berthollet's essays do not relate only on the velocity of the reaction but also on its equilibrium. Today these considerations may appear obvious but at the time they received fierce critics. These theories and the embryonic conception of equilibrium was favourably considered by some important Chemists as Berzelius, Davy and Gay-Lussac, but most of the scientific community did not considered them being incompatible with Proust's and Dalton's Laws that monopolized the attention of the scientific community in the period.

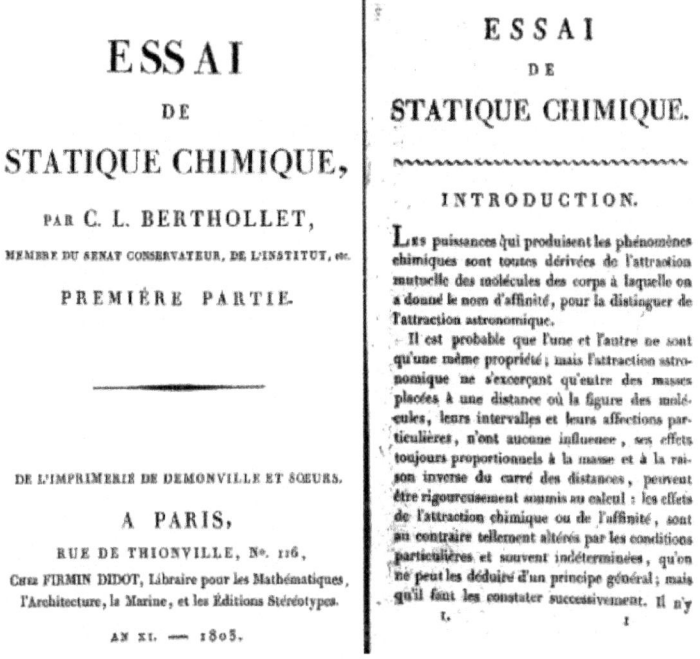

Figure: 4. Title and first page of Berthollet Essay

Berthollet made other significant considerations, for example the fact that for solids the Affinity remain costant. So Affinities are not absolute but become dependant on the quantities of reactants (except solids), but how those quantities was defined? In the Essai he defined the Affinity $A=a/E$, where a is a constant dependant on the substance and E its equivalent weight. Multiplying the mass of the substance for unit of volume w by the precedent expression he defined the Active Mass of the reactant equal to the concentration , (numbers of equivalent per unit volume: w/E).

The reasons of this rejection depended also by the fact that most of the conclusions of Berthollet and his predecessors was qualitative and not supported by adequate analytical data. To get the first quantitative observations and thermodynamic interpretations of reacting systems we have to wait the second half of 19th Century thanks to the development of analytical chemistry.

Time: A New Quantitative Observable

It is difficult today arguing about Chemical Kinetics without Thermodynamic but this branch of our science was established originally by simple chronological measurements of chemical processes (King 1981).

V. *Ueber das Gesetz, nach welchem die Einwir-kung der Säuren auf den Rohrzucker stattfindet; von Ludwig Wilhelmy in Heidelberg.*

Der die Polarisationsebene des durch seine Auflösung ge-henden Lichts nach rechtsdrehende Rohrzucker wird be-kanntlich durch Einwirkung von Säuren in linksdrehenden Schleimzucker verwandelt. Da man nun mit Hülfe eines Polarisationsapparats, namentlich unter Anwendung der So-leil'schen Doppelplatte, mit grofser Leichtigkeit und Sicher-heit der Ablesung in jedem Augenblicke bestimmen kann, wie weit diese Umwandlung vorgeschritten ist, so schien mir hierdurch die Möglichkeit gegeben, die Gesetze des in Rede stehenden Vorgangs zu ermitteln, andererseits aber die Aufgabe von Interesse, festzustellen, in welcher Weise die chemische Action, wenigstens in diesem speciellen Falle, der aber gewifs nur ein einzelner Repräsentant einer grö-fseren Reihe von Erscheinungen seyn wird, — denn in der Natur folgt Alles allgemeinen Gesetzen — abhängig sey von sämmtlichen ihr Eintreten bedingenden Umständen.

Ich glaubte, dafs man auf diesem Wege werde feststel-len können, in welcher Weise diese Action — ähnlich wie der Dampfdruck und die Ausdehnung der Körper — eine Function der Temperatur sey, in wiefern sie — analog der elektrischen und magnetischen Anziehung und Absto-fsung — ihrem Werth nach abhängig sey von dem Ab-

Figure: 5. Wilhelmy paper Title page

The development of quantitative relations and laws derived from the use of advanced analytical techniques but these did not give real contributions until the end of 19th Century thanks to a suitable mathematical construct. Initially analytical observations was used to collect a multitude of data from many different systems thinking in this way to get universal laws in the optic of Natural Philosophy. It is the passage from the many experiments to the good experiment that made the true change. The intense experimental phase around the half of 19th Century may be efficiently described by Wilhelmy and Gladstone works. Ludwig Ferdinand Wilhelmy (1812-1864), a German physicist published in 1850 an important paper on the kinetic on the inversion of sugar with acids (Wilhelmy, 1850, Fig. 5).

He used a new technique, Polarimetry, for evaluating the dependence of reaction velocity on the quantity of reactants and temperature. In this paper probably appeared the first differential equation used in chemistry

$$-\frac{dZ}{dt} = MZS$$

(1)

Reaction velocity is the negative derivative of the sugar quantity Z in time t, S the acid quantity and M the quantity of inverted sugar in the infinitesimal time dt. Considering an excess of acid S is constant and supposing also M constant the solution results:

$$Z = Z_0 e^{-MSt}$$

(2)

Wilhelmy verified that M remains almost constant in time and observed the dependance of that constant with temperature. Wilhemly's paper results impressive for its anticipations, it was written forty years before Arrhenius work on the same topic. Unfortunately, although written in a prestigious Journal, (the Poggendorffs Annalen der Physik und Chemie, later Annalen der Physik), the paper passed unnoticed by contemporary scholars. It will be rediscovered only in 1884 by Ostwald.

Not only Polarimetry but also other techniques useful for kinetic studies was developed in this period. Colorimetric titrations was used by John Hall Gladstone (1827-1902), Fullerian Professor of Chemistry in London, to get precise measurements of equilibrium and to investigate the effect of salts on reaction dynamic. We will quote the conclusions of Gladstone about the action of thiocyanate on iron salts to notice the evolution of the language and concepts on the topic (Gladstone, 1855):

1. Where two or more binary compounds are mixed under such circumstances that all the resulting bodies are free to act and react, each electro-positive element arrang-es itself in combination with each electro-negative element in certain constant proportions.

2. These proportions are independent of the manner in which the different elements were originally combined.

3. These proportions are not merely the resultant of the various strengths of affinity of the several substances for one another, but are dependent also on the mass of each of tie substances in the fixture.

4. An alteration in the mass of any one of the binary compounds present alters the amount of every one of the other binary compounds, and that in a regularly pro- gressive ratio; sudden transitions only occurring where a substance is present which is capable of combining with another in more than one proportion.

5. This equilibrium of affinities arranges itself in most cases in an inappreciably short space of time, but in certain instances the elements do not attain their final state of combination for hours, or even days.

6. The phenomena that present themselves where precipitation, volatilization, crystallization, and perhaps other actions occur, are of an opposite character, simply because one of the substances is thus removed from the field of action, and the equilibrium that was first established is thus destroyed.

7. There is consequently a fundamental error in all attempts to determine the relative strength of affinity by precipitation; in all methods of quantitative analysis founded on the colour of a solution in which colourless salts are also present; and in all conclusions as to what compounds exist in a solution drawn from such empirical rules as that " the strongest base combines with the strongest acid."

From Gladstone experiments Chemists on the field begun to use extensively optical methods verifying Berthollet's statements and two facts emerged clearly: the presence of the equilibrium conditions in contrast with Proust's Law, the hypothetical achieving of a complete reaction after an infinite time

CLOCKWORK STOICHIOMETRY

We will see that many exemplary experiments survived the second half of 19th Century and results still dominating in chemical didactics. The acid esterification of alcohols is emblematic in this sense: it is difficult nowadays to find an introductory textbook that does not explain this reaction as basic example.

In most of cases nevertheless the origin of that example is not cited. It comes from a series of experiments done by a couple of Parisian chemists around the 1860: Pierre Eugene Marcelin Berthelot (1827-1907) and Leon Peon de Saint-Gilles (1832-1863), the first one full professor of chemistry at the Ecole de Pharmacie, the second a wealthy dilettante.

Their work (Berthelot & Saint Gilles, 1862) will be extensively used by other two important couples of scholars: Guldberg and Waage, Harcourt and Esson. We will speak of them later in the chapter. In the title they referred to etherification but in the paper they speak about esterification, probably a misprint. Arising from his interest in esterification, Berthelot studied the kinetics of reversible reactions. Working with Saint Gilles, he produced an equation for the reaction velocity depending on reactants concentrations. This was incorrect because they did not considered in the expression the inverse reaction. Other interesting considerations was the hypothesis of an exponential dependence on temperature and the fact that equilibrium position is independent from the kind of alcohols and acids used. The conclusions of Berthelot and Sait Gilles was not particularly new compared to those of Wilhelmy. They found similar expressions and both esterification and sugar inversion are good systems for the study of kinetic and equilibrium. May be that the use of differential equations was not usual for the chemists at the time. Significantly Guldberg and Esson was mathematicians that helped later the chemists Waage and Harcourt. The fact that chemists used mathematics so late after decades of data collections may surprise the actual reader, but we have to consider that a systematic study of mathematics was not considered in chemistry courses until after the second world war. This delay did not regarded only Chemical Kinetics but chemistry in general. We have to wait Physical Chemistry, other developments in Analytical Chemistry and a general evolution of Chemistry equipment and instruments to free chemists from very difficult and hardworking experimental praxis for the development of theoretical reflections and laws.

The law of Mass Action Again

The first quantitative expression of the law of mass action was presented by Cato Maximilian Guldberg (1833-1902) and Peter Waage (1839-1900) two years later (Guldberg & Waage, 1864). They was Norwegian Professors of Mathematics and Chemistry at the Christiania University of Oslo and brothers in law (fig. 6). It is interesting to consider the blessed situation of chemistry in Scandinavia coming from the necessities of mining industry and from the large number of eminent chemists like Scheele, Bergman, Berzelius, the same Guldberg and Waage, Arrhenius and Nobel later. Scandinavian insulation and advanced knowledge promoted many autonomous researches and caused often independent contemporary discoveries with other European groups. Despite being isolated as the use of the Laurent-Gherard notation demonstrate, (that notation was diffused more than ten years before), from 1862 and 1864 they repeated and examined experiments and results of Berthelot and Saint Gilles on esterification, Rose's work on Barium salts and that of Scherer on heterogeneous reaction between silica and soda. The study of so different processes derived from the will of the authors to get a new law, universal for all chemical processes. The style of the 1864 paper was polemical against the precedent theories of affinity that the authors considered inconclusive or erroneous.

Figure: 6. Guldberg & Waage

Guldberg and Waage preferred a less speculative and more direct approach simply enunciating the formula for the definition of action of mass and volume:

$$F_{chem} = \alpha \left(\frac{M}{V}\right)^a \left(\frac{N}{V}\right)^b$$

(3)

Where F_{chem} represents the chemical force, M and N the quantity of the reacting substances, V the volume, α, a and b constants wich, other conditions being equal, depends only from the nature of the substances. If one begins with a general system containing four active substances pairwise interacting, (direct and opposite reactions between two reactants and two products), and considering the balance of chemical forces Guldberg and Waage obtained the expression for chemical equilibrium:

$$\alpha (p-x)^a (q-x)^b = \alpha' (p'+x)^{a'} (q'+x)^{b'}$$

(4)

where p, q, p', q' are the initial concentrations of reactants and products, x the amount of transformed reactants at equilibrium reaching, α, a, b, α', a', b', constants with the previous meaning that can be calculated from the initial concentrations, the amount x and experimental data. We can quote a passage of the 1864 paper because it resolves the apparent contradiction between affinities and equilibrium theories towards Proust's and Dalton's Laws.

That a chemical process, as so often is the case in chemistry, seems to occur in only one direction, so that either complete or no substitution takes place, arises easily from our formula. Since the active forces do not increase proportionally to the masses, but according to a power of the same, the relationship of the exponent does not have to be particularly large before the unchanged or changed amount becomes so small that it does not let itself be revealed by our usual analytical methods. Timidly in this paper and in the following works of the two scientists there is the consciousness that the expressions derives from microscopic processes between atoms and molecules. They present a progressive clarification and distinction of the concept of chemical forces, initially considered in a Newtonian way to get the balance-equilibrium expression and in their last paper (Guldberg & Waage, 1879) assimilated to bond strength and reaction velocity, the macroscopic kinetic observable. In this last work are present many interesting intuitions, there is an hypothesis of the microscopic interaction mechanism and from that and the stoichiometric coefficients a try to explain theoretically the exponential coefficients, previously arbitrary or purely phenomenological and there is a mild use of thermodynamic data. Guldberg and Waage went a step further in the correct direction introducing suitable formulas for equilibrium and velocity expressions but do not have the theoretical instruments to justify and interpret it correctly. They examined a huge number of different chemical systems falling in the old trap of getting general laws from the many experiments rather than the good experiment.

THERMODYNAMICS REVOLUTION

The first non systematic introduction of thermodynamics in Chemical Kinetics is due to the second couple of scientists previously cited: Augustus George Vernon Harcourt (1834-1919) and William Esson (1839-1916). Harcourt was an important chemist, member of the Royal society and president of Chemical society, Esson a mathematician and Savilian professor of Geometry. They worked at University of Oxford in a period particularly fruitful for Sciences in Britain. It is the peak of positivism and at the time different sciences, included chemistry, got clearly distinct university courses. Their activity covered a period of fifty years and represented the main passage from natural philosophy speculations to modern scientific reasoning. Influenced by Van't Hoff they will definitively abandon ambiguous terms like Affinity and Chemical Forces. In 1864 Harcourt presented his first publications contemporary to Guldberg and Waage paper. In this work only the name of Harcourt appears but Esson asked the collaboration of a chemist around six years before to applying his mathematical methods to experimental chemistry. In 1865 it was Harcourt that asked Esson to collaborate and their partnership will continue for the rest of their lives (Harcourt & Esson 1865). In the first part of their studies they searched chemical processes suitable for kinetics measurements. Harcourt found an initial valid system: the oxidation of oxalic acid with potassium permanganate. He supposed a two step mechanism:

1. $K_2Mn_2O_8 + 3MnSO_4 + H_2O = K_2SO_4 + 2H_2SO_4 + 5MnO_2$
2. $MnO_2 + H_2SO_4 + H_2C_2O_4 = MnSO_4 + 2CO_2 + 2H_2O$

Verifying it, like today, by the presence of the intermediate manganese oxide and by the acceleration of reaction if manganese sulphate was present at the beginning of the reaction. Examining again the data around 1866 with Esson they plotted a curve of time vs quantity of reactants verifying a logarithmic trend. To get a better plot they needed to interrupt the reaction at will and to analyze the quantity of substances reacted at time of interruption. So they considered another reaction: the oxidation of hydroiodic acid with hydrogen peroxide in presence of definite quantities of thiosulfate and a starch indicator. They measured the time passed before the appearance of blue solutions after the consumption of thiosulfate. In this way they confirmed precisely the logarithmic trend and published their results (Harcourt & Esson, 1867). They extended also an interesting comparison about the energetic of chemical processes. A chemical reaction is like the fall of bodies: the initial activity of reactants is converted in reaction transformation as the potential energy of a falling body in kinetic energy. Reaction velocity so does not remain constant depending on reactants activity. To get a function of velocity they need to consider an infinitesimal time interval introducing, again in analogy with Mechanics, an instantaneous reaction velocity. The velocity of change, equal to the negative time derivative of reactants quantity, is assumed to be proportional to their original quantity y and a constant a depending on the considered system:

$$\frac{dy}{dt} = -ay$$

(5)

Results and methods of this system was published again and better described in other two important papers: the Bakerian Lecture (Harcourt & Esson, 1895) and the last paper written by the authors Harcourt & Esson, 1912, fig. 7). These publications represents probably the most important works for the beginning of modern Chemical Kinetics. They introduced the today common symbol for reaction rate constant k and evaluated formally its dependence on temperature. There is a clear conception of the microscopic nature of chemical processes, they supposed for example that rate constant nullifies at absolute zero, considering that inert atoms and molecules could not encounter and interact each other. To describe temperature dependence of rate constant we will consider the theoretical explanation done by Esson from the experimental ratio between two rate constants at different temperatures found by Harcourt:

$$\frac{k}{k'} = \left(\frac{T}{T'}\right)^m$$

(6)

Where k, k' are the rate constants, T, T' the absolute temperatures and m an experimental pure number. Expressing the equation (6) in differential form m becomes a proportionality constant of infinitesimal changes of the temperatures and rate constants:

$$\frac{dk}{k} = m\frac{dT}{T}$$

(7)

The value of m resulted constant for all temperatures, depending only on the chemical system considered. Considering a big excess of one reactant its chemical activity, (Esson used the term potential but we will use activity to avoid confusion with chemical potential μ), may be supposed constant during reaction course because its quantity remain nearly the same, so the variation in rate constant may be caused only by temperature variation (the precedent argumentation for equation (5) is difficult to use in this case). In this conditions Esson talk about stable conversion of thermal energy to chemical energy with m a constant of proportionality between the different energies. Reconsidering equation (5) and integrating we can obtain an expression for chemical potential energy, (in Esson's terms), whose variation remain constant for the same variation of reactant concentration at different temperatures:

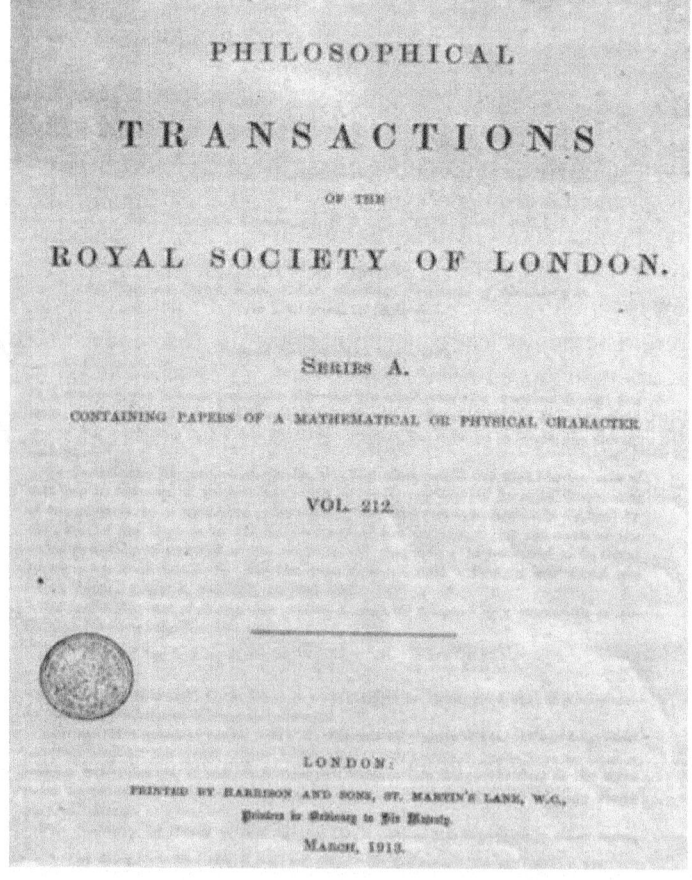

Figure: 7. Front page of the Phil. Trans. volume with the last paper of Harcourt and Esson

$$f(y_2) - f(y_1) = f'(y_2) - f'(y_1) = kt = k't'$$

$$\text{(8)}$$

Where the terms with asterisk derives from a reaction at temperature T'. From (6), (7) and (8) equations we can obtain the following expression:

$$\frac{k}{k'} = \frac{t'}{t} = \left(\frac{T}{T'}\right)^m$$

$$\text{(9)}$$

Contrary to rate constant the reaction times and temperatures are measurable directly, and from the equation (9) we can obtain the value of the exponent m. Once found this relationship from the careful examination of suitable reaction systems Harcourt and Esson checked its validity for a vast number of different reactions: organic, inorganic, biological, in gas phase and so on. In all cases they obtained the value of m different from case to case but constant for different temperatures intervals. Many experiment of Harcourt and Esson was also considered by Van't Hoff and they correlated their law with his thermodynamic hypothesis. They found confirmation also of Van't Hoff parametric formula for m:

$$m = bT^{-1} + a + cT$$

$$\text{(10)}$$

The dependence of m with temperature is for example m=a for dissolution of metals with acids or the action of drugs in muscles, m=cT for decomposition of dibromosuccinic acid, $m=bT^{-1}$ for ethyl acetate hydrolysis with sodium hydroxide. In most of the cases m results constant, but Harcourt and Esson admitted that in some cases this does not happen, contradicting their hypothesis.

For the resolution of this and other problems we have to wait Van't Hoff and Arrhenius but thermodynamics got his entrance into Chemical Kinetics thanks to Harcourt and Esson extensive work, even if it is less famous than that of Guldberg and Waage.

The Birth of Physical Chemistry

The fundamental passage for the development of modern Chemical Kinetics was done when stages of reaction was associated to definite thermodynamic states. This passage was done by a tern of important names: Svante August Arrhenius (1859-1927), Jacobus Hendricus Van't Hoff (1852-1911) and Wilhelm Ostwald (1853-1932), fig. 8.

Figure: 8. From left to right: Arrhenius, Van't Hoff and Ostwald

In the first years of its construction Physical Chemistry practically corresponded to Chemical Kinetics. We will see before the contributions of Van't Hoff and Ostwald the "founders" of Physical chemistry and creators of its first journal: the "Zeitschrift fur Physikalische Chemie", published for the first time in 1887 at Liepzig, fig. 9.

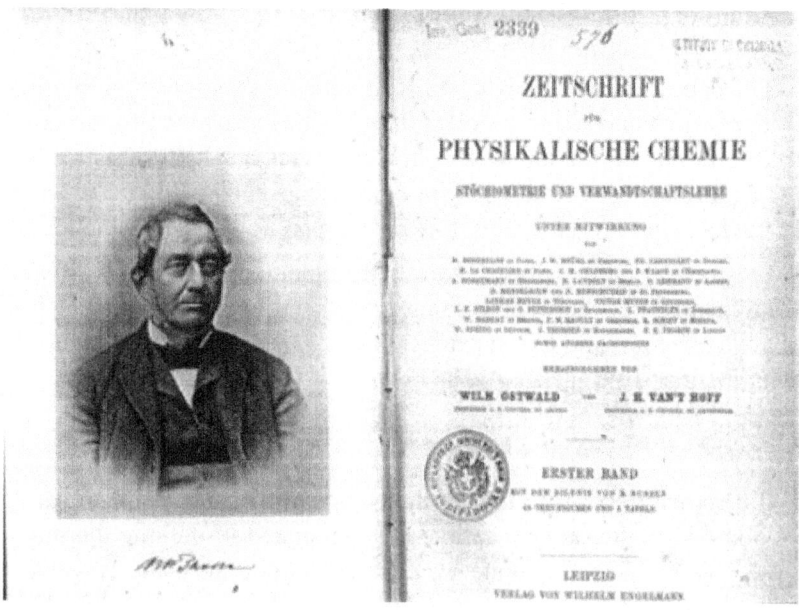

Figure: 9. Title page of the first number of the Zeitschrift

Ostwald and Catalysis

Ostwald contributed directly on Chemical Kinetics less than Van't Hoff and Arrhenius, indeed he is more known for his position in the debate about atoms and for his contributions for the comprehension of Catalysis. His main activity was done at Liepzig, where he became professor of Phisical Chemistry in 1887, after an important academic career at Riga Polytechnic where he wrote his major Opera: "Lehrbuch der Allgemainen Chemie", Treatise on General Chemistry, a reference book for chemistry for many years later. In 1909 he won the Nobel Price for Chemistry thanks to his work on Catalysis. In this period he partially accepted the existence of atoms after the results of Perrin and Einstein on Brownian motion. Even if his direct contribution on Chemical Kinetics was limited it was a field that interested him for all his academic career. His first publications regarded the verification of the Law of Mass Action on different salts hydrolysis reactions (Ostwald, 1879-1884). He later rediscovered also the work of Wilhelmy on the inversion of sugar supposing erroneously that the acids do not react directly but act as accelerator (Ostwald, 1884). That erroneous interpretation was the origin of his interest on catalytic phenomena that we will treat briefly being only partially related at the scope of the chapter. Catalysis was discovered in the first half of 19th Century and initially was considered only as a physical action. After Berzelius studies in this field the phenomenon was considered as a chemical one and its action extended for all chemical reactions. Liebig, a pupil of Berzelius, viewed the phenomenon in terms of the radical theory: Catalysis manifests when the forces of attraction between radicals, (activated species in modern terms), are changed due to the contact with a third body that does not combine with the original reacting species.

The explanations of Catalysis was also considered from an energetic point of view: Mitschelich, Mayer and others thought it as a sort of trigger that discharged an hidden chemical energy by physical contact. Ostwald merged the two approaches: there is not a direct physical catalytic force or action nor a direct modification of the chemical bonds but the thermodynamic of the whole system is changed with new ways of lower free energy in the chemical transformation (Ostwald, 1902). Substantially the actual conception of the phenomena. He pointed out that the development of the new theory of Catalysis was not possible without the development of Chemical Kinetics because it was deeply involved with the velocity of reaction (Ostwald, 1909). The correct interpretation of Catalysis was one of the first big success and confirmation of the Kinetic Theory. We are in debt with Ostwald also for the popularization of Gibbs work, not very known until the end of the 19th Century.

The link between K and k: Van't Hoff

Jacobus Hendricus Van't Hoff (1852-1911), was a Dutch Chemist that worked in Holland and France before joining Ostwald in Germany. He gave essential contribution to many fields of chemistry and physics: from the conception of Stereochemistry (Van't Hoff, 1875), to the thermodynamic explanations of Osmosis and solutions dynamics (Van't Hoff. 1885). For his studies on solutions he won the Nobel Prize for Chemistry in 1901. His essential contributions to Chemical Kinetics, besides the part previously cited in the first part of this chapter, culminated in the discovery of the relation between the rate constant and the equilibrium constant (Van't Hoff, 1884). He interpreted the Chemical Equilibrium as the balance between opposite reactions so he related equilibrium constant to the ratio of the rate constants of the direct and reverse reaction. From an application of Clausius-Clapeyron equation Van't Hoff found the dependence of the equilibrium constant K from the absolute temperature T:

$$\ln K = -\frac{Q}{RT} + C$$

(11)

Where Q represents the isochoric heat of reaction, C an arbitrary integration constant and R the gas constant. Equilibrium constant dependence on temperature is different for exothermic and endothermic reactions, Van't Hoff called this conclusion mobile equilibrium, a principle that Le Chatelier generalized in the same period. From the equation (11) and the relation of K with the rate constants he obtained the phenomenological equation for the dependence of the rate constant k with temperature:

$$\frac{d\ln k}{dT} = \frac{A}{T^2} + B$$

(12)

Where A is related to some not specified heat and B remain indeterminate. In his later works he determined experimentally the values of these constants for many reactions but did not obtain a theoretical interpretation of them.

Van't Hoff classified chemical reactions at microscopic level as mono-bi and poly molecular processes, interpreting the polymolecular processes from their stoichiometry as a sequence of mono and/or bimolecular steps. From these conclusions and the equation (12) Arrhenius will get the basis for his studies.

The Arrhenius Equation

The first hypothesis on the conductibility of ions in electrolytic solutions and on the electrolyte dissociation of acid and basis of the young Swedish chemist Svante August Arrhenius (1859-1927) was not well accepted in his own country. He searched abroad a support for his studies and obtained it from Ostwald and Van't Hoff. He worked with them for six years between 1885 and 1891 and wrote an important paper in 1887 (Arrhenius, 1887). From thereafter his theories on ionic mobility received attention and acceptance and he won the Nobel Prize for chemistry in 1903. After the german period he returned to Sweden and studied the application of Physical chemistry to biology processes giving the basis for Biochemistry (Arrhenius, 1915). With Ostwald and Van't Hoff he worked also on the Kinetics of electrolyte solutions and exposed his most important conclusions in a fundamental paper (Arrhenius, 1989) where he reconsidered the classical case of inversion of sugar with acids. Arrhenius wanted to obtain the phenomenological coefficients of the precedent formulas from the number of ions in solution but found discrepancies between excepted and experimental data at high temperatures. Considering also the contributions due to more frequent collisions with the help of kinetic theory of gases applied to liquid phase he estimated a variation of 2% but the discrepancies was higher, around 15%. Moreover the acidity of the solution, or the number of H+ ions, vary very slowly with temperature (around 0.05% for K°). What really react therefore to justify a so big dependence with temperature? Arrhenius assumed the existence of a new specimen in the reaction: the active sugar. It is the number of molecules of active sugar that determine the velocity of reaction, they are the true reacting species. There is another subordinate equilibrium inside the reaction between sugar and active sugar that determinate its kinetic. He reinterpreted the rate constant as the ratio between the quantities of active and total sugar and evaluated its dependence in function of the temperature:

$$\frac{d\ln k}{dT} = \frac{q}{2T^2}$$

(13)

It is no more necessary to define the constant B from (12) and A is now $q/2$ the half heat of activation of sugar. Arrhenius valuated also successfully the question of the activated part of the acid adding different electrolytes to solution.

The equation for the dependence of velocity of reaction with temperature results:

$$V_1 = V_0 e^{\frac{q(T_1 - T_0)}{2T_1 T_0}} = v_0 e^{-q\left(\frac{1}{T_1} - \frac{1}{T_0}\right)}$$

(14)

Where v_1 and v_0 are the velocities at temperatures T_1 and T_0. From equations (13) and (14) he obtained directly his famous formula for the rate constant:

$$k = Ae^{-\frac{E}{RT}}$$

(15)

Where A is a frequency factor and E the energy of activation. The essential is the introduction of the concept of activation, but the physical explanation of the constants remained vague.

GENESIS AND DEVELOPMENT OF TRANSITION STATE THEORY

Van't Hoff, Arrhenius and Ostwald put the foundation for a formal systematization of Chemical Kinetics but did not achieve a self-consistent theory. Thermodynamics alone was able to treat the reactions from a macroscopic point of view, but results insufficient to fully interpret the microscopic processes. To get a exhaustive picture of the mechanisms from at atomic or molecular scale we will need the application of Statistical Mechanics and the development of Quantum Physics. This is the mainly reason why we have to wait around forty years before a new for a new breakthrough in Chemical Kinetics. Anyway this forty years are characterized by many debates and other discoveries in this field (Laidler & King 1983). First of all the Arrhenius equation, mainly welcomed, created some perplexities in the researchers that studied particular class of reactions where its use was really problematic. Max Bodenstein (1871-1942), a German physical chemist from Heidelberg that collaborated with Walter Nernst in Gottingen and took his chair at the Berlin University after his retirement, was one of these researchers. Bodenstein worked on gas reactions dynamics at the end of 19th Century (Bodenstein, 1899). Reactions in gas phase presents more difficulties and peculiar behaviors respect to liquid ones. Bodenstein accepted the hypothesis of activated species but supposed apparent or false equilibria between them and stable reactants especially for the particular systems he examined. Bodenstein intuited a fully new class of phenomena, what we now call nonequilibrium processes, and initially provoked some interest, but this concept was too early to get a development at the time. Theoretical basis for Transition State Theory, (hereafter called TST), needed a true equilibrium state and this approach become dominant. Other important contributions due to Bodenstein was in clarifying mechanisms of many heterogeneous and catalyzed reactions and the discovery of the mechanism of Chain Reactions around 1920, a field that we will reconsider later analyzing Christiansen work. In this period there was a great attention about the molecularity of mechanisms and of particularly interest was a debate about unimolecular reactions. The debate was that about the so called Radiative Theory (King & Laidler, 1984), proposed mainly by Jean Baptiste Perrin (1870-1942), around 1917. Perrin proposed that unimolecular processes was activated only by blackbody radiation. The hypothesis, fallacious, continued for nearly ten years involving many and important figures as Einstein for example. Even being wrong Radiative Theory represents an interesting case study and boosted the research on different activation causes other than thermal collisions.

Between 1920 and 1930 many scholars like Wigner, Pelzer, Polanyi and Eyring at the Haber Laboratory of the Kaiser Wilhelm Institut of Berlin established a rigorous statistical approach to Chemical Kinetics (Polanyi & Wigner, 1928; Wigner & Pelzer, 1932). Important contributions in this sense was done also by Marcelin that introduced the modern terminology and the Gibbs standard energy of activation and Kramers & Christiansen (Kramers & Christiansen, 1923).

Quantum Mechanical Interpretation

After the achievement of the wave equation for the hydrogen molecule due to Heitler and London the Hungarian Michael Polanyi (1891-1976), director of the Haber Laboratory in Berlin, and his host, the young Mexican American, Henry Eyring (1901-1981) wanted to apply it to the quantum mechanical description of the reaction of atom exchange between ortho and para hydrogen molecules: $H + H_2(orto) = H_2(para) + H$. They generalized that description for others bimolecular gas reactions between trhee atoms in a fundamental paper in 1931 (Eyiring & Polanyi 1931). The energy of the molecular and atom states during reactions was eventually exactly calculated thanks to Quantum Mechanics. They obtained the bond and activation energies plotting the energy in function of the distance between the atoms and molecules involved in the process and built the first diagrams of potential surfaces vs reaction coordinates. You can see an example in fig. 10.

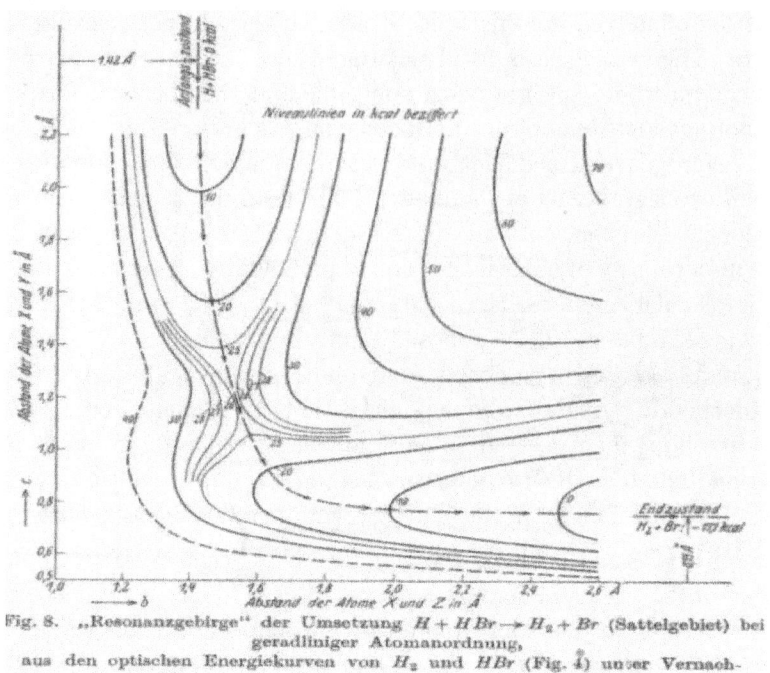

Fig. 8. „Resonanzgebirge" der Umsetzung $H + HBr \longrightarrow H_2 + Br$ (Sattelgebiet) bei geradliniger Atomanordnung, aus den optischen Energiekurven von H_2 und HBr (Fig. 4) unter Vernachlässigung des Coulombschen Anteils berechnet.

Figure: 10. Potential surface for the reaction H + HBr H₂ + Br

TST Presentation

The energetic description of all the configuration states of a chemical system was applied to Chemical Kinetics independently by the two researchers four years later, Polanyi from Manchester (Polanyi & Evans 1935) and Eyring from Princeton (Eyring 1935). The primacy is traditionally given to the most famous of the two, Eyring: the publications had some month of difference but the work was contemporary and a natural consequence of their previous joint work. Absolute reaction rates are obtained statistically from the probability of rising of the reactants molecules from their fundamental state to the saddle of the maximum of the potential surface diagram (the activated complex). Evaluating the ratio of the partition functions of the activated and fundamental state of the reactants and the limitation of the degrees of freedom due to the particular geometry of the reaction surface (the saddle point of the activated complex reduce the degrees of freedom to one) Eyring obtained an Arrhenius type equation with a clear and definite value of the pre-exponential and exponent factors:

$$k = \left(\frac{k_b T}{h}\right) e^{-\frac{\Delta G^{\ddagger}}{RT}}$$

(16)

Where G$\ddagger$ is the Gibbs energy of activation, kB is Boltzmann's constant, and h is Planck's constant. Eyring considered also the possible variations of the equation (16) due to the molecularity of the reaction.

The paper of Evans and Polanyi (Polanyi & Evans 1935) presented similar conclusions to that of Eyring but moreover tried to evaluate the interactions and energy exchanges between the reactants and the other actors of the chemical system (the solvent for example). The investigations of Evans and Polanyi are not a simple detail, because they make evident the limits of the TST. The most known of them are the appearance of unexpected products due to particular form of the saddle surface, the tunnel effect through low energy barriers, the population of higher energy states rather than the only saddle state for high temperature reactions. A methodological limit is the vision of the process as a "big" isolated molecule where all the actors: reactants, activated complexes and products are contemporary presents and in equilibrium. This picture is valid when the main process is the establishment of the equilibrium between fundamental and activated states but results fallacious when other processes, as the interaction with solvent in diffusion controlled reactions for example become dominant. The other picture, less known, that sees the reaction as a process of diffusion will be examined in the next and last part of the chapter.

GENESIS AND DEVELOPMENT OF DIFFUSIVE-STOCHASTIC THEORIES

The diffusion description, elaborated by Christiansen around 1935 (Christiansen 1936) and fully systematised in 1940 by Kramers (Kramers 1940), was an interesting and successful method complementary to transition state theory (TST). It received, however, little or no attention in chemistry circles for a long time (Zambelli 2010).

Hendrik Anthony Kramers (1894 –1952) was a Dutch physicist. He worked mainly in Germany and Denmark and was one of the most important collaborator of Bohr in the famous Copenhagen Institute of Theoretical Physics. His interest in Chemical Kinetics derived from the collaboration with Jens Anton Christiansen (1888–1969), later full professor of Physical Chemistry at the Copenhagen University, around 1922.

Christiansen visited the Bohr Institute after his PhD graduation for a period of nearly one year. It is possible that he already came to Copenhagen with the hope of finding some mathematical-physical assistance for his studies of chemical reactions. Christiansen's studies treated the dynamics of specific chemical reactions: in this PhD Thesis he introduced for the first time the term chain reactions (ketten reaction in Danish). His developments in this field together with that of Bodenstein previously cited resulted fundamental for the work of Nikolay Semenov (1896-1986) and Cyril Norman Hinshelwood (1897-1967) that will produce a definitive theory on chain reactions around 1950.

Christiansen's Approach

Christiansen tried to apply the description and the model of chain reactions to different mechanisms (Christiansen 1922) and wrote a paper with Kramers in 1923, cited previously, about unimolecular reactions confronting the activation mechanism due to thermal collisions and radiation absorption. They treated the radiation mechanism with the fundamental Einstein's quantum theory about matter-radiation interaction (Einstein 1917). Other work of Einstein and Smoluchowski will be necessary later for Christiansen-Kramers approach. After the paper the collaboration probably ended and the two researchers will reconsider separately these arguments around fifteen years later. Christiansen developed the model of a chemical reaction as an intra-molecular diffusion process in the half of the thirties. He published two papers in 1935 (Christiansen 1935) and 1936 (Christiansen 1936) on this research. The paper of 1936 is particularly significant. Christiansen confronted Arrhenius's theory of activated states with a little known theory (Nernst 1893) of Walther Hermann Nernst (1864–1941). In Nernst's theory, the reaction velocity is obtained, by analogy with Ohm's law, as the ratio between a chemical potential and a chemical resistance. Christiansen intended the chemical potential as the difference of the chemical activities of the beginning and the final states and the chemical resistance was represented by a particular integral depending on temperature and diffusion constant. The purpose of Christiansen was to demonstrate, extending Arrhenius's conception, that the methods of Nernst and Arrhenius are analogous. The generalization of Arrhenius's theory is obtained by supposing an open, possibly infinite, sequence of many consecutive steps, thus gaining an expression consistent with that of Nernst. Christiansen discretized a chemical reaction considering not only one activated state, as in Arrhenius's model, but a series of consecutive n stages which result in reciprocal virtual equilibrium. The equilibria between reactants, products and intermediates are supposed to be valid because Christiansen considered the quantity of intermediates constant during the slow stages of reaction, so the process is stationary or quasi-stationary. These are a group of assumptions similar to those made in the theory of diffusion. In fact, according to Christiansen's hypothesis, the equilibrium quantity of the activated complexes may be put in relationship to the concentrations of a diffusing substance along the sections of a column. From this diffusive description, he obtained an expression for the reciprocal reaction rate which was consistent with that obtained on a thermodynamic basis. Christiansen expressed the velocity of reaction v in the form of a diffusion equation:

$$v = -\frac{D}{\varphi}\frac{\partial c\varphi}{\partial x}$$

(17)

Where D and φ are the diffusion and activity coefficients, c the concentration and x a reaction coordinate. This expression implies that the transport of molecules is produced by the concentration gradient and by molecular forces (their contribution represented by the activities φ). From (17) and other assumptions about the activity coefficient Christiansen obtained another equation analogous to that of Einstein and Smoluchowski about Brownian motion:

$$v = -D\frac{\partial c}{\partial x} + \frac{D}{RT}cK$$

(18)

The generalization of the Arrhenius conception brings us naturally to consider the transformation of a molecule during a reaction as an intra-molecular diffusion. To demonstrate this generalization Christiansen made some fundamental assumptions. He considered the case of a simple potential barrier, a symmetrical bi-stable one as shown in Fig. 11. This case will be examined better thanks to Kramers work of 1940.

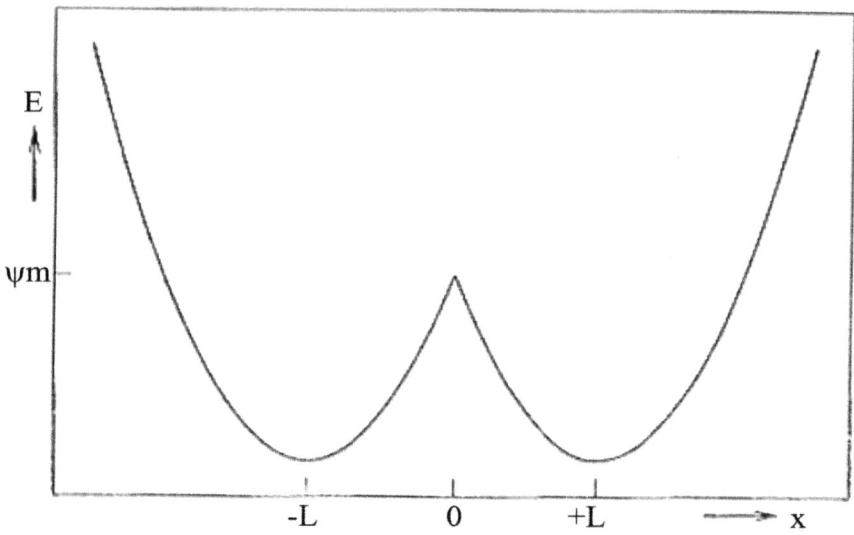

Figure: 11. Bi-stable potential barrier, the figure is taken from Christiansen's paper

The Application of Klein-Kramers Equation to Chemical Kinetics

The main biographers of Kramers, in particular Dirk ter Haar, claim that his interest in Chemical kinetics is a simple mathematical exercise of style. This may be partially true but Kramers's work would be impossible without Christiansen's previous contribution and his collaboration with Oskar Benjamin Klein (1894–1977) a Swedish theoretical physicist student of Arrhenius that during the years from 1917 to 1921 travelled many times back and forth between Copenhagen and Stockholm to complete his PhD thesis in which he examined the forces between ions in strong electrolyte solutions. The result was a generalized description of liquid dynamics and the formulation of what we call today the Klein– Kramers equation (Klein 1922).

Kramers paper of 1940 presents what today we call the "Kramers problem": the dynamics of a particle moving in a bi-stable external field of force, subject to the irregular forces of a surrounding medium in thermal equilibrium. The particle, originally caught in a potential well, may escape by passing over a potential barrier. Constructing a diffusion equation for the density distribution of particles in phase space it is possible to calculate the probability and the escape rate as a function of the temperature and viscosity of the medium. Kramers considered the one dimensional motion of a particle of unit mass starting from a Langevin equation of the system for the time derivative of the velocity v:

$$\frac{dv}{dt} = -\xi v + f(t) - \frac{dV(x)}{dx}$$

(19)

where V(x) is the potential field, x the friction and f(t) a time dependent stochastic force. Searching for the distribution law of the particle in phase space on the basis of a given distribution of the random forces he obtained the diffusion equation for the particle distribution from the statistical moments of the random force:

$$\frac{\partial P(x, x_0, v, v_0, t)}{\partial t} = -\Gamma P(x, x_0, v, v_0, t)$$

(20)

Where the operator Γ is defined as:

$$\Gamma = v \frac{\partial}{\partial x} - \frac{dV(x)}{dx} \frac{\partial}{\partial v} - \xi \frac{\partial}{\partial v} \left(v + k_B T \frac{\partial}{\partial v} \right)$$

(21)

This express the Klein-Kramers equation of the system. Kramers found solutions of equation (21) in the stationary cases in a good range of viscosity. We remark that the stationarity condition is not strictly an approximation, but rather the simplest case of a non-equilibrium state of a system. If we consider the potential surface of a chemical system and assume the role of the solvent in the viscosity coefficient Kramers approach describes efficiently the course of chemical reactions.

CONCLUSIONS

Kramers found the Eyring-Polanyi equation (16) as a particular case of medium-small viscosity. It may seem impressive to see that the results of TST, based on quantum mechanics, come out as a particular case of Kramers pure classical method. But there are precise limitations to the use of Kramers method. Methodologically, even if the diffusive stochastic approach has some theoretical advantages, it is more difficult to adapt and apply to the description of chemical reactions than TST. It requires notable mathematical knowledge and physical concepts that are not so familiar in chemistry. TST on the other hand, relying on the powerful means of quantum mechanics, produces more predictive results, although we have to apply phenomenological coefficients in some cases and make some arbitrary assumptions. Diffusion approach, although less known, and TST represents the basic theories for contemporary studies on Chemical Kinetics, a disciple that now is a part of Physical Chemistry but that before Quantum Mechanics corresponded practically to it and that contributed so deeply, as we have seen, to the whole construction of Chemical Science.

REFERENCES

1. Arrhenius, S.A. (1887). Über die Dissociation der in Wasser gelösten Stoffe, Z. Phys. Chem., 1, pp. 631-648, Available in English from http://www.chemteam.info/Chem-History/Arrhenius-dissociation.html

2. Arrhenius, S.A. (1915). Quantitative laws in biological chemistry, London, U.K.

3. Arrhenius, S.A. (1889). Uber die Reaktionsgeschwindigkeit bei der Inversion von

4. Rohrzucker durch Säuren, Z. Phys. Chem., 4, 1889, pp. 226-248

5. Berthelot, P.E.M. & Saint Gilles, L.P. (1862). Recherches sur les affinités de la formation et de la decomposition des ethers, Annales de chimie et physique (3) 65 1872, 66 1872, 68 1873, pp. 385-422, pp. 5-110, pp. 225-359, Available from http://gallica.bnf.fr/ark:/12148/bpt6k34806t/f383.image.langEN

6. Bodenstein, M. (1899). Gasreaktionen in der chemischen Kinetik, I, II, III, Z. Phys. Chem. 29, pp. 147-158, pp. 295-314, pp. 315-333

7. Christiansen, J.A. & Kramers, H.A. (1923). Über die Geschwindigkeit chemische Reaktionenen, Z. Phys. Chem. 104, pp. 451-471

8. Christiansen, J.A. (1922). Über das Geschwindigkeitsgesetz monomolekularer Reactionen, Z. Phys. Chem. 103, pp. 91-98

9. Christiansen, J.A. (1935). Einige Bemerkungen zur Anwendung der Bodensteinschen Methode der stationären Konzentrationen der Zwischenstoffe in der Reakionskinetik. Z. Phys. Chem. 28B, pp. 303-310

10. Christiansen, J.A. (1936). Über eine Erweiterung der Arrheniusschen Auffassung der chemischen Reaction, Z. Phys. Chem. 33B, pp.145-155

11. Einstein, A. (1917). Zur Quantentheorie der Strahlung, Phys. Z. 18, pp 121-128, Available in English from http://books.google.com/books?id=8KLMGqnZCDcC&pg=PA63&ots=h9g_x_ptx u&dq=%22the+formal+similarity+between+the+chromatic%%2022%22&sig=rrtVd32Es QTQUsYa e3hRnkunW0I#v=onepage&q&f=false

12. Evans, M.G. & Polanyi M. (1935). Some applications of the transition state method to the calculation of reaction velocities, especially in solution, Trans. Faraday Soc. 31, pp. 875-893

13. Eyring, H. & Polanyi M. (1931). Uber einfache Gasreaktionen, Z. Phys. Chem. B 12, pp. 279- 311

14. Eyring, H. (1935). The Activated Complex in Chemical Reactions, J. Chem. Phys. 3, pp. 107- 115

15. Geoffroy, E.F. (1718). Table des différents rapports observés en chimie entre différentes substances, Mémoires de l'Academie Royale des Sciences, pp. 202-212, Available from http://gallica.bnf.fr/ark:/12148/bpt6k3519v/f330

16. Gladstone, J.H. (1855). On Circumstances Modifying the Action of Chemical Affinity, Phil. Trans. Roy. Soc. London 175, pp. 179-223, Available from http://www.jstor.org/stable/108516

17. Guldberg, C. M. & Waage, P. (1864). Etudes sur l'Affinité, Forhandlinger: Videnskabs-Selskabet i Christiana, 35, Available in English from http://www.nd.edu/~powers/ame.50531/guldberg.waage.1864.pdf

18. Guldberg, C.M. & Waage, P. (1879). Über Die Chemische Affinität, Journal Prakt. Chem., 127, pp. 69-114

19. Harcourt, A. G. V. & Esson W. (1865). On the Laws of Connexion between the Conditions of a Chemical Change and Its Amount, Phil. Trans. London 156, 1866, pp. 193-221, Available from http://www.jstor.org/stable/108945

20. Harcourt, A.G.V. & Esson W. (1867). On the Laws of Connexion between the Conditions of a Chemical Change and Its Amount. II. On the Reaction of Hydric Peroxide and

21. Hydric Iodide. Proc. Roy Soc. 15, 1867, pp. 262-265, Available from http://www.jstor.org/stable/112633

22. Harcourt, A.G.V. & Esson W. (1895). Bakerian Lecture: On the Laws of Connexion between the Conditions of a Chemical Change and Its Amount. III. Further Researches on the Reaction of Hydrogen Dioxide and Hydrogen Iodide, Phil. Trans. London 186A, 1895, 817-895

23. Harcourt, A.G.V. & Esson W. (1912). On the Variation with Temperature of the Rate of a Chemical Change, Phil. Trans. London 212A, 1913, 187-204

24. King, M.C. (1981). Experiments with time, Ambix 23, pp. 70-82

25. King, M.C. & Laidler, K.J. (1983). The Development of Transition-State Theory, J. Phys. Chem. 87, pp. 2657-2664, Available from http://www.qi.fcen.uba.ar/materias/qf2/TCA.pdf

26. King, M.C. & Laidler, K.J. (1984). Chemical kinetics and the radiation hypothesis, Archive for History of Exact Sciences 30, 1, pp. 45-86

27. Klein, O. (1922). Zur statistischen Theorie der Suspensionen und Lösungen, Arkiv Mat. Astr. Fys. 16(5), pp 1-51, Available from http://

su.diva-portal.org/smash/record.jsf?pid=diva2:440187

28. Kramers, H.A. (1940). Brownian Motion in a Field of Force and the Diffusion Model of Chemical Reactions, Physica 7, pp. 284-304, Available from   http://www-lpmcn.univ-lyon1.fr/~barrat/phystat-he/kramers1940.pdf

29. Lavoisier, A.L. (1782). Considerations sur la dissolution des metaux dans les acides, Mémoires de l'Académie des sciences 1782, pp. 492-527, Available from http://www.lavoisier.cnrs.fr/ice/ice_book_detail-fr-text-lavosier-Lavoisier-49- 5.html

30. Nernst, W. (1893). Über die Beteiligung eines Lösungmittels an chemischen Reaktionen. Z. Phys. Chem..11, pp 345-359

31. Ostwald, W. (1879-84). Chemische Affinitätsbestimmungen, J. Prakt. Chem., 19, 1879, pp. 468- 484; 22, 1880, pp. 251-260; 23, 1881, pp. 209-225, 517-536; 24, 1881, pp. 486-497; 29, 1884, pp. 49-52

32. Ostwald, W. (1884). Studien zur chemischen Dynamik. III. Die Inversion des Rohrzuckers, J. Prakt. Chem. 29, pp. 385-408

33. Ostwald, W. (1902). Uber Katalyse, Liepzig, 1902

34. Ostwald, W. (1909). Grundriss der allgemeinen Chemie, Liepzig, 1909

35. Pelzer, H. & Wigner, E. (1932). Über die Geschwindigkeitskonstante von Austauschreaktionen, Z. Phys. Chem. 15B, pp. 445-471

36. Polanyi, M. & Wigner, E. (1928). Über die Interferenz von Eigenschwingungen als Ursache von Energieschwankungen und chemischer Umsetzungen, Z. Phys. Chem. 139, pp. 439-452

37. Van't Hoff, J.H. (1875). Chimie dans l'espace, Rotterdam, 1875

38. Van't Hoff, J.H. (1884). Etudes de dynamique chimique, Amsterdam, 1884

39. Van't Hoff, J.H. (1885). L›Équilibre chimique dans les Systèmes gazeux ou dissous à I›État dilué, Recueil des Travaux Chimiques des Pays-Bas 4, 12, pp. 424-427

40. Wilhelmy, L. (1850). Über das Gesetz, nach welchem die Einwirkung der Säuren auf den Rohrzucker stattfindet, Pogg. Ann. 81, pp. 413-433, Available from http://gallica.bnf.fr/ark:/12148/bpt6k15166k/f427.table

41. Zambelli, S. (2010), Chemical kinetics and diffusion approach: the history of the KleinKramers equation, Arch. Hist. Exact Sci. 64, 4, pp. 395-428

Chapter 7

A GREATLY UNDER-APPRECIATED FUNDAMENTAL PRINCIPLE OF PHYSICAL ORGANIC CHEMISTRY

Robin A. Cox

Formerly Department of Chemistry, University of Toronto, 80 St. George St., Toronto, ON, M5S 3H6, Canada

ABSTRACT

If a species does not have a finite lifetime in the reaction medium, it cannot be a mechanistic intermediate. This principle was first enunciated by Jencks, as the concept of an enforced mechanism. For instance, neither primary nor secondary carbocations have long enough lifetimes to exist in an aqueous medium, so S_N1 reactions involving these substrates are not possible, and an S_N2 mechanism is enforced. Only tertiary carbocations and those stabilized by resonance (benzyl cations, acylium ions) are stable enough to be reaction intermediates. More importantly, it is now known that neither H_3O^+ nor HO^- exist as such in dilute aqueous solution. Several recent high-level calculations on large proton clusters are unable to localize the positive charge; it is found to be simply "on the cluster" as a whole. The lifetime of any ionized water species is exceedingly short, a few molecular vibrations at most; the best experimental study, using modern IR instrumentation, has the most probable hydrated proton structure as $H_{13}O_6^+$, but only an estimated quarter of the protons are present even in this form at any given instant. Thanks to the Grotthuss mechanism of chain transfer along hydrogen bonds, in reality a proton or a hydroxide ion is simply instantly available anywhere it is needed for reaction. Important mechanistic consequences result. Any charged oxygen species (e.g., a tetrahedral intermediate) is also not going to exist long enough to be a reaction intermediate, unless the charge is stabilized in some way, usually by resonance. General acid catalysis is the rule in reactions in concentrated aqueous acids. The Grotthuss mechanism also means that reactions involving neutral water are favored; the solvent is already highly structured, so the entropy involved in bringing several solvent molecules to the reaction center is unimportant. Examples are given.

INTRODUCTION

In recent years, the study of the mechanisms of organic reactions has been considerably enhanced by the study of putative reaction intermediates [1], often under conditions in which the species are stable enough for spectroscopic examination. For instance, carbocations and other species have been studied extensively in superacid media by Olah and his colleagues [2–4]. However, if a species is to be a reaction intermediate, it has to be stable enough to have a lifetime of at least a few molecular vibrations under the reaction conditions, say greater than 10^{-13}–10^{-14} s [5]. Jencks pointed this out a number of years ago now [6], as the concept of an "enforced mechanism"; if a species cannot exist under the reaction conditions a mechanism involving it is impossible, and an alternate one is "enforced".

At the time Jencks wrote his review [6] not a lot was known about the lifetimes of putative reaction intermediates. However, more is known now, and although it is still not easy to apply, the author believes that much more attention has to be paid to what I might call the "Jencks Principle". For instance, it is certain that primary carbocations cannot exist in a primarily aqueous medium [7], although mechanisms involving them are still occasionally proposed [8]. It is now apparent that this is true of secondary carbocations too [9,10]. In some (but not all) textbooks one still sees mention of "mixed S_N1 and S_N2" mechanisms involving secondary substrates [11], due primarily to the early work of the Hughes and Ingold school [12,13], which has since been discredited [13]. It is now well established that secondary substrates react by an S_N2 process [14], for instance as shown in Scheme I, although for the example shown [15,16] the specific mechanism given is still speculative. The scheme is drawn this way in consequence of the observation that hydroxide ion does not add to carbonyl groups directly, but instead attacks a water molecule which does the actual addition [17–19]. Enough kinetic evidence to prove or disprove this probably exists [15,16], and work to do this is underway [20]. Hydroxide ion is not very reactive. It is less solvated, and hence much more reactive, in alcohol solvents, and in pure DMSO its reactivity is increased by some twelve orders of magnitude [21].

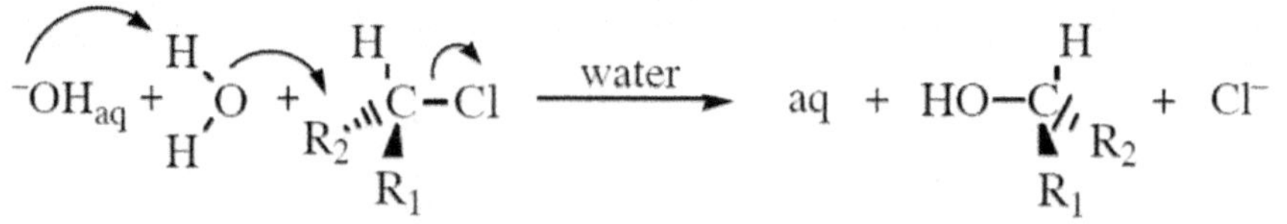

$$\text{$^-$OH}_{aq} + \overset{H}{\underset{H}{O}} + \underset{R_1}{\overset{H}{R_2\text{\textbackslash\textbackslash}C-Cl}} \xrightarrow{\text{water}} aq + HO-\underset{R_1}{\overset{H}{C/\!/R_2}} + Cl^-$$

Scheme I: S_N2 substitution of a secondary alkyl halide by hydroxide ion.

For the mechanisms of reactions in aqueous media, far more important is the observation that species such as H_3O^+ (usually called the Eigen cation [22]), $H_5O_2^+$ (usually called the Zundel cation [23,24], although also strongly preferred by the school of Vinnik and Librovich at the Institute of Physical Chemistry in Moscow [25]), $H_9O_4^+$ (first postulated by Bell [26], but often (mistakenly) also called the Eigen cation) and the many others which have been proposed [27] (not that there has ever been any believable experimental evidence for any of them [28,29]) do not have lifetimes long enough to exist. Although far less work has been done, recent studies show that HO^- cannot exist as such in water either [30–32]. Recent very high-quality IR measurements on acid solutions [33,34] show that the only structure that has any kind of real existence in them is the proposed $H_{13}O_6^+$ [35], shown in Scheme II [34], but even this has a very short lifetime; the authors state [36]: "The lifetime of the five central protons is close to the time of their vibrational transitions. In ~70% of these cations it is shorter than the time of normal vibrations and the IR spectrum degenerates to a continuum absorption". In addition, in several modern theoretical calculations on proton clusters containing many water molecules it is found not to be possible to isolate the positive charge, it is simply "on the cluster" as a whole [37].

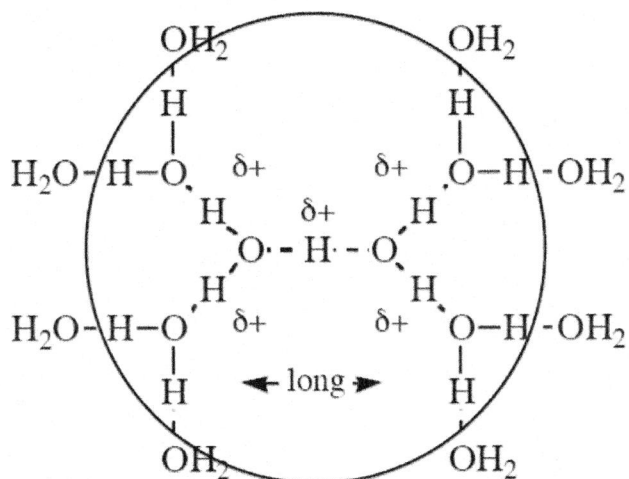

Scheme II: Structure of the only solvated proton species detected in water.

Consequently, we may only speak of "H_{aq}^+" and "HO_{aq}^-" as being reactants [28–34]. The Grotthuss chain transfer process along hydrogen bonds in water simply ensures that a proton or a hydroxide ion is available instantaneously where or when it is needed. (This is such a widely accepted transport mechanism in water that specific references to it are difficult to find. The original is [38]). This has all kinds of consequences for reaction mechanisms in predominantly aqueous acidic and basic media. For instance, we can no longer speak of "general" and "specific" acid and base catalysis of reactions. Far better to speak of "pre-equilibrium proton transfer", in the case of reactions that involve the formation of a stable ionized intermediate (usually by resonance), and of "proton transfer as part of the rate-determining step", in the other cases. Several examples follow. The highly structured nature of liquid water [39] also ensures that reaction mechanisms involving several water molecules acting in concert are also favored. The entropy involved in bringing water molecules into the right positions is not a concern as the structure is already there, and the Grotthuss process ensures that all proton transfers are essentially instantaneous. Several examples of reactions of this type will be given as well.

RESULTS AND DISCUSSION

General Acid Catalysis in Strong Acid Media

As far as the common strong acids HCl, $HClO_4$ and H_2SO_4 are concerned, the only acid species present is "H_{aq}^+" under normal conditions, and reactions in all of them therefore ought to proceed at the same rate at the same acid concentration [40].

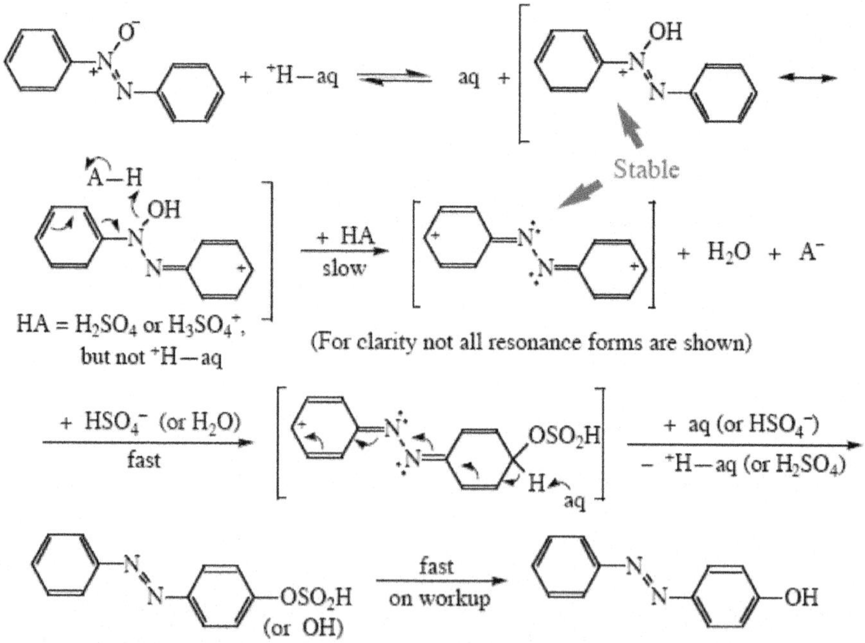

Scheme III: Wallach rearrangement of azoxybenzene in sulfuric acid.

Sulfuric acid is the only one that can be used from 0 wt% to 100 wt%, the dilute solution containing H_{aq}^+. Above the 1:1 $H_2O:H_2SO_4$ molecular ratio (84.48 wt%) there is, of course, no free water present, but the solution now contains catalytically active undissociated sulfuric acid molecules. Above 99.5 wt% autoprotolysis becomes important, with the very strong acid species $H_3SO_4^+$ present as a possible catalyst as well [41]. I found catalysis by both of the latter species as far back as 1974 in the Wallach rearrangement of azoxybenzene, Scheme III [41–43]. This reaction has been extensively reviewed [44,45], so I will not say much about it here. The species which are stable enough to exist in the reaction solution are indicated in the Scheme; interestingly, both of them have been observed experimentally under stable ion conditions [4]. Theoretical calculations have shown the dicationic species to have the structure shown, with little communication between the two halves of the molecule [42].

Interestingly H_{aq}^+ is not a strong enough acid species to catalyze the reaction, only catalysis by H_2SO_4 and by $H_3SO_4^+$ being observed [41,44]. The reaction does not work in $HClO_4$, a stronger acid system in H_0 terms but only containing H_{aq}^+ with no undissociated $HClO_4$ molecules present [45,46]. It does go in pure FSO_3H and $ClSO_3H$, both being quite strong acid species [46].

Another case of general acid catalysis was observed in the hydrolysis of several ethyl thiolbenzoates in sulfuric acid at concentrations above 60 wt%, where catalysis by H_{aq}^+ was observed, catalysis by undissociated H_2SO_4 molecules taking over above 80 wt% in concentration [47], Scheme IV.

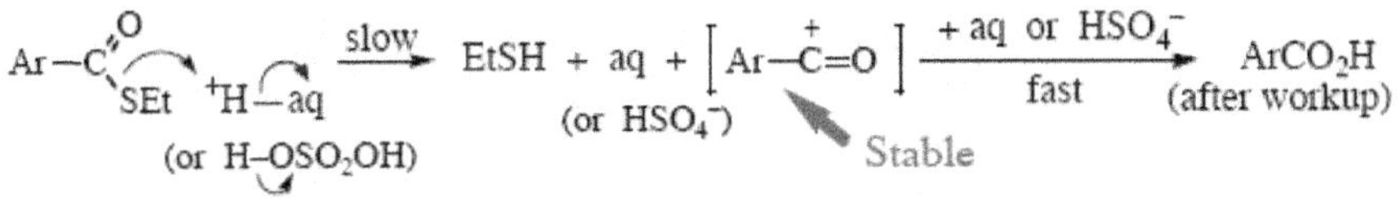

Scheme IV: Hydrolysis of ethyl thiolbenzoates in sulfuric acid.

Ether Hydrolyses

The hydrolyses of trioxane and similar molecules in dilute acid have been taken by many authors (even by myself [48]) to be typical A1 processes, protonation followed by rate-determining breakup of the protonated intermediate. However, if H_3O^+ cannot exist in water, other species with positive charge on oxygen which is not resonance-stabilized are not going to be capable of existence either. This means that the mechanism of the hydrolysis of trioxane is going to be that given inScheme V. (Scheme V shows the breakup to three formaldehyde molecules taking place all at once, but a similar stepwise breakup is of course also possible.)

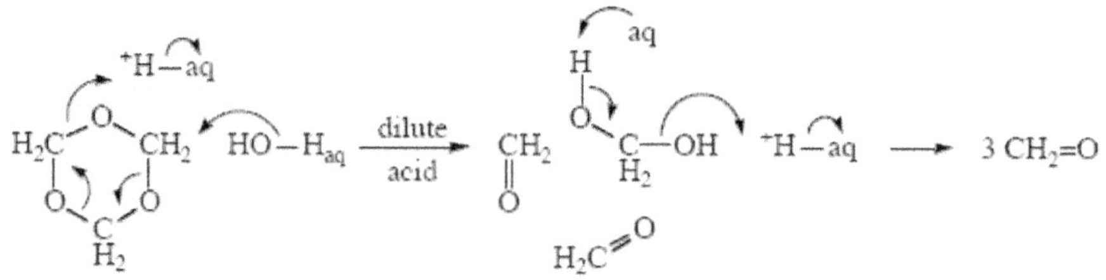

Scheme V: Hydrolysis of trioxane in dilute acid.

There is plenty of kinetic data on this reaction in several different acid media available for analysis [49]. The preferable method to use for this is the excess acidity correlation analysis [48], which is used here. The applicable rate equation is shown as Equation 2.

$$k_\psi C_S = k_0 a_S a_{H_2O} a_{H^+_{aq}} / f_\ddagger = k_0 C_S a_{H_2O} C_{H^+_{aq}} \frac{f_S f_{H^+_{aq}}}{f_\ddagger}$$

(1)

$$\log k_\psi - \log C_{H^+_{aq}} - \log a_{H_2O} = \log k_0 + m^\ddagger m^* X$$

(2)

Here the observed rate constants are k_ψ [49], the medium-independent rate constant (*i.e.*, the rate constant in the aqueous standard state) is k_0, the proton concentration is $C_{Haq}{}^+$, the water activity is $a_{H_2}O$ and the excess acidity is X, all available data for all three acid systems [48]. The slope parameters m^* and $m^\ddagger$ describe the behavior of the protonated substrate and the transition state as the acidity changes, necessarily combined here [48]. Plots according to Equation 2 are given in Figure 1.

As can be seen, the plots for all three acids are accurately linear. For illustration purposes a thick line is given for all of the data combined, slope 1.333 ± 0.022, intercept -9.198 ± 0.018, correlation coefficient 0.993 over 54 points. However, the points for the three individual acids fall (very accurately, correlation coefficients 0.9990 in HCl, 0.9994 in $HClO_4$, 0.9994 in H_2SO_4) on slightly different lines, which undoubtedly reflects the fact that the water activities for the three acids are not known equally well. Water activities in the aqueous sulfuric acid medium [50] are very accurately known [51], but this is not the case for HCl [52–54] and, particularly, $HClO_4$ [55–58]. All of the plots fit the appropriate lines more closely than was previously found by treating the process as a traditional A1 reaction [48].

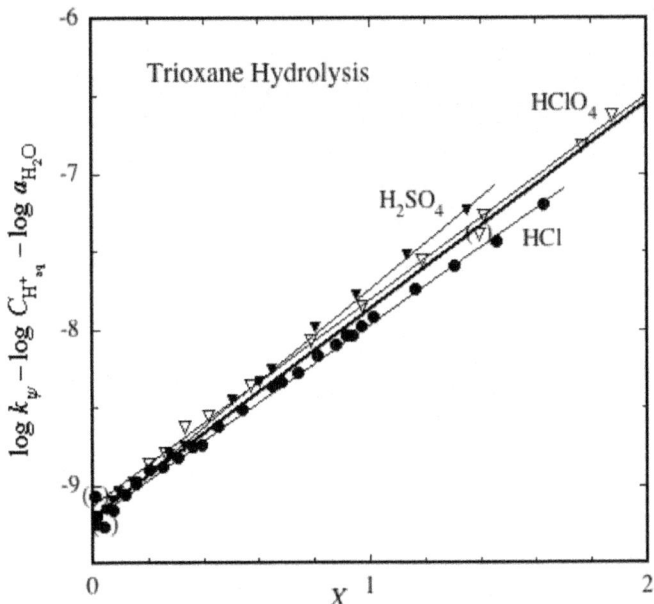

Figure 1: Excess acidity plot for trioxane hydrolysis in dilute H_2SO_4, HCl and $HClO_4$.

If this process is really a case of general acid catalysis, rates measured in aqueous buffer systems should show this. Trioxane hydrolysis is too slow a reaction to have been studied in this way, but the closely related hydrolysis of paraldehyde (the acetaldehyde trimer) is much faster [48], and evidence for general acid catalysis has indeed been found [59,60], although this fact does not seem to be widely known (or has been ignored). A plot like Figure 1 can also be drawn for paraldehyde, but the kinetics cover a much smaller acidity range, and the scatter is bad.

Another ether system for which kinetic results are available [61] is the hydrolysis of diethyl ether at high temperatures and high acidities in aqueous sulfuric acid. The mechanism proposed here is shown in Scheme VI.

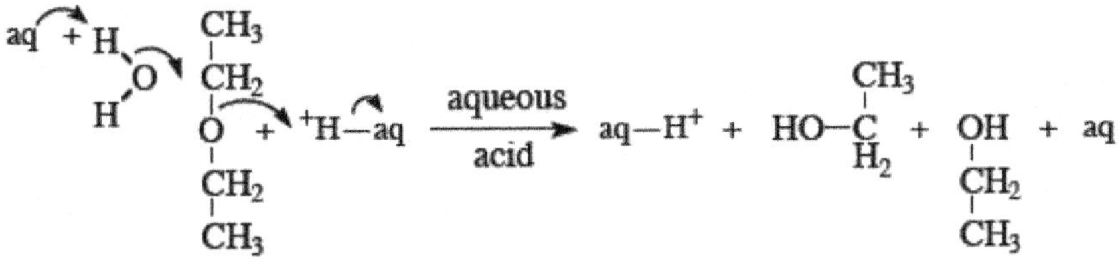

Scheme VI: Acid hydrolysis of diethyl ether.

This is essentially the same mechanism as that shown in Scheme V, and the same excess acidity rate equation, Equation 2, applies. In sulfuric acid this mechanism is only going to apply as long as there is free water available, *i.e.*, not above a concentration of 85.48 wt%. Above this acidity another well-characterized mechanism takes over [61], involving a much faster direct reaction between the diethyl ether and SO_3, which is available for reaction above this acidity. Thus in an excess acidity plot one would expect linearity below 85.48 wt%, and an upward deviation above this point. This is exactly what is observed, as Figure 2 illustrates.

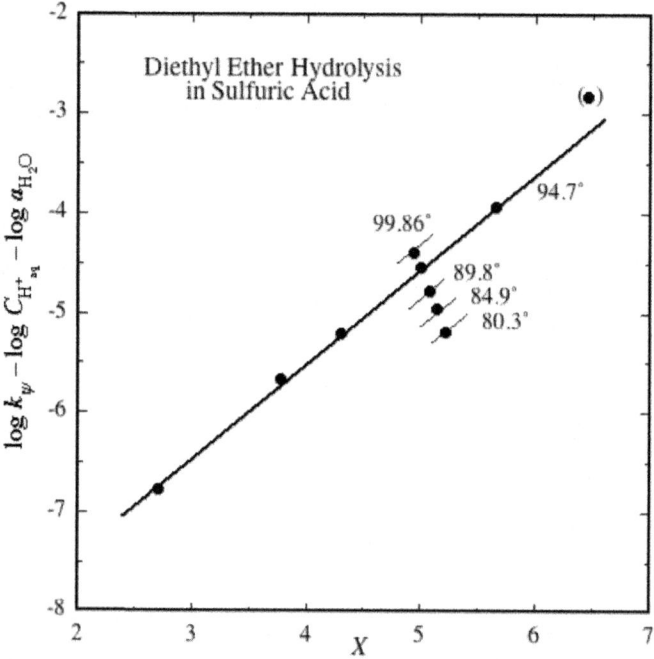

Figure 2: Excess acidity plot for the hydrolysis of diethyl ether in relatively concentrated H_2SO_4, at several temperatures.

The topmost point in Figure 2 is at an acidity of 90 wt%, and deviates upwards as expected. (In the original paper [61] a plot of log rate constant against acidity curves downward over the acidity region which gives linearity here.) The $m^*m^{\ddagger}$slope is 0.949 ± 0.015, and as different temperatures are available, the activation parameters for the reaction can be calculated: $\Delta H^{\ddagger}$ = 32.8 ± 1.4 kcal·mol^{-1}; $\Delta S^{\ddagger}$ = −12.4 ± 4.7 cal·deg^{-1}·mol^{-1}, both perfectly reasonable numbers. (They only concern the substrate, as X, log $C_{H_{aq}}{}^{+}$and log $a_{H2}O$ have all been corrected to the reaction temperature, as before [48].) The correlation coefficient is 0.9993.

Figures 1 and 2 constitute strong evidence in favor of the mechanisms given here. Interestingly, it does not matter whether the substrate can be considered to be primarily protonated at the acidity of the reaction or not; oxygen-protonated species in which the charge cannot be delocalized are not going to be reaction intermediates as their lifetimes are too short! When the charge **can** be delocalized, intermediate lifetimes are much longer. For instance, the methoxymethyl cation, where the charge is delocalized over carbon and oxygen, is calculated to have a lifetime of about 1 ps [62], which, although short, is quite long enough for it to be a reaction intermediate.

Amide Hydrolyses

Benzamides, and presumably other suitable amides, have two hydrolysis mechanisms [63]. In weakly acidic aqueous H_2SO_4 media, a pre-equilibrium proton transfer gives a stable delocalized protonated amide intermediate, to which water adds; see Scheme VII. From this a neutral tetrahedral intermediate is formed directly; charged ones cannot exist in an aqueous medium. (Log rate constants, corrected for incomplete amide protonation, are linear in the log water activity, slope two. Molarity-based water activities must be used for consistency with the other species concentrations, rather than the listed mole-fraction-based ones [48].)

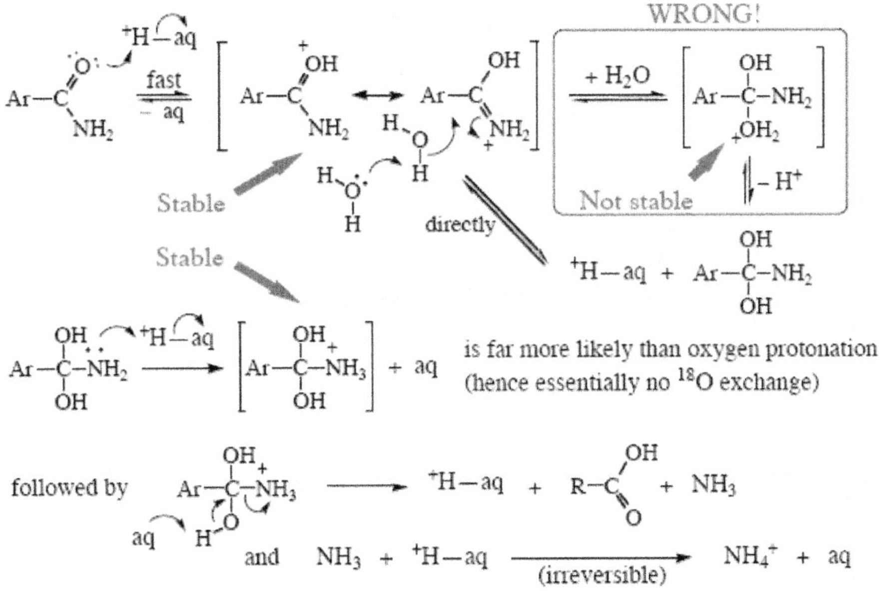

Scheme VII: Acid hydrolysis of benzamides in <60 wt% H_2SO_4.

In more strongly acidic media the mechanism changes [63]; the kinetics show a second, concerted, proton transfer taking place, giving an acylium ion which is stable under the reaction conditions, and that two water molecules are involved [63]. This mechanism is a bit tricky to draw, but I have made an attempt in Scheme VIII. Since an acylium ion is involved, this mechanism would only occur for those amides capable of giving stable ones, primarily benzamides. For other types of amide evidence is lacking; amides are particularly stable and their acid hydrolysis is very slow and quite difficult to study. The catalyzing acid is given as H_{aq}^{+}; presumably in H_2SO_4 media stronger than ~85 wt% the catalyst would be undissociated H_2SO_4, see above [63].

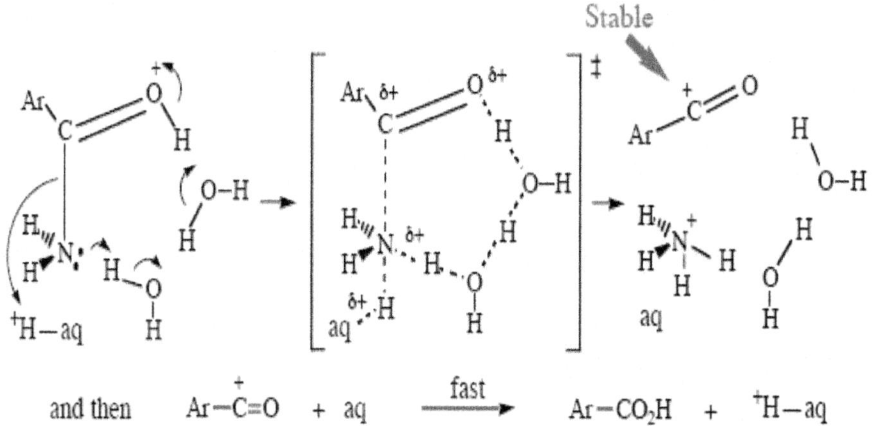

Scheme VIII: Acid hydrolysis of benzamides in >60 wt% H_2SO_4.

Ester Hydrolyses

At acidities below ~85 wt% the mechanisms of these processes are similar to those for benzamides [63] (and benzimidates [64]) as shown in Scheme IX [64], which differs from Scheme VII for amides in that the neutral tetrahedral intermediate does not contain a nitrogen atom, and so it is susceptible to ^{18}O-exchange, which is observed [65]; it is essentially not found in amide hydrolysis [66].

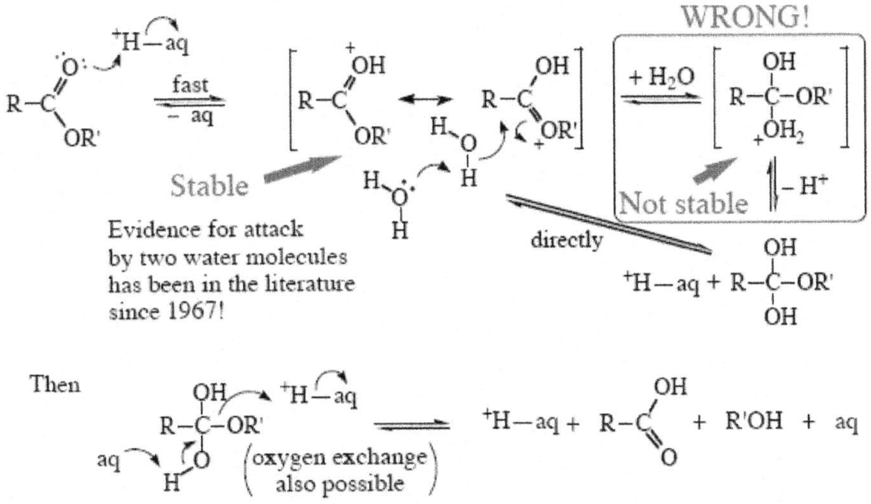

Scheme IX: Acid hydrolysis of esters in <85 wt% H_2SO_4.

In the strong acid region, above ~85 wt% H_2SO_4, other mechanisms take over. If the substrate contains a group capable of forming a stable carbocation, e.g., a benzylic or a tertiary group, this can leave directly from the protonated ester, and this can be the preferred mechanism at acidities much lower than 85 wt% H_2SO_4 [67,68]. This is shown in Scheme X.

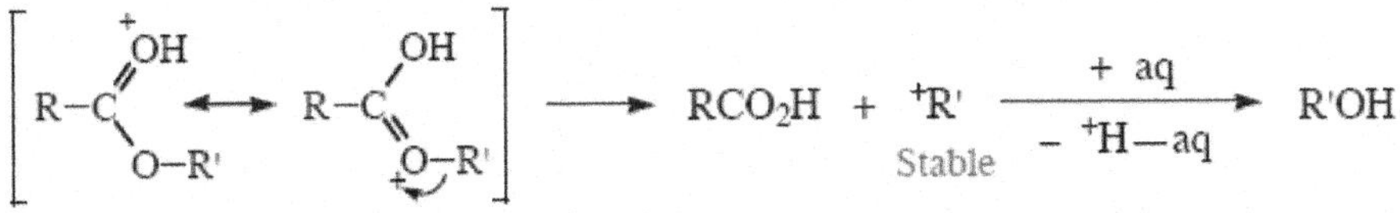

Scheme X: Acid hydrolysis of esters capable of forming carbocations.

For other esters in strong acid an additional proton transfer is probably involved, to give an acylium ion; the previously proposed [67,68] "proton switch" mechanism is probably wrong. This again is quite difficult to draw, but I have made an attempt in Scheme XI. This mechanism is not yet established, but work is underway to do this [20].

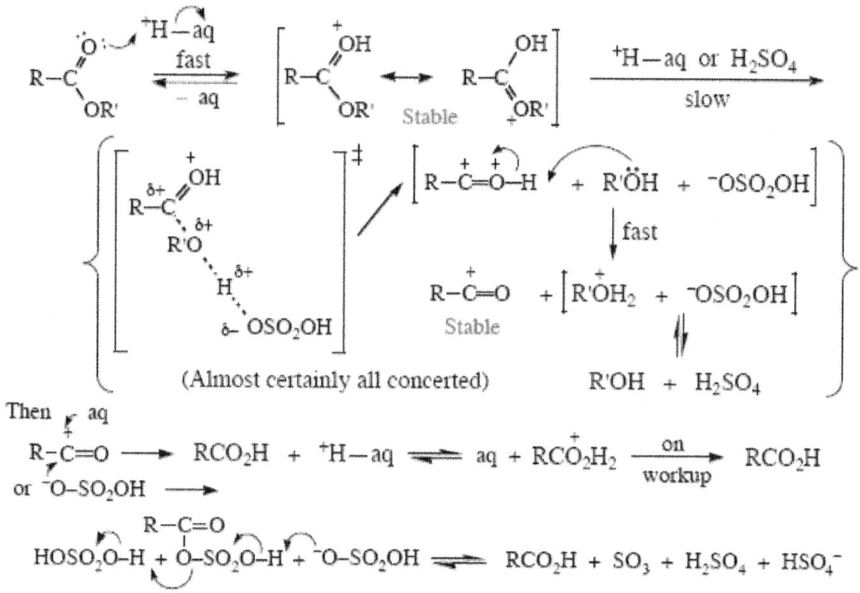

Scheme XI: Acid hydrolysis of other esters in >85 wt% H_2SO_4.

In basic media, it is becoming increasingly apparent that hydroxide ions do not themselves add directly to carbonyl groups, but that HO_{aq}^- removes a proton from a water molecule which then adds to the carbonyl, the result being a neutral tetrahedral intermediate [17–19]. Heavy-atom isotope effect studies make this appear even more likely [69]. Since the process is reversible, extensive oxygen exchange into the substrate is observed as well [70,71]. The most probable mechanism is given here as Scheme XII. Formation of a neutral intermediate ensures that the negative charge is dispersed into the solvent. Electronegative oxygen is certainly more able to support a negative charge than a positive one, but the principle of having any charge, positive or negative, dispersed as widely as possible ensures that all tetrahedral intermediates formed in either acidic or basic processes would be neutral. Species that are represented by various authors as T^+, T^-, $T^\pm$ and, especially, T^{2-} do not exist in aqueous media.

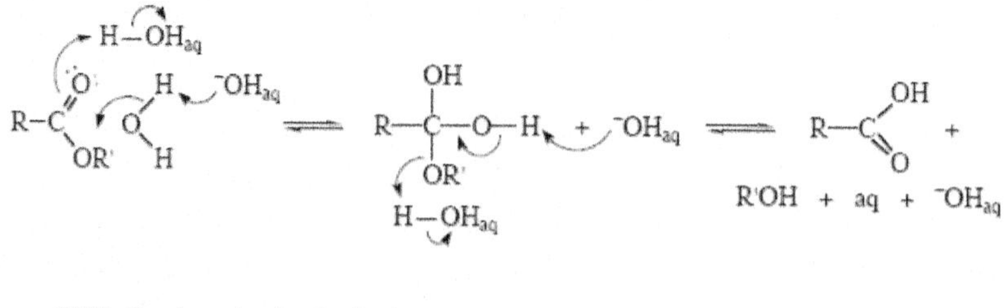

Scheme XII: Basic ester hydrolysis.

Mechanisms Involving Chains of Water Molecules

There are quite a number of these known now. The principles seem to be that if a reaction can be achieved without any charge transfer taking place it is favored, and that reactions involving chains of water molecules are favorable because the structure necessary for reaction essentially already exists; water molecules do not have to be moved into position, which is unfavorable entropically. For instance, acylimidazoles hydrolyze by forming a tetrahedral intermediate directly, Scheme XIII [72]. Incidentally, this work showed that the excess acidity correlation analysis works well even for reactions that are not acid-catalyzed [72].

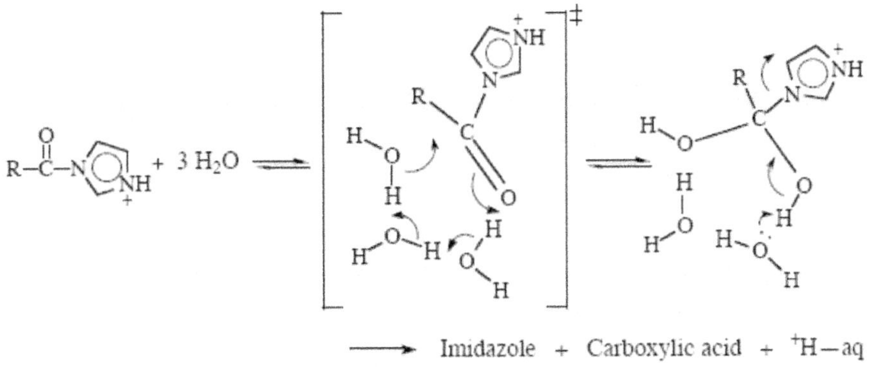

Scheme XIII: The mechanism of hydrolysis of acylimidazoles in water.

I proposed a mechanism for the hydrolysis of nitramide in neutral water on the basis of nothing but its elegance [73], and was gratified that detailed modern theoretical calculations, in the gas-phase and also in solution [74], showed that it was in fact correct. This is shown in Scheme XIV.

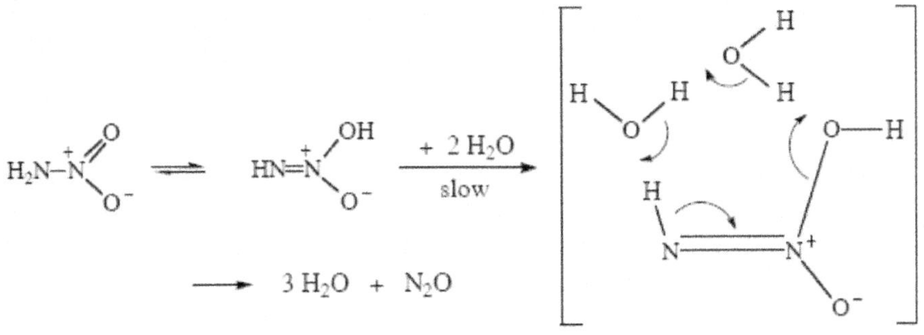

Scheme XIV: Nitramide hydrolysis in neutral water.

The hydrolyses of acid chlorides and acid anhydrides are fast reactions which have not received a lot of attention. Several mechanisms have been proposed [75–78], but the latest research would indicate that the actual mechanism may well be a simple cycle involving water as well, Scheme XV [78].

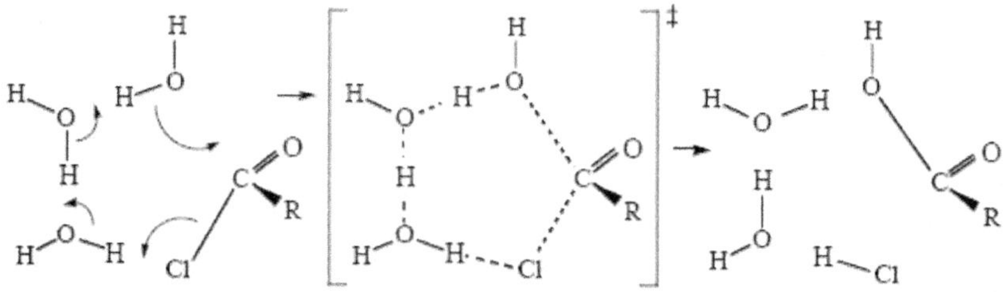

Scheme XV: A possible mechanism for acid chloride hydrolysis in water.

CONCLUSIONS

- If a species does not have a finite lifetime in the solution in which the reaction is performed it cannot be a reaction intermediate. No primary or secondary carbocations in aqueous media; only T^0, no T^+, T^-, $T^\pm$ or T^{2-} tetrahedral intermediates.

- Positive or negative charge, if present, will be as delocalized as possible during the reaction, especially in reaction intermediates, often into the aqueous solvent. A highly electronegative atom like oxygen is simply not going to support a positive charge all by itself. O^+ is almost as unlikely as F^+!

- Also, reactions will be unimolecular, as far as possible, for entropic reasons (S_N1 favored over S_N2); however, mechanisms involving chains of water molecules are favored in aqueous media thanks to the highly structured nature of water and the Grotthuss process.

There are a number of philosophical implications. Many years ago chemists weaned themselves from using "H^+" as a reactant, once it was pointed out that free protons are only stable in a hard vacuum. Now we are going to have to wean ourselves from using "H_3O^+" or "HO^-" as reactants in aqueous solution as well. Of course these species do exist, under special circumstances. In sulfuric acid above the 1:1 mole ratio point (~85 wt%) all the remaining water is present in the form H_3O^+. The perchloric acid hydrate sold as a solid in glass vials is $H_3O^+ \cdot ClO_4^-$ (and is pretty dangerous stuff!). The terms to use are "H_{aq}^+" and "HO_{aq}^-".

We are going to have to cease using the terms "general" and "specific" acid and base catalysis. Much to be preferred, I think, is to refer to "pre-equilibrium proton transfer" when an intermediate that is stable under the reaction conditions is formed in a first fast step, and to "concerted with proton transfer", or something similar, when the proton transfer is involved in the rate-determining step, as in many of the examples discussed above.

Very recently some common organic reactions have begun to be studied in liquid ammonia as a solvent, rather than in water [79,80]. It is going to be very interesting to compare the mechanisms of the same reaction in the two different solvents.

ACKNOWLEDGMENTS

Valuable correspondence with Tina Amyes, Bill Bentley, John Marlier, Howard Maskill, Chris Reed and Evgenii Stoyanov is gratefully acknowledged, and I thank those who saw the poster I presented on this subject at a recent conference for their (mostly!) useful comments.

REFERENCES AND NOTES

1. For instance, the many reviews given in the series *Reviews of Chemical Intermediates*; Elsevier: Amsterdam, The Netherlands., commencing in1981.

2. Olah, G.A.; White, A.M. Stable carbonium ions. XCI. Carbon-13 nuclear magnetic resonance spectroscopic study of carbonium ions. *J. Am. Chem. Soc* **1969**, *91*, 5801–5810.

3. Olah, G.A. My search for carbocations and their role in chemistry. *Angew. Chem. Int. Ed. Engl* **1995**, *34*, 1393–1405, and references therein.

4. Olah, G.A.; Dunne, K.; Kelly, D.P.; Mo, Y.K. Stable carbocations. CXXIX. Mechanism of the Benzidine and Wallach rearrangements based on direct observation of dicationic reaction intermediates and related model compounds. *J. Am. Chem. Soc* **1972**, *94*, 7438–7447.

5. Moore, W.J. *Physical Chemistry*, 4th ed.; Prentice Hall: Englewood Cliffs, NJ, USA, 1972; p. 769.

6. Jencks, W.P. When is an intermediate not an intermediate? Enforced mechanisms of general acid-base catalyzed, carbocation, carbanion, and ligand exchange reactions. *Acc. Chem. Res* **1980**, *13*, 161–169.

7. Richard, J.P.; Amyes, T.L.; Toteva, M.M. Formation and stability of carbocations and carbanions in water and intrinsic barriers to their reactions. *Acc. Chem. Res* **2001**, *34*, 981–988, and references therein.

8.   Vorob'eva, E.N.; Kuznetsov, L.L.; Gidaspov, B.V. Kinetics of decomposition of primary aliphatic *N*-nitroamines in aqueous sulfuric acid. *Zh. Org. Khim* **1983**, *19*, 698–704. *Russ. J. Org. Chem* **1983**, *19*, 615–620.

9.   Dietze, P.E.; Jencks, W.P. Oxygen exchange into 2-butanol and hydration of 1-butene do not proceed through a common carbocation intermediate. *J. Am. Chem. Soc* **1987**, *109*, 2057–2062.

10. Dietze, P.E.; Wojciechowski, M. Oxygen scrambling and stereochemistry during the trifluoroethanolysis of optically active 2-butyl 4-bromobenzenesulfonate. *J. Am. Chem. Soc* **1990**, *112*, 5240–5244.

11. Bruice, P.Y. *Organic Chemistry*, 3rd ed.; Prentice Hall: Upper Saddle River, NJ, USA, 2001; pp. 380–381.

12. Ingold, C.K. *Structure and Mechanism in Organic Chemistry*, 2nd ed.; Cornell University Press: Ithaca, NY, USA, 1969; p. 430.

13. Murphy, T.J. Absence of S_N1 involvement in the solvolysis of secondary alkyl compounds. *J. Chem. Educ* **2009**, *86*, 519–524.

14. Bentley, T.W.; Schleyer, P.v.R. The S_N2-S_N1 spectrum. 1. Role of nucleophilic solvent assistance and nucleophilically solvated ion pair intermediates in solvolyses of primary and secondary arenesulfonates. *J. Am. Chem. Soc* **1976**, *98*, 7658–7666.

15. Bunton, C.A.; Konasiewicz, A.; Llewellyn, D.R. Oxygen exchange and the Walden inversion in *sec*-butyl alcohol. *J. Chem. Soc* **1955**, 604–607.

16. Bunton, C.A.; Llewellyn, D. R. Tracer studies on alcohols. Part II. The exchange of oxygen-18 between *sec*-butyl alcohol and water. *J. Chem. Soc* **1957**, 3402–3407.

17. Mata-Segreda, J.F. Hydroxide as a general base in the saponification of ethyl acetate. *J. Am. Chem. Soc* **2002**, *124*, 2259–2262.

18. Haeffner, F.; Hu, C.-H.; Brinck, T.; Norin, T. The catalytic effect of water in basic hydrolysis of methyl acetate: A theoretical study. *J. Mol. Struct. (Theochem.)* **1999**, *459*, 85–93.

19. Hori, K.; Hashitani, Y.; Kaku, Y.; Ohkubo, K. Theoretical study on oxygen exchange accompanying alkaline hydrolysis of esters and amides. *J. Mol. Struct. (Theochem.)* **1999**, *461–462*, 589–596.

20. Cox, R.A. Scarborough, ON, Canada, Unpublished work; 2011.

21. Dolman, D.; Stewart, R. Strongly basic systems. VIII. The H_- function for dimethyl sulfoxide-water-tetramethylammonium hydroxide. *Can. J. Chem* **1967**, *45*, 911–924.

22. Eigen, M. Proton transfer, acid-base catalysis, and enzymatic hydrolysis. Part 1. elementary processes. *Angew. Chem. Int. Ed. Engl* **1964**, *3*, 1–19.

23. Zundel, G. Hydrogen bonds with large proton polarizability and proton transfer processes in electrochemistry and biology. *Adv. Chem. Phys* **2000**, *111*, 1–217, and many earlier papers.

24. Niedner-Schatteburg, G. Infrared spectroscopy and ab initio theory of isolated $H_5O_2^+$: from buckets of water to the Schrödinger equation and back. *Angew. Chem. Int. Ed. Engl* **2008**, *47*, 1008–1011.

25. Librovich, N.B.; Maiorov, V.D.; Savel'ev, V.A. The $H_5O_2^+$ ion in the vibrational spectra of aqueous solutions of strong acids.

26. Bascombe, K.N.; Bell, R.P. Properties of concentrated acid solutions. *Discuss. Faraday Soc* **1957**, *24*, 158–161.

27. Robertson, E.B.; Dunford, H.B. The state of the proton in aqueous sulfuric acid. *J. Am. Chem. Soc* **1964**, *86*, 5080–5089.

28. Ault, A. Telling it like it is: Teaching mechanisms in organic chemistry. *J. Chem. Educ* **2010**, *87*, 922–923.

29. Silverstein, T.P. The solvated proton is NOT H_3O^+! *J. Chem. Educ* **2011**, *88*, 875.

30. Roberts, S.T.; Ramasesha, K.; Petersen, P.B.; Mandal, A.; Tokmakoff, A. Proton transfer in concentrated aqueous hydroxide visualized using ultrafast infrared spectroscopy. *J. Phys. Chem. A* **2011**, *115*, 3957–3972.

31. Marx, D.; Chandra, A.; Tuckerman, M.E. Aqueous basic solutions: hydroxide solvation, structural diffusion, and comparison to the hydrated proton. *Chem. Rev* **2010**, *110*, 2174–2216.

32. Tuckerman, M.E.; Chandra, A.; Marx, D. Structure and dynamics of HO_{aq}^-. *Acc. Chem. Res* **2006**, *39*, 151–158.

33. Stoyanov, E.S.; Stoyanova, I.V.; Reed, C.A. The structure of the hydrogen ion (H_{aq}^+) in water. *J. Am. Chem. Soc* **2010**, *132*, 1484–1486.

34. Stoyanov, E.S.; Stoyanova, I.V.; Reed, C.A. The unique nature of H^+ in water. *Chem. Sci* **2011**, *2*, 462–472.

35. Jiang, J.-C.; Wang, Y.-S.; Chang, H.-C.; Lin, S.H.; Lee, Y.T.; Niedner-Schatteburg, G.; Chang, H.-C. Infrared spectra of $H^+(H_2O)_{5-8}$ clusters: evidence for symmetric proton hydration. *J. Am. Chem. Soc* **2000**, *122*, 1398–1410.

36. Stoyanov, E.S.; Reed, C.A. Private communication, Department of Chemistry, University of California: CA, USA, 2011.

37. Shevkunov, S.V. Computer simulation of molecular complexes $H_3O^+(H_2O)_n$ under conditions of thermal fluctuation. II. Work of formation and structure. *Zh. Obshch. Khim* **2004**, *74*, 1585–1592. *Russ. J. Gen. Chem* **2004**, *74*, 1471–1477.

38. Grotthuss, C.J.T. Sur la décomposition de l'eau et des corps qu'elle tient en dissolution à l'aide de l'électricité galvanique. *Ann. Chim* **1806**, *LVIII*, 54–74.

39. Marcus, Y. Effect of ions on the structure of water: structure making and breaking. *Chem. Rev* **2009**, *109*, 1346–1370, and references therein.

40. Aqueous HCl is only usable up to about 38 wt%, when the water is saturated with gaseous HCl, and aqueous perchloric acid only up to 78 wt% or so, when the solution solidifies at 25 °C. Nitric acid has problems and is not normally used; it is considerably weaker, it is an oxidizing agent, as is strong perchloric acid, and it can give NO_2^+ and related species at higher concentrations. Aqueous HF is not often used; it is very weak at high dilution, and if concentrated it can dissolve glassware. Trifluoromethanesulfonic acid would probably be useful, but it is very expensive. Methanesulfonic acid is not used much. Trifluoroacetic and the other carboxylic acid variants are too weak to be useful.

41. Cox, R.A. Mechanistic studies in strong acids. I. General considerations. Catalysis by individual acid species in sulfuric acid. *J. Am. Chem. Soc* **1974**, *96*, 1059–1063.

42. Cox, R.A.; Fung, D.Y.K.; Csizmadia, I.G.; Buncel, E. An *ab initio* molecular orbital study of the geometry of the dicationic Wallach rearrangement intermediate. *Can. J. Chem* **2003**, *81*, 535–541.

43. Buncel, E.; Keum, S.-R.; Rajagopal, S.; Cox, R.A. Rearrangement mechanisms for azoxypyridines and axoxypyridine *N*-oxides in the 100% H_2SO_4 region—the Wallach rearrangement story comes full circle. *Can. J. Chem* **2009**, *87*, 1127–1134.

44. Cox, R.A.; Buncel, E. Rearrangements of Hydrazo, Azoxy and Azo Compounds*The Chemistry of the Hydrazo, Azo and Azoxy Groups*; Patai, S., Ed.; Wiley: London, UK, 1975; Volume 1, pp. 775–859.

45. Cox, R.A.; Buncel, E. Rearrangements of Hydrazo, Azoxy and Azo Compounds: Kinetic, Product and Isotope Studies*The Chemistry of the Hydrazo, Azo and Azoxy Groups*; Patai, S., Ed.; Wiley: London, UK, 1997; Volume 2, pp. 569–602.

46. Cox, R.A.; Buncel, E.; Bolduc, R. Department of Chemistry, Queen's University: Kingston, Canada, Unpublished observations; 1971.

47. Cox, R.A.; Yates, K. Mechanistic studies in strong acids. VIII. Hydrolysis

mechanisms for some thiobenzoic acids and esters in aqueous sulfuric acid, determined using the excess acidity method. *Can. J. Chem* **1982**, *60*, 3061–3070.

48. Cox, R.A. Excess acidities. *Adv. Phys. Org. Chem* **2000**, *35*, 1–66.

49. Bell, R.P.; Bascombe, K.N.; McCoubrey, J.C. Kinetics of the depolymerization of trioxane in aqueous acids, and the acidic properties of aqueous hydrogen fluoride. *J. Chem. Soc* **1956**, 1286–1291.

50. Giauque, W.F.; Hornung, E.W.; Kunzler, J.E.; Rubin, T.R. The thermodynamic properties of aqueous sulfuric acid solutions and hydrates from 15 to 300 K. *J. Am. Chem. Soc* **1960**, *82*, 62–70.

51. Zeleznik, F.J. Thermodynamic properties of the aqueous sulfuric acid system to 350 K. *J. Phys. Chem. Ref. Data* **1991**, *20*, 1157–1200.

52. Randall, M.; Young, L.E. The calomel and silver chloride electrodes in acid and neutral solutions. The activity coefficient of aqueous hydrochloric acid and the single potential of the deci-molal calomel electrode. *J. Am. Chem. Soc* **1928**, *50*, 989–1004.

53. Åkerlöf, G.; Teare, J.W. Thermodynamics of concentrated aqueous solutions of hydrochloric acid. *J. Am. Chem. Soc* **1937**, *59*, 1855–1868.

54. Liu, Y.; Grén, U.; Theliander, H.; Rasmuson, A. Simultaneous correlation of activity coefficient and partial thermal properties for electrolyte solutions using a model with ion-specific parameters. *Fluid Phase Equilibria* **1993**, *83*, 243–251.

55. Pearce, J.N.; Nelson, A.F. The vapor pressures and activity coefficients of aqueous solutions of perchloric acid at 25°. *J. Am. Chem. Soc* **1933**, *55*, 3075–3081.

56. Robinson, R.A.; Baker, O.J. The vapor pressures of perchloric acid solutions at 25°. *Trans. Proc. R. Soc. N. Z* **1946**, *76*, 250–254.

57. Wai, H.; Yates, K. Determination of the activity of water in highly concentrated perchloric acid solutions. *Can. J. Chem* **1969**, *47*, 2326–2328.

58. Bidinosti, D.R.; Biermann, W.J. A redetermination of the relative enthalpies of aqueous perchloric acid solutions from 1 to 24 molal. *Can. J. Chem* **1956**, *34*, 1591–1595.

59. Bell, R.P.; Brown, A.H. Kinetics of the depolymerization of paraldehyde in aqueous solution. *J. Chem. Soc* **1954**, 774–778.

60. Hamer, D.; Leslie, J. The Hammett acidity function in reactions catalyzed by carboxylic acids. The hydrolysis of methylal and the depolymerization of trioxane. *J. Chem. Soc* **1960**, 4198–4202.

61. Jaques, D.; Leisten, J.A. Acid-catalysed ether fission. Part II. Diethyl ether in aqueous acids. *J. Chem. Soc* **1964**, 2683–2689.

62. Ruiz Pernía, J.J.; Tuñón, I.; Williams, I.H. Computational simulation of the lifetime of methoxymethyl cation in water. A simple model for a glycosyl cation: When is an intermediate an intermediate? *J. Phys. Chem. B* **2010**, *114*, 5769–5774.

63. Cox, R.A. Benzamide hydrolysis in strong acids—the last word. *Can. J. Chem* **2008**, *86*, 290–297.

64. Cox, R.A. A comparison of the mechanism of hydrolysis of benzimidates, esters, and amides in sulfuric acid media.*Can. J. Chem* **2005**, *83*, 1391–1399.

65. Bender, M.L. Oxygen exchange as evidence for the existence of an intermediate in ester hydrolysis. *J. Am. Chem. Soc***1951**, *73*, 1626–1629.

66. McClelland, R.A. Benzamide oxygen exchange concurrent with acid hydrolysis. *J. Am. Chem. Soc* **1975**, *97*, 5281.

67. Yates, K.; McClelland, R.A. Mechanisms of ester hydrolysis in aqueous sulfuric acids. *J. Am. Chem. Soc* **1967**, *89*, 2686–2692.

68. Yates, K. Kinetics of ester hydrolysis in concentrated acid. *Acc. Chem. Res* **1971**, *4*, 136–144.

69. Marlier, J.F. Heavy-atom isotope effects on the alkaline hydrolysis of methyl formate. The role of hydroxide ion in ester hydrolysis. *J. Am. Chem. Soc* **1993**, *115*, 5953–5956.

70. Bender, M.L.; Ginger, R.D.; Unik, J.P. Activation energies of the hydrolysis of esters and amides involving carbonyl oxygen exchange. *J. Am. Chem. Soc* **1958**, *80*, 1044–1048.

71. Shain, S.A.; Kirsch, J.F. Absence of carbonyl oxygen exchange concurrent with the alkaline hydrolysis of substituted methyl benzoates. *J. Am. Chem. Soc* **1968**, *90*, 5848–5854.

72. Cox, R.A. The mechanism of the hydrolysis of acylimidazoles in aqueous mineral acids. The excess acidity method for reactions that are not acid catalyzed. *Can. J. Chem* **1997**, *75*, 1093–1098.

73. Cox, R.A. The acid catalyzed decomposition of nitramide. *Can. J. Chem* **1996**, *74*, 1779–1783.

74. Eckert-Maksic, M.; Maskill, H.; Zrinski, I. Acidic and basic properties of nitramide, and the catalyzed decomposition of nitramide and related compounds; an *ab initio* theoretical investigation. *J. Chem. Soc. Perkin Trans* **2001**, *2*, 2147–2154.

75. Bentley, T.W.; Harris, H.C. Solvolyses of *para*-substituted benzoyl chlorides in trifluoroethanol and in highly aqueous media. *J. Chem. Soc. Perkin Trans* **1986**, *2*, 619–624.

76. Williams, A. Concerted mechanisms of acyl group transfer reactions in solution. *Acc. Chem. Res* **1989**, *22*, 387–392.

77. Bentley, T.W.; Ebdon, D.N.; Kim, E.-J.; Koo, I.S. Solvent polarity and organic reactivity in mixed solvents: Evidence using a reactive molecular probe to assess the role of preferential solvation in aqueous alcohols. *J. Org. Chem* **2005**, *70*, 1647–1653.

78. Ruff, F.; Farkas, Ö. Concerted S_N2 mechanism for the hydrolysis of acid chlorides: comparisons of reactivities calculated by the density functional theory with experimental data. *J. Phys. Org. Chem.* **2011**, *24*, 480–491.

79. Ji, P.; Atherton, J.; Page, M.I. Liquid ammonia as a dipolar aprotic solvent for aliphatic nucleophilic substitution reactions. *J. Org. Chem* **2011**, *76*, 1425–1435.

80. Ji, P.; Atherton, J.H.; Page, M.I. The kinetics and mechanisms of aromatic nuclear substitution reactions in liquid ammonia. *J. Org. Chem* **2011**, *76*, 3286–3295

Chapter 8

EFFECTS OF CHEMICAL REACTIONS ON UNSTEADY FREE CONVECTIVE AND MASS TRANSFER FLOW FROM A VERTICAL CONE WITH HEAT GENERATION/ABSORPTION IN THE PRESENCE OF VWT/VWC

Bapuji Pullepu,[1] P. Sambath,[1] and K. K. Viswanathan[2]

[1]Department of Mathematics, SRM University, Kattankulathur, Tamil Nadu 603203, India

[2]UTM Centre for Industrial and Applied Mathematics and Department of Mathematical Sciences, Faculty of Science, Universiti Teknologi Malaysia, 81310 Johor Bahru, Johor, Malaysia

ABSTRACT

A mathematical model for the effects of chemical reaction and heat generation/absorption on unsteady laminar free convective flow with heat and mass transfer over an incompressible viscous fluid past a vertical permeable cone with nonuniform surface temperature $T'_w(x) = T'_\infty + ax^n$ and concentration $C'_w(x) = C'_\infty + bx^m$ is considered here. The dimensionless governing boundary layer equations of the flow that are transient, coupled, and nonlinear partial differential equations are solved by an efficient, accurate, and unconditionally stable finite difference scheme of Crank-Nicholson type. The velocity, temperature, and concentration profiles have been studied for various parameters, namely, chemical reaction parameter λ, the heat generation and absorption parameter Δ, Schmidt number Sc, Prandtl number Pr, buoyancy ratio parameter N, surface temperature power law exponent n, and surface concentration power law exponent m. The local as well as average skin friction, Nusselt number, and Sherwood number are discussed and analyzed graphically. The present results are compared with available results in open literature and are found to be in excellent agreement.

INTRODUCTION

The problem of two-dimensional axisymmetric free convective flow past a vertical cone with different boundary conditions has attracted the attention of many researchers in recent years. When a heated surface is in contact with the fluid, the result of temperature difference causes buoyancy force, which induces the natural convection. Free convection flows under the influence of gravitational force have been studied in detail because they occur frequently in nature. Simultaneous heat and mass transfer in natural convection flows on a vertical cone has a wide range of applications in the field of science and technology. Also it plays an important role in manufacturing industries for the design of reliable equipment for nuclear power plants, gas turbines, and various propulsion devices for aircraft, missiles, satellites, and space vehicles. The flow of a fluid is caused not only by the temperature differences, but also by concentration differences. These concentration differences affect the flow and temperature near the surface of a body embedded in a fluid. In engineering applications, the concentration differences are created either by injecting the foreign gases or by coating the surface with evaporating material which evaporates due to the heat of the surface. The presence of foreign masses in air and water like hydrogen H_2 and water H_2O causes some kind of chemical reaction. Heat is generated due to this chemical reaction. A common example of heat and mass transfer is the evaporation of lake water into the wind flowing over it. In some cases, mass transfer is predominant and heat transfer may be negligible; in other cases, both remain equally predominant. Mass transfer proceeds as long as there is a difference in concentrations of some chemical species in the mixture. Hence, the concentration gradient acts as a driving potential in mass transfer, just as the temperature gradient does in heat transfer. Since 1953 several authors have developed similarity/nonsimilarity solutions for two-dimensional axisymmetrical problems for natural convection laminar flow over vertical cone in steady state (see [1–4]). Kafoussias [5] analyzed the effects of mass transfer on a free convective flow past a vertical cone surface embedded in an infinite, incompressible, and viscous fluid. Yih [6, 7] studied in saturated porous media the combined heat and mass transfer effects over a full cone with uniform wall temperature/concentration or heat/mass flux and for truncated cone with nonuniform wall temperature/variable wall concentration or variable heat/mass flux using the Keller box implicit difference method. Chamkha [8] considered the problem of steady state laminar heat and mass transfer by natural convection boundary layer flow around a permeable truncated cone in the presence of magnetic field and thermal radiation effects; nonsimilar solutions were obtained and solved numerically by an implicit finite difference methodology. Later Chamkha and Quadri [9] solved the problem of combined heat and mass transfer by hydromagnetic natural convection over a cone embedded in a non-Darcian porous medium with heat generation/ absorption effects; a nonsimilar form of the solution was solved numerically by an implicit, iterative, and finite difference method. Afify [10] studied the effects of radiation and chemical reaction on a steady free convective flow and mass transfer of an optically dense viscous, incompressible, and electrically conducting fluid past a vertical isothermal cone in the presence of a magnetic field; the resulting similarity equations were solved numerically using a fourth-order Runge-Kutta scheme with the shooting technique. Chamkha and Al-Mudhaf [11] focused on the study of unsteady heat and mass transfer by mixed convection flow over a vertical permeable cone rotating in an ambient fluid with a time-dependent angular velocity in the presence of a magnetic field and heat generation or absorption effects with the cone surface which is maintained at variable temperature and concentration and obtained numerical solutions by solving the governing partial differential equations using an implicit, iterative finite difference scheme. Chamkha et al. [12] studied the effects of coupled heat and mass transfer by boundary layer free convection over a vertical flat plate embedded in a fluidsaturated porous medium in the presence of thermophoretic particle deposition and heat generation or absorption effects; the governing partial differential equations are transformed into ordinary differential equations by using special transformations and the resulting similarity equations are solved numerically by an efficient implicit tridiagonal finite difference method. EL-Kabeir et al. [13] used perturbation method to study the effect of heat and mass transfer on free convection flow with a uniform suction and injection over a cone in a micropolar fluid. EL-Kabeir and Abdou [14] studied the effects of chemical reaction and heat and mass transfer on MHD flow over a vertical isothermal cone surface in micropolar fluids with heat generation/absorption effects and obtained numerical solutions by using the fourthorder Runge-Kutta method with shooting technique. Also ElKabeir et al. [15] discussed the linear transformation group approach to simulate the problem of heat and mass transfer in steady, two-dimensional, laminar, and boundary layer flow of a viscous, incompressible, and electrically conducting fluid over a vertical permeable cone surface saturated porous medium in the presence of a uniform transverse magnetic field and thermal radiation effects. Cheng [16] presented a boundary layer analysis about the natural convection heat and mass transfer near a vertical cone with variable wall temperature and concentration in a porous medium saturated with non-Newtonian power law fluids; coordinate transform is used to obtain the nonsimilar governing equations, and the transformed boundary layer equations are solved by the cubic spline collocation method. Cheng [17, 18] analyzed the Soret and Dufour effects on the boundary layer flow due to natural convection heat and mass transfer over a downward-pointing vertical cone and truncated cone in a porous medium saturated with Newtonian fluids with constant wall temperature and concentration, similarity analysis is performed, and similarity equations are solved by cubic spline collocation method. Murti et al. [19] discussed the radiation and chemical reaction effects on heat and mass transfer in non-Darcy nonNewtonian fluid over a vertical surface; the governing boundary layer equations and boundary conditions are simplified by using similarity transformations and are solved numerically by means of fourth-order Runge-Kutta method coupled with double-shooting technique. Kishore et al. [20] studied viscoelastic buoyancy driven MHD free convective heat and mass transfer past a vertical cone with thermal radiation and viscous dissipation and obtained numerical solutions for the governing equations using Crank-Nicholson method. Mahdy [21] focused on the study of combined heat and mass transfer on double-diffusive convection near a vertical truncated cone in a fluid-saturated porous medium in the presence of a firstorder chemical reaction and heat generation or absorption with variable viscosity. Viscosity of the fluid is assumed to be an inverse linear function of the temperature; the nondimensional nonsimilar governing equations are solved numerically using the fourth-order Runge-Kutta integration scheme with Newton-Raphson shooting technique

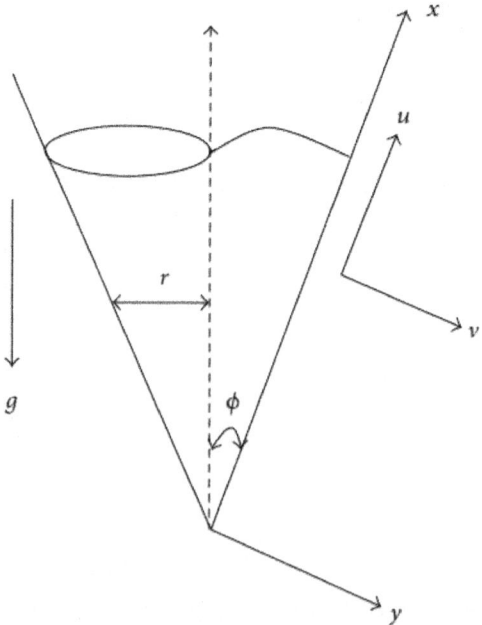

Figure 1: Physical model and coordinate system.

Mohiddin et al. [22, 23] discussed the combined effects of thermal radiation and viscous dissipation on unsteady, laminar, and free convective flow with heat and mass transfer over an incompressible viscous fluid past vertical cone with variable surface temperature and concentration in the presence of a transverse magnetic field applied normal to the surface, heat and mass transfer in a Walters-B viscoelastic fluid along a vertical cone using Crank-Nicholson finite difference scheme. Patil and Pop [24] considered the unsteady mixed convection boundary layer flow over a vertical cone to investigate the combined effects of the buoyancy force, thermal and mass diffusion in the presence of the first-order chemical reaction and surface mass transfer. The governing boundary layer equations are transformed into a nondimensional form by a group of nonsimilar transformations and the resulting system of coupled nonlinear partial differential equations is solved numerically by the combination of quasilinearization technique and an implicit finite difference scheme. Recently El-Kabeir and El-Sayed [25] studied the problem of heat and mass transfer by free convection of a viscoelastic fluid past a vertical isothermal cone surface in the presence of transverse uniform magnetic field and chemical reaction effect taking into account the effects of viscous dissipation, Joule heating, and thermal radiation.The cone surface is maintained at constant temperature and constant species concentration. The governing partial differential equations are transferred into a system of ordinary differential equations, which are solved numerically using a fourth-order Runge-Kutta scheme with the shooting method. Awad et al. [26] studied the Soret and Dufour effects on the skin friction coefficient, the heat and the mass transfer from an inverted cone in a porous medium. Numerical solutions for the governing momentum, energy, and concentration equations were found using a shooting method together with a sixth-order Runge-Kutta method. The results were validated by using a linearization method. Also Narayana et al. [27] studied the Soret and Dufour effects on free magneto hydrodynamic convection from a vertical spinning cone. They discussed two different types of boundary heating, namely, linear surface temperature (LST), where the surface of the cone is maintained at a temperature that varies linearly with the distance from origin, and linear surface heat flux (LSHF). The nonlinear coupled governing equations were solved using a shooting technique together with a Runge-Kutta method of four slopes. Basiri Parsa et al. [28] discussed steady laminar magneto hydrodynamic boundary layer flow past a stretching surface with uniform free stream and internal heat generation or absorption in an electrically conducting fluid. The governing boundary layer and temperature equations for this problem are first transformed into a system of ordinary differential equations using similarity variables and then solved by using a new analytical method and numerical method, by using a fourthorder Runge-Kutta and shooting method.

The objective of the present investigation, namely, transient free convective flow from a nonisothermal vertical cone with heat generation/absorption and chemical reaction, has not received any attention in literature. Also it has a wide range of applications in the field of nuclear reactor safety, solar energy plants, drying, and dehydration process in chemical and food process, design of space crafts and steam generators, and so forth. Hence, the present work studies and deals with the transient free convective flow from a nonisothermal vertical cone with the above said effects. The governing boundary layer equations are solved by an implicit finite difference scheme of Crank-Nicolson type for various values of parameters λ, Δ, Sc, Pr, N, n, and m. In order to check the accuracy of the numerical results, the present results are compared with the available results of Chamkha [8] and they are found to be in excellent agreement.

MATHEMATICAL FORMULATION

An axisymmetric unsteady, laminar free convection flow of a viscous incompressible fluid past a vertical cone with nonuniform surface temperature and concentration under the influence of chemical reaction and heat generation/absorption is considered. It is assumed that the effects of viscous dissipation and pressure gradient along the boundary layer are negligible.

It is also assumed that there exists first-order chemical reaction between the fluid and the species concentration. The concentration C' of the diffusing species is assumed to be very small in comparison to the other chemical species far away from the surface of the cone C'_∞. Hence the Soret and Dufour effects are neglected. It is also assumed that the cone surface and the surrounding fluid which is at rest are at the same temperature T'_∞ and concentration C'_∞. Then at time $t' > 0$, the temperature of the cone surface is suddenly raised to $T'_w(x) = T'_\infty + ax^n$ and the concentration near the cone surface is also raised to $C'_w(x) = C'_\infty + bx^m$ and both are maintained at the same level, where a, b are the positive constants and n, m are the exponents in power law variation in surface temperature and concentration, respectively. The coordinate system is chosen (as shown in Figure 1) such that x measures the distance along surface of the cone from the apex ($x=0$) and y measures the distance normally outward. The fluid properties are assumed to be constant except the density variations causing a body force in the momentum equation. The governing boundary layer equations of continuity, momentum, energy, and concentration under Boussinesq approximation are as follows:

equation of continuity:

$$\frac{\partial}{\partial x}(ru) + \frac{\partial}{\partial y}(rv) = 0;$$ (1)

equation of momentum:

$$\frac{\partial u}{\partial t'} + u\frac{\partial u}{\partial x} + v\frac{\partial u}{\partial y}$$

$$= g\beta\left(T' - T'_\infty\right)\cos\phi + v\frac{\partial^2 u}{\partial y^2}$$

$$+ g\beta^*\left(C' - C'_\infty\right)\cos\phi;$$ (2)

equation of energy:

$$\frac{\partial T'}{\partial t'} + u\frac{\partial T'}{\partial x} + v\frac{\partial T'}{\partial y} = \alpha\frac{\partial^2 T'}{\partial y^2} + \frac{Q_o}{\rho c_p}\left(T' - T_\infty\right);$$ (3)

equation of concentration:

$$\frac{\partial C'}{\partial t'} + u\frac{\partial C'}{\partial x} + v\frac{\partial C'}{\partial y} = D\frac{\partial^2 C'}{\partial y^2} - k_1\left(C' - C_\infty\right).$$ (4)

The initial and boundary conditions are

$$t' \le 0: \quad u = 0, \quad v = 0,$$

$$T' = T'_\infty, \quad C' = C'_\infty \quad \forall x, y,$$

$$t' > 0: \quad u = 0, \quad v = 0,$$

$$T'(x) = T'_\infty + ax^n,$$

$$C'(x) = C'_\infty + bx^m \quad \text{at} \quad y = 0,$$

$$u = 0, \quad T' = T'_\infty, \quad C' = C'_\infty \quad \text{at} \quad x = 0,$$

$$u \longrightarrow 0, \quad T' \longrightarrow T'_\infty, \quad C' \longrightarrow C'_\infty \quad \text{as} \quad y \longrightarrow \infty.$$ (5)

Local skin friction, local Nusselt number, and local Sherwood number are given by

$$\tau_x = \mu\left(\frac{\partial u}{\partial y}\right)_{y=0},$$

$$Nu_x = \frac{-x\left(\partial T'/\partial y\right)_{y=0}}{T'_w - T'_\infty},$$

$$Sh_x = \frac{-x\left(\partial C'/\partial y\right)_{y=0}}{C'_w - C'_\infty}.$$

$$(6)$$

Using the following nondimensional quantities:

$$X = \frac{x}{L}, \qquad Y = \frac{y}{L}(Gr_L)^{1/4}, \qquad R = \frac{r}{L},$$

where $r = x \sin\phi,$

$$V = \frac{vL}{v}(Gr_L)^{-1/4}, \qquad U = \frac{uL}{v}(Gr_L)^{-1/2},$$

$$t = \frac{vt'}{L^2}(Gr_L)^{1/2},$$

$$T = \frac{\left(T' - T'_\infty\right)}{\left(T'_w - T'_\infty\right)}, \qquad Gr_L = \frac{g\beta\left(T'_w - T'_\infty\right)L^3 \cos\phi}{v^2},$$

$$Pr = \frac{v}{\alpha},$$

$$T = \frac{\left(C' - C'_\infty\right)}{\left(C'_w - C'_\infty\right)}, \qquad Gr^* = \frac{g\beta^*\left(C'_w - C'_\infty\right)L^3 \cos\phi}{v^2},$$

$$Sc = \frac{v}{D}, \qquad N = \frac{Gr^*}{Gr_L}, \qquad \Delta = \frac{Q_o L^2}{C_p \mu}(Gr_L)^{-1/2},$$

$$\lambda = \frac{k_1 L^2}{v}(Gr_L)^{-1/2}.$$

$$(7)$$

Equations (1), (2), (3), (4), and (5) can then be written in the following nondimensional form:

$$\frac{\partial(UR)}{\partial X} + \frac{\partial(VR)}{\partial Y} = 0,$$

$$\frac{\partial U}{\partial t} + U\frac{\partial U}{\partial X} + V\frac{\partial U}{\partial Y} = T + NC + \frac{\partial^2 U}{\partial Y^2},$$

$$\frac{\partial T}{\partial t} + U\frac{\partial T}{\partial X} + V\frac{\partial T}{\partial Y} = \frac{1}{Pr}\frac{\partial^2 T}{\partial Y^2} + \Delta T,$$

$$\frac{\partial C}{\partial t} + U\frac{\partial C}{\partial X} + V\frac{\partial C}{\partial Y} = \frac{1}{Sc}\frac{\partial^2 C}{\partial Y^2} - \lambda C.$$

$$(8)$$

The corresponding nondimensional initial and boundary conditions are

$$t \leq 0: \quad U = 0, \quad V = 0,$$
$$T = 0, \quad C = 0 \quad \forall X, Y,$$
$$t > 0: \quad U = 0, \quad V = 0, \quad T = X^n,$$
$$C = X^m \quad \text{at } Y = 0,$$
$$U = 0, \quad T = 0, \quad C = 0 \quad \text{at } X = 0,$$
$$U \longrightarrow 0, \quad T \longrightarrow 0, \quad C \longrightarrow 0 \quad \text{as } Y \longrightarrow \infty.$$

$$(9)$$

Local skin friction, local Nusselt number, and local Sherwood number in nondimensional quantities are

$$\tau_X = \text{Gr}_L^{3/4} \left(\frac{\partial U}{\partial Y} \right)_{Y=0},$$

$$\text{Nu}_X = \frac{X}{T_{Y=0}} \left(\frac{-\partial T}{\partial Y} \right)_{Y=0} \text{Gr}_L^{1/4},$$

$$\text{Sh}_X = \frac{X}{C_{Y=0}} \left(\frac{-\partial C}{\partial Y} \right)_{Y=0} \text{Gr}_L^{1/4}.$$

$$(10)$$

Average skin friction, average Nusselt number, and average Sherwood number in nondimensional quantities are

$$\bar{\tau} = 2\text{Gr}_L^{3/4} \int_0^1 X \left(\frac{\partial U}{\partial Y} \right)_{Y=0} dX,$$

$$\overline{\text{Nu}} = 2\text{Gr}_L^{1/4} \int_0^1 \frac{X}{T_{Y=0}} \left(\frac{-\partial T}{\partial Y} \right)_{Y=0} dX$$

$$\overline{\text{Sh}} = 2\text{Gr}_L^{1/4} \int_0^1 \frac{X}{C_{Y=0}} \left(\frac{-\partial C}{\partial Y} \right)_{Y=0} dX.$$

$$(11)$$

SOLUTION PROCEDURE

The unsteady nonlinear coupled partial differential equations (8) with the initial and boundary conditions (9) are solved by employing a finite difference scheme of Crank-Nicholson type which is rapidly convergent and unconditionally stable as discussed by Soundalgekar and Ganesan [29], Bapuji et al. [30], and Muthucumaraswamy and Ganesan [31, 32]. The region of integration is considered as a rectangle with $X_{max}(=1)$ and $Y_{max}(=20)$, where Y_{max} corresponds to $Y=\infty$ which lies very well outside both the momentum and thermal boundary layers. The maximum of Y was chosen as 20, after some preliminary investigation so that the last two boundary conditions of (9) are satisfied within the tolerance limit of $10-5$. The mesh sizes have been fixed as $\Delta X = 0.05$, $\Delta Y = 0.05$ with time step $\Delta t = 0.01$. The computations are carried out first by reducing the spatial mesh sizes by 50% in one direction and later in both directions by 50%. The results are compared. It is observed in all cases that the results differ only in the fifth decimal place. Hence, the choice of the mesh sizes seems to be appropriate. The scheme is unconditionally stable as described by Bapuji et al. [33]. The local truncation error is $(\Delta t2 + \Delta Y2 + \Delta X)$ and it tends to zero as Δt, ΔY, and ΔX tend to zero. Hence, the scheme is compatible. Stability and compatibility ensures the convergence.

RESULT AND DISCUSSION

In order to prove the accuracy of our numerical results, the present results for the steady state flow at $X = 1.0$ are compared with available solutions from the open literature. The numerical values of the local skin friction τ_x and the local Nusselt number Nu_x for different values of the Prandtl number with $M=0$, $N=0$, $n=0$, and $R_A = 0$ are compared with the results of Chamkha [8] in Table 1, where $f''(\infty, 0)$ and $-\theta'(\infty, 0)$ are the steady state local skin friction and the local Nusselt number for a full cone. It is observed that the results are in good agreement with each other.

Table 1: Comparison of steady-state local skin friction and local Nusselt number values at $X = 1.0$ with those of Chamkha [8] for full cone, for various values of Pr when $n = 0$, $M = 0$, $N = 0$ and $R_d = 0$.

| Pr | Local skin friction | | Local Nusselt number | |
	Chamkha [8] $f''(\infty, 0)$	Present values $\tau_X / Gr_L^{3/4}$	Chamkha [8] $-\theta'(\infty, 0)$	Present results $Nu_X / Gr_L^{1/4}$
0.001	1.5135	1.4149	0.0245	0.0294
0.01	1.3549	1.3356	0.0751	0.0797
0.1	1.0962	1.0911	0.2116	0.2115
1	0.7697	0.7688	0.5111	0.5125
10	0.4877	0.4856	1.0342	1.0356
100	0.2895	0.2879	1.9230	1.9316
1000	0.1661	0.1637	3.4700	3.5186

Velocity, temperature, and concentration profiles for different values of Prandtl number Pr, heat generation/absorption parameter Δ, are shown in Figures 2(a)–2(c). The positive values of Δ represent the presence of heat generation and the negative values correspond to heat absorption. It is noted from Figure 2(a) that the effect of the heat generation/ absorption parameter Δ on the velocity distribution. It is seen from the figure that when heat is generated the buoyancy force increases which induces the flow rate to increase and giving rise to the velocity profiles but the momentum boundary layer decreases as Δ and Pr increases. Figure 2(b) depicts the temperature increases for higher values of Δ and lower values of Pr; the thermal boundary layer decreases for the larger value of Δ and Pr. From Figure 2(c) it is seen that the concentration decreases and the time taken to reach the steady state is increased when Δ and Pr increase. Also the concentration boundary layer becomes thin for higher values of Δ and Pr

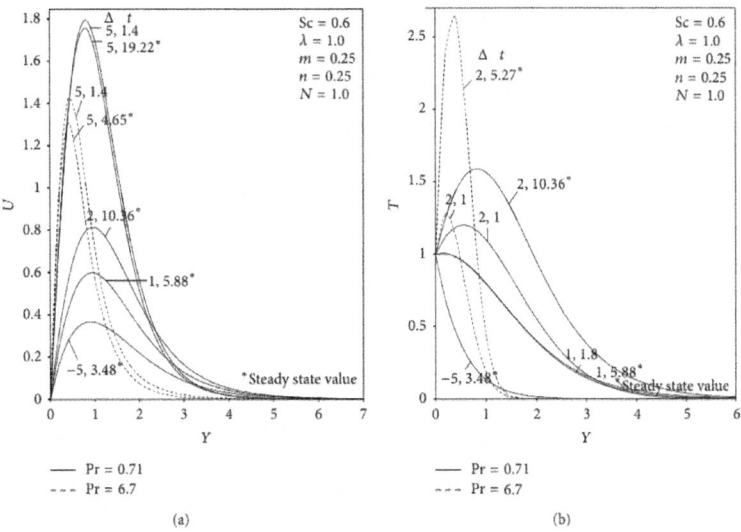

(a)   (b)

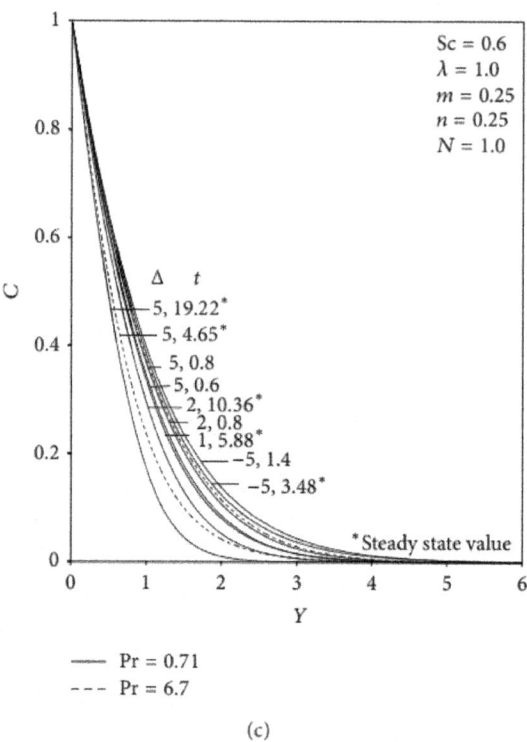

(c)

Figure 2: (a) Transient velocity profiles at $X = 1.0$ for different values of Δ and Pr. **(b)** Transient temperature profiles at $X = 1.0$ for different values of Δ and Pr. **(c)** Transient concentration profiles at $X = 1.0$ for different values of Δ and Pr.

Figures 3(a)–3(c) depict the transient velocity, temperature, and concentration profiles for various values of chemical reaction parameter λ and Schmidt number Sc. As the Schmidt number increases the concentration decreases. This causes the concentration buoyancy effects to decrease. The velocity and boundary layer thickness decreases with an increase in λ and Sc (Figure 3(a)), whereas the temperature increases for larger values of Sc and smaller values of λ and the thermal boundary layer becomes thin for smaller values of Sc and λ (Figure 3(b)). Figure 3(c) shows that the concentration decreases for lower values of λ and higher values of Sc. With increasing Sc the velocity is depressed through the boundary layer; that is, the flow is retarded. Higher Sc values will physically correspond to a decrease of molecular diffusivity of the primary fluid causing a decrease in the rate of species diffusion. Lower Sc values will exert the reverse influence since they correspond to higher molecular diffusivities. Concentration boundary layer thickness is therefore considerably greater for Sc = 0.6 than for Sc = 10.

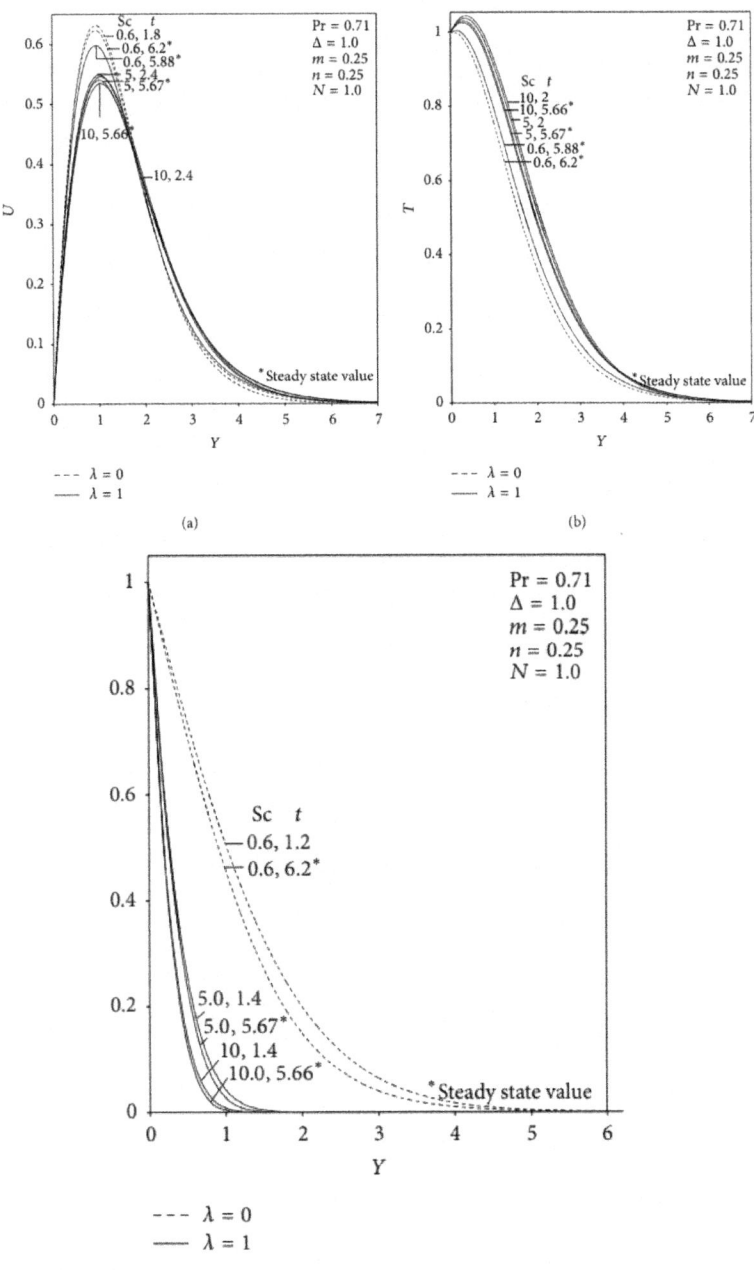

Figure 3: (a) Transient velocity profiles at $X = 1.0$ for different values of λ and Sc. (b) Transient temperature profiles at $X = 1.0$ for different values of λ and Sc. (c) Transient concentration profiles at $X = 1.0$ for different values of λ and Sc.

Figures 4(a)–4(c) show the influence of the surface concentration power law exponent m on velocity, temperature, and concentration distributions. We observe in Figure 4(a) that the velocity is maximized throughout the boundary layer with a decrease in m. Figure 4(b) indicates that the temperature increases for smaller values of m. As such increasing power law exponents in the cone surface concentration variations serve to decelerate the flow in the boundary layer. Concentration of the species increases for smaller values of m and larger values of n is observed from Figures 4(c) and 5. The effects of the buoyancy ratio parameter N on the transient velocity, temperature, and concentration profiles are shown in Figures 6(a)–6(c). The velocity increases steadily with time reaches a temporal maximum and consequently it reaches the steady state. However, time required to reach the steady state depends upon buoyancy ratio parameter N. An increase in N leads to an increase in the velocity; that is, as N increases, the combined buoyancy force also increases; therefore, the velocity increases near the surface of the cone (Figure 6(a)). As we move away from the surface of the cone, the temperature decreases for all the values of N (Figure 6(b)); thus for higher value of buoyancy ratio parameter N the fluid cools rapidly and concentration field decreases with increasing value of buoyancy ratio parameter N (Figure 6(c))

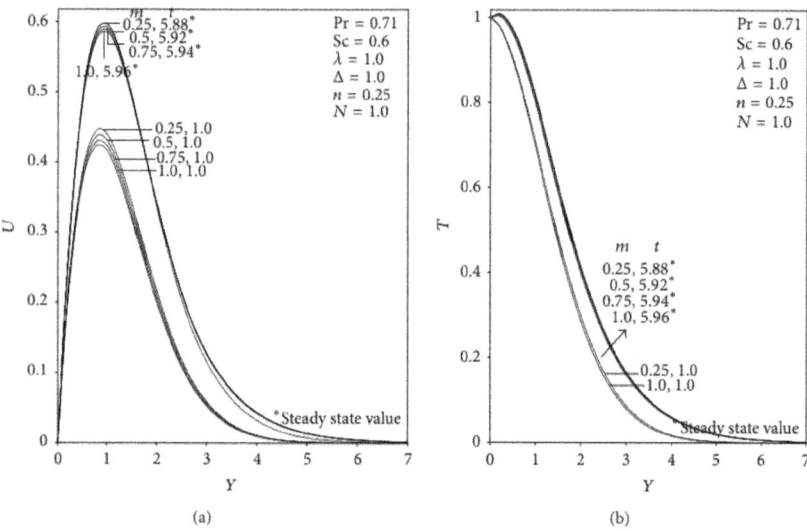

(a)   (b)

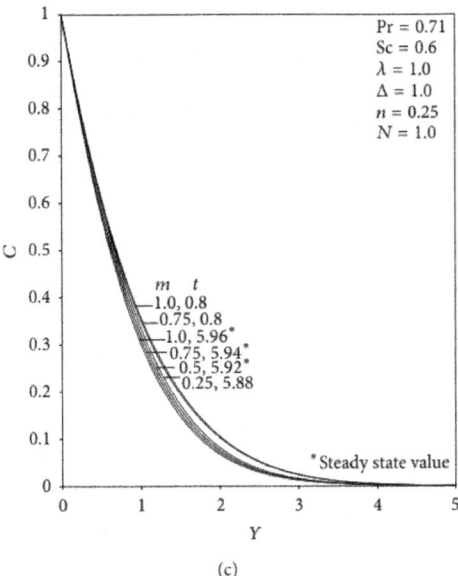

(c)

Figure 4: (a) Transient velocity profiles at $X = 1.0$ for different values of m. (b) Transient temperature profiles at $X = 1.0$ for different values of m. (c) Transient concentration profiles at $X = 1.0$ for different values of m.

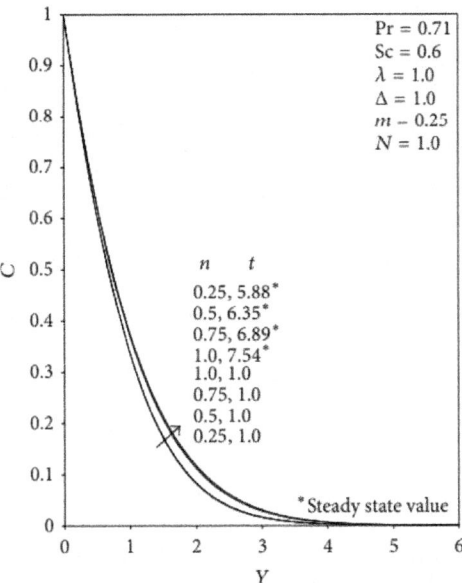

Figure 5: Transient concentration profile at $X = 1.0$ for different values of n.

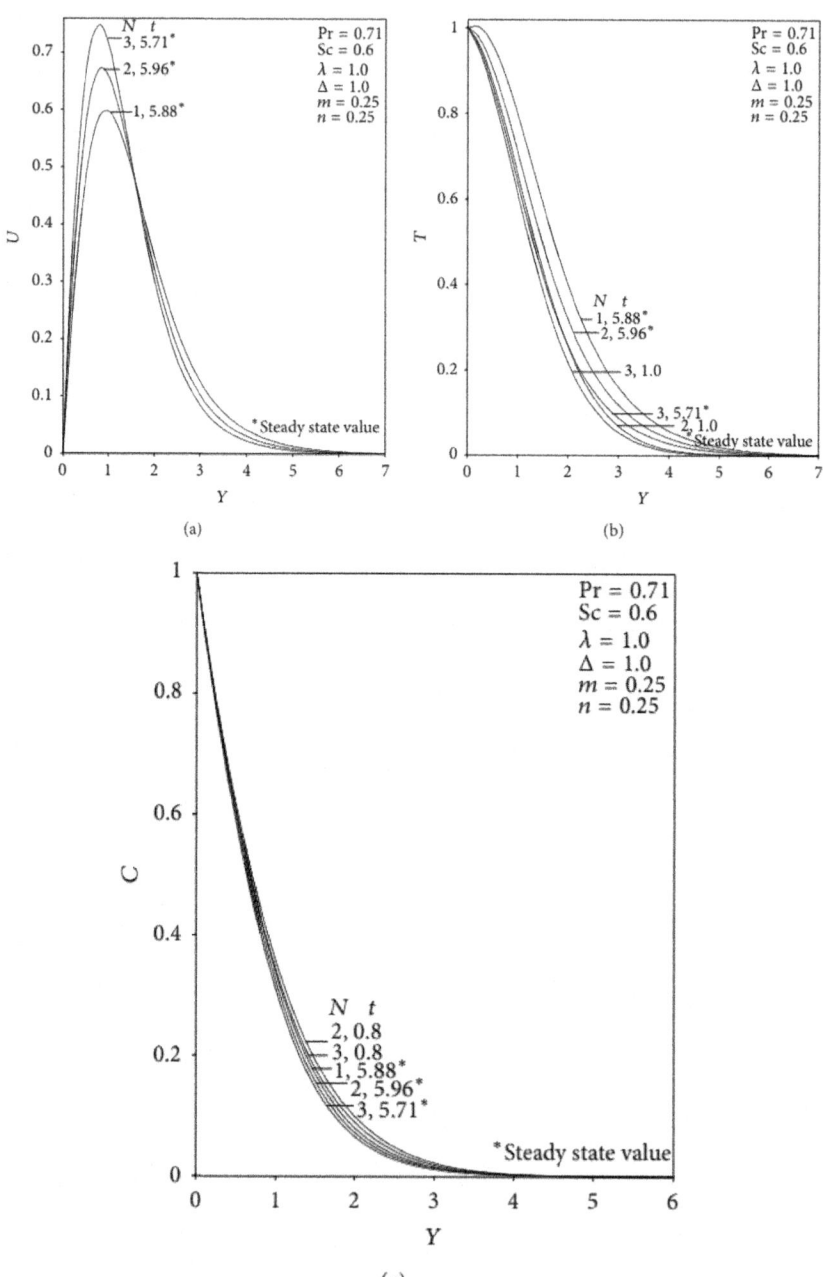

Figure 6: (a) Transient velocity profiles at $X = 1.0$ for different values of N. (b) Transient temperature profiles at $X = 1.0$ for different values of N. (c) Transient concentration profiles at $X = 1.0$ for different values of N.

Local values of the skin friction τ_x, Nusselt number Nu_x, and Sherwood number Sh_x for different parameters Pr, Δ, λ, Sc, m, n, and N are plotted through Figures 7(a)– 7(c) to Figures 11(a)–11(c). Figures 7(a)–7(c) indicate that the local skin friction increases for larger values of Δ and Pr (Figure 7(a)). The local Nusselt number increases for smaller values of Δ and larger values of Pr; that is, the heat generation/absorption parameter Δ has the tendency to increase the magnitude of the local Nusselt number for $\Delta < 0$ (Figure 7(b)), whereas the local Sherwood number increases for larger values of Δ and smaller values of Pr (Figure 7(c)). Figures 8(a)–8(c) depict the effects of chemical reaction parameter λ and Schmidt number Sc on the local skin friction, local Nusselt number, and local Sherwood number. The local skin friction and local Nusselt number increase for smaller values of λ and Sc. Increasing λ and Sc clearly boosts the wall skin friction (Figure 8(a)). With increasing Sc, the local Nusselt number (Figure 8(b)) is consistently reduced. The surface species gradient, that is, mass transfer rate at the cone surface, is strongly elevated with a rise in λ and Sc is observed from Figure 8(c). Figures 9(a)–9(c) show the effect of the surface concentration power law exponent m on the local skin friction, local Nusselt number, and local Sherwood number. The local skin friction and local Nusselt number increase for smaller values of m, while the local Sherwood number increases for higher values of m. Figure 10 indicates the effect of surface temperature power law exponent n on the local Sherwood number; it increases for larger values of n. It is observed from Figures 11 (a)–11(c) the effect of buoyancy ratio parameter N on the local skin friction, local Nusselt number, and local Sherwood number. Figure 11(a) illustrates a rise in

N accompanying a stronger increase in assisted buoyancy force; it strongly accelerates the flow and enhances the shear stress. The time required to attain the steady state is decreased with this increase in N. Inspection of Figures 11(b) and 11(c) shows that an increase in N strongly boosts both NuX and Sh_x; that is, it enhances the heat transfer gradient and mass transfer gradient at the cone surface. Further the Sherwood number increases with the increase in N. The physical reason is that positive force produced remarkable overshoot near the surface within the boundary layer for low Prandtl number fluid (Pr = 0.71) but for high Prandtl number fluid (Pr = 6.7) the velocity overshoot is not significant. Also, the buoyancy force enhanced the skin friction coefficient as well as the local Nusselt number or local heat transfer rate. Simultaneously the time required to attain the steady state is reduced with an increase in N for both Nu_x and Sh_x

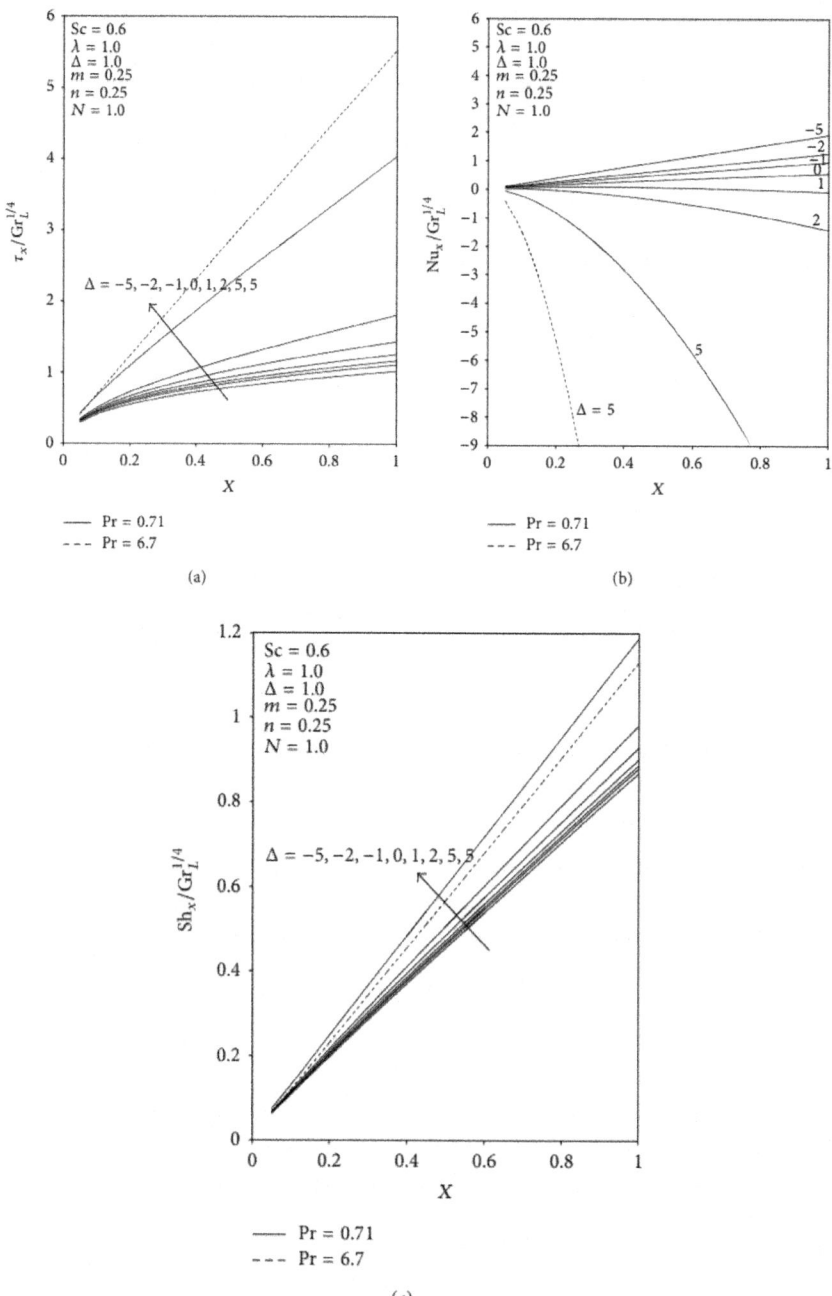

Figure 7: (a) Local skin friction for different values of Δ and Pr. (b) Local Nusselt number for different values of Δ and Pr. (c) Local Sherwood number for different values of Δ and Pr

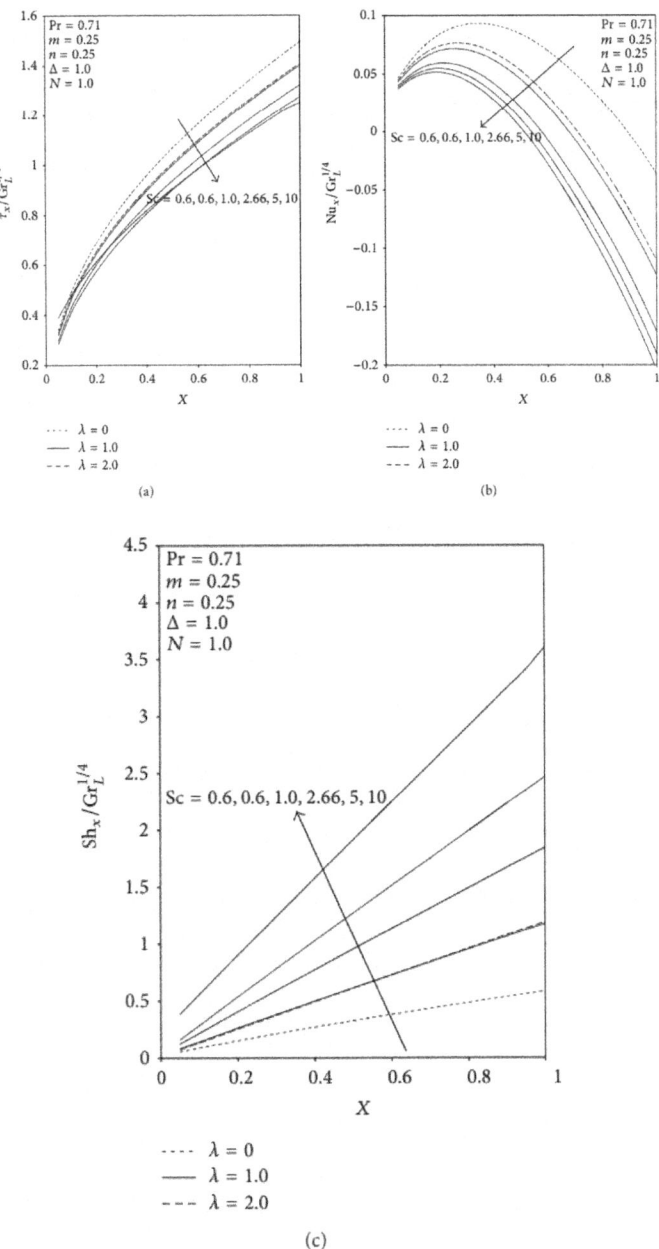

Figure 8: (a) Local skin friction profiles for different values of λ and Sc. (b) Local Nusselt number profiles for different values of λ and Sc. (c) Local Sherwood number profiles for different values of λ and Sc.

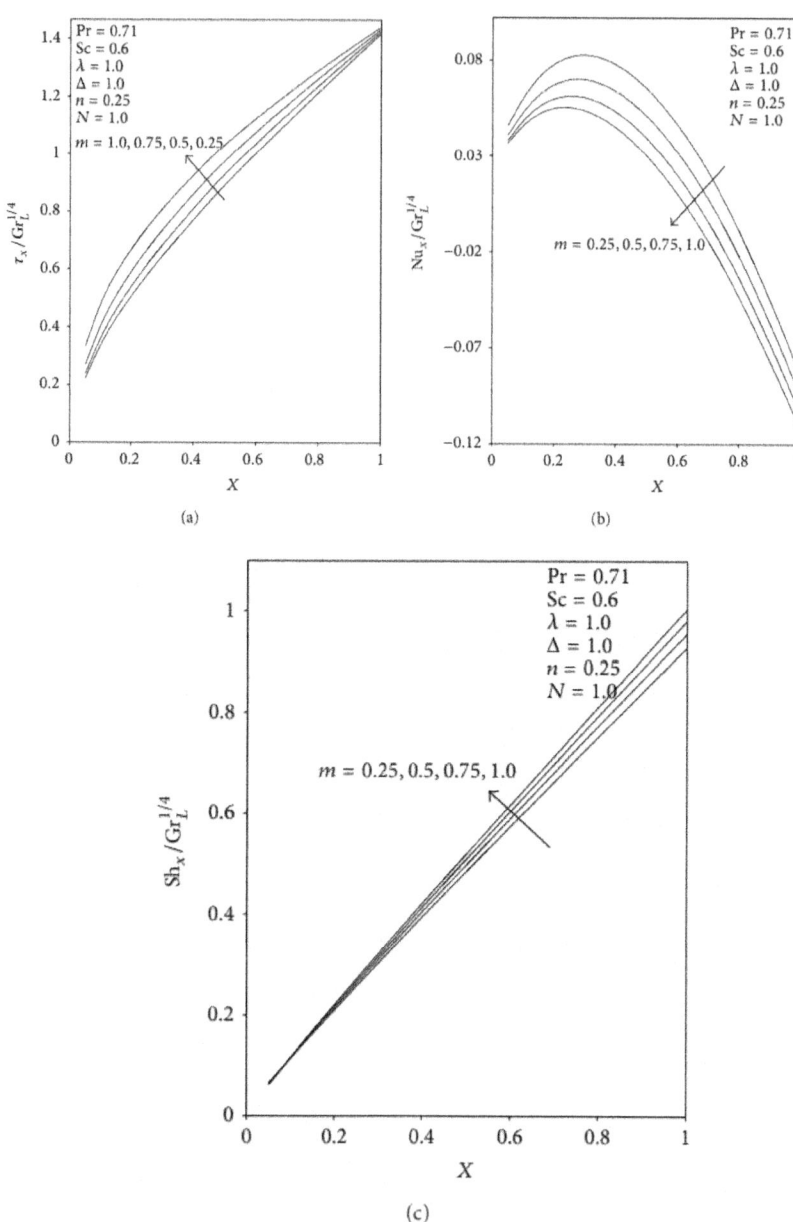

Figure 9: (a) Local skin friction profiles for different values of m. (b) Local Nusselt number profiles for different values of m. (c) Local Sherwood number profiles for different values of m.

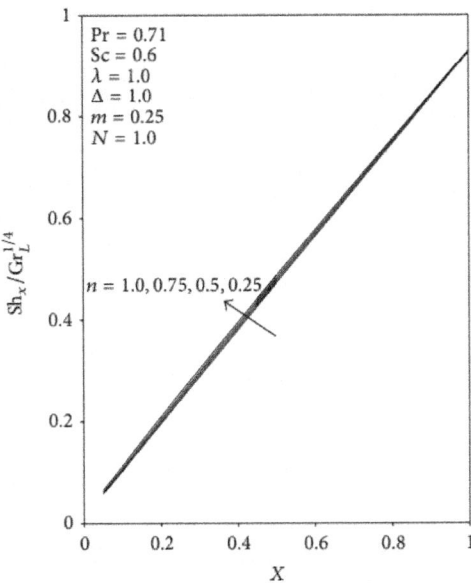

Figure 10: Local Sherwood number profiles for different values of n.

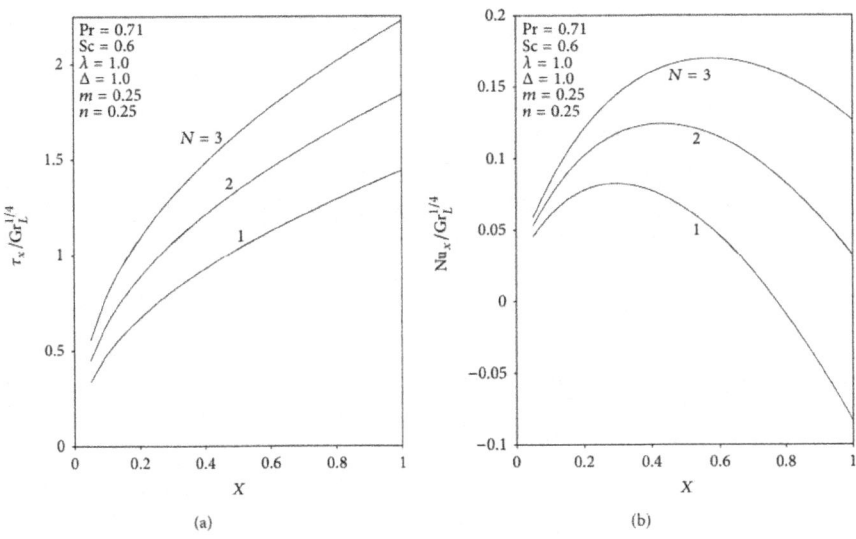

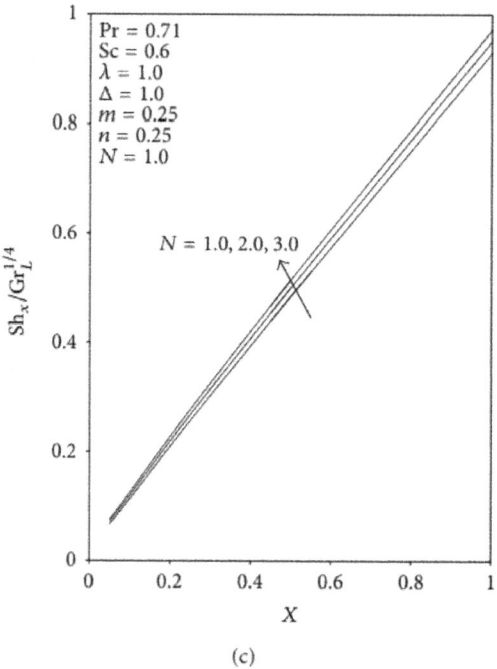

(c)

Figure 11: (a) Local skin friction profiles for different values of N. (b) Local Nusselt number profiles for different values of N. (c) Local Sherwood number profiles for different values of N.

Time dependences of the average values of skin friction $\bar{\tau}$, Nusselt number $\overline{Nu}$, and Sherwood number $\overline{Sh}$ for various parameters Pr, Δ, λ, Sc, m, n, and N are plotted through Figures 12(a)–12(c) to Figures 16(a)–16(c). From Figures 12(a)–12(c), it is noticed that the effects of heat generation/absorption parameter Δ and Pr on the average skin friction. It increases for larger values of Δ and Pr, whereas the average Nusselt number increases for smaller values of Δ and higher values of Pr. The average Sherwood number increases for larger values of Δ and smaller values of Pr. Figures 13(a)– 13(c) depict the effects of the chemical reaction parameter λ and Schmidt number Sc on the average skin friction, average Nusselt number, and average Sherwood number. The average skin friction and Nusselt number increase for smaller values of λ and Sc but the average Sherwood number increases for larger values of λ and Sc. Figures 14(a)–14(c) show the effect of the surface concentration power law exponent m on the average skin friction, average Nusselt number, and average

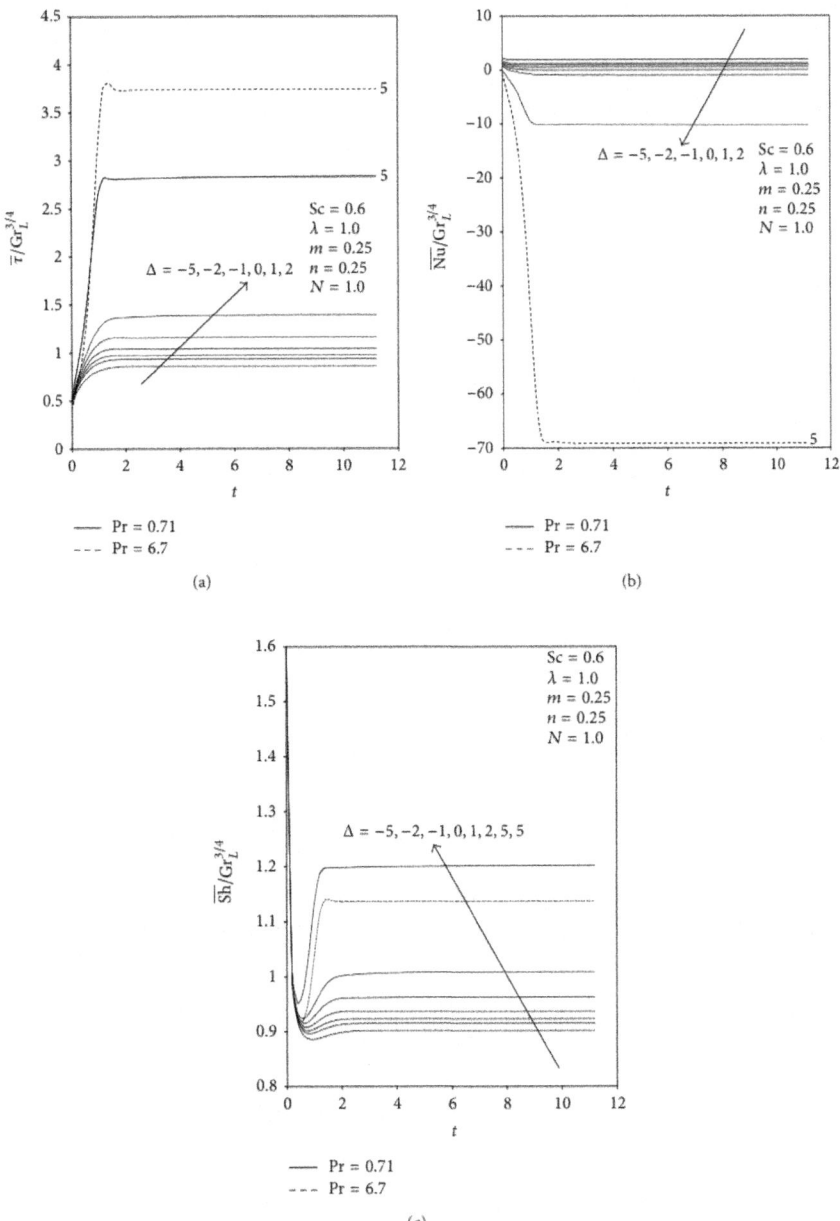

Figure 12: (a) Average skin friction profiles for different values of Δ and Pr in transient state. (b) Average Nusselt number profiles for different values of Δ and Pr in transient state. (c) Average Sherwood number profiles for different values of Δ and Pr in transient state.

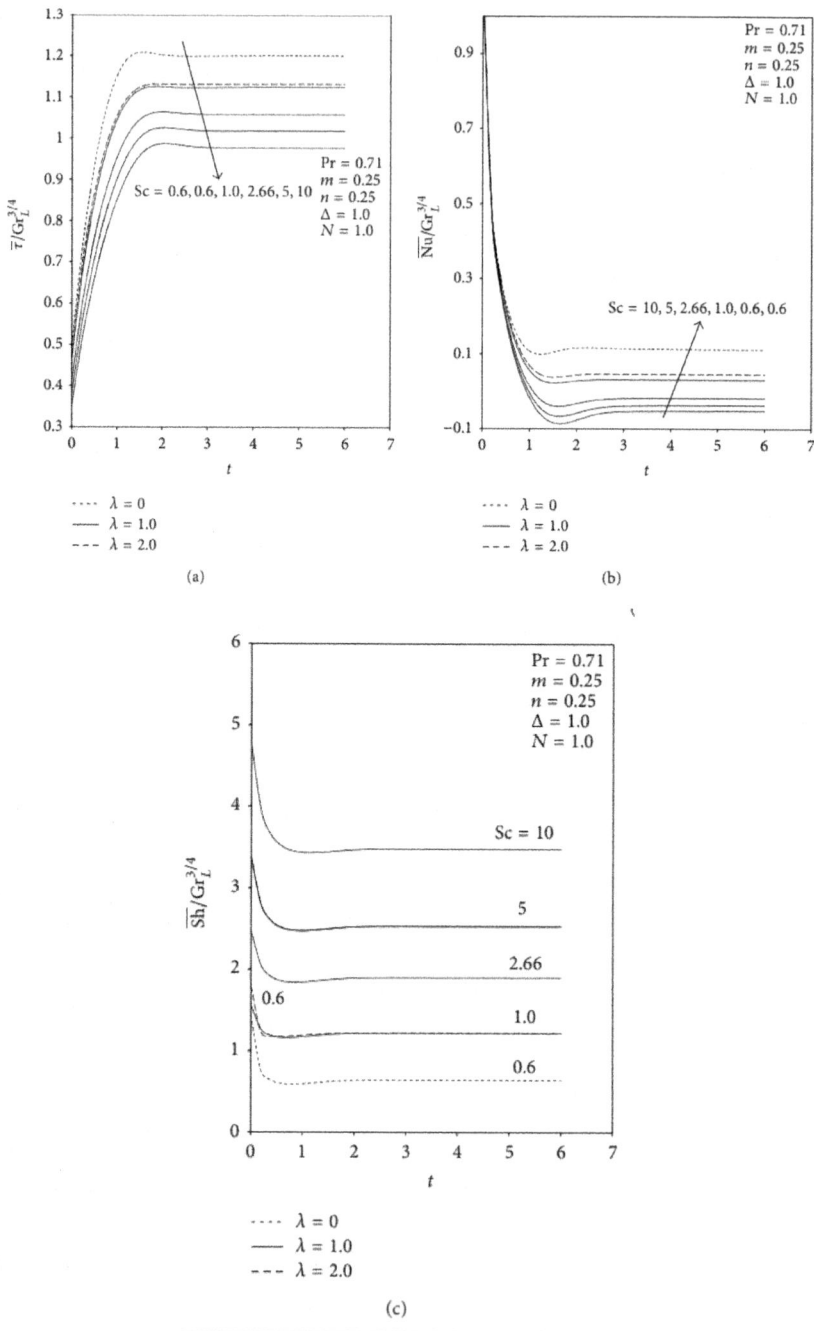

Figure 13: (a) Average skin friction profiles for different values of λ and Sc in transient state. (b) Average Nusselt number profiles for different values of λ and Sc in

transient state. (c) Average Sherwood number profiles for different values of λ and Sc in transient state.

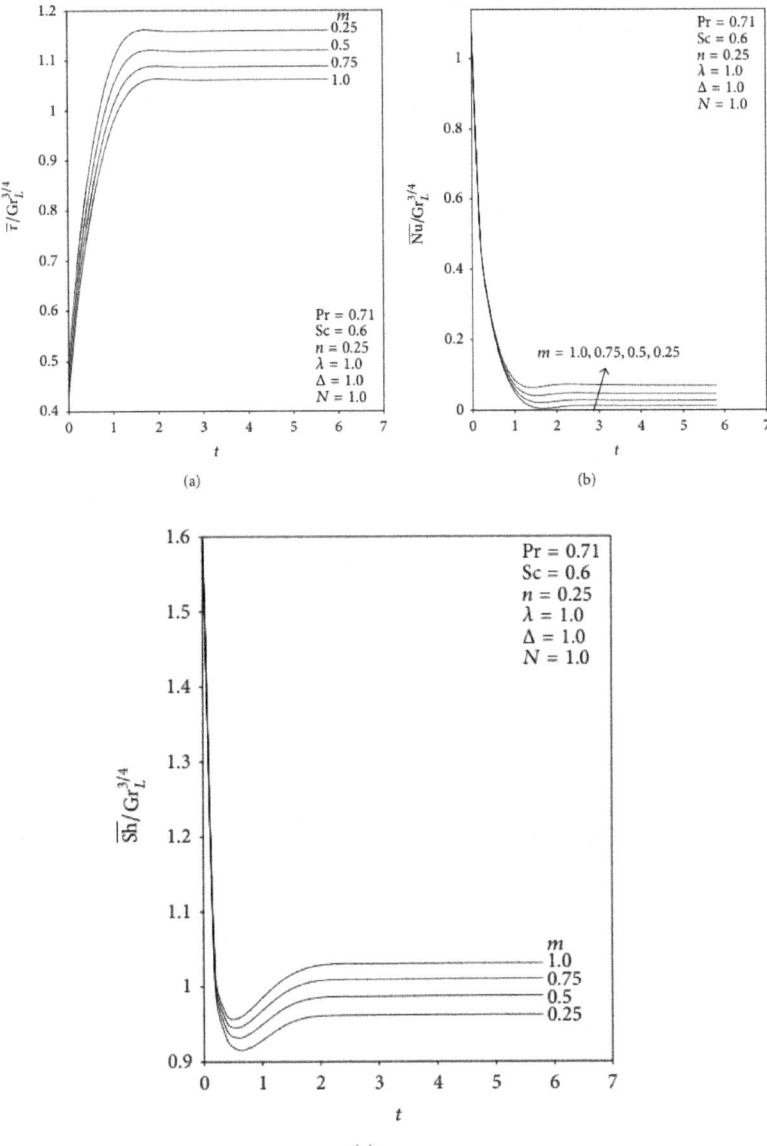

(a)

(b)

(c)

Figure 14: (a) Average skin friction profiles for different values of m in transient state. (b) Average Nusselt number profiles for different values of m in transient state. (c) Average Sherwood number profiles for different values of m in transient state.

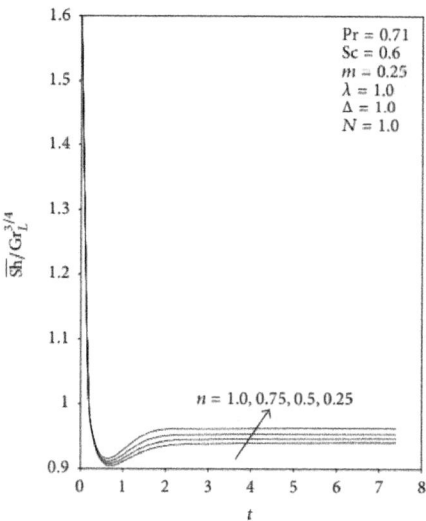

Figure 15: Average Sherwood number profiles for different values of n in transient state.

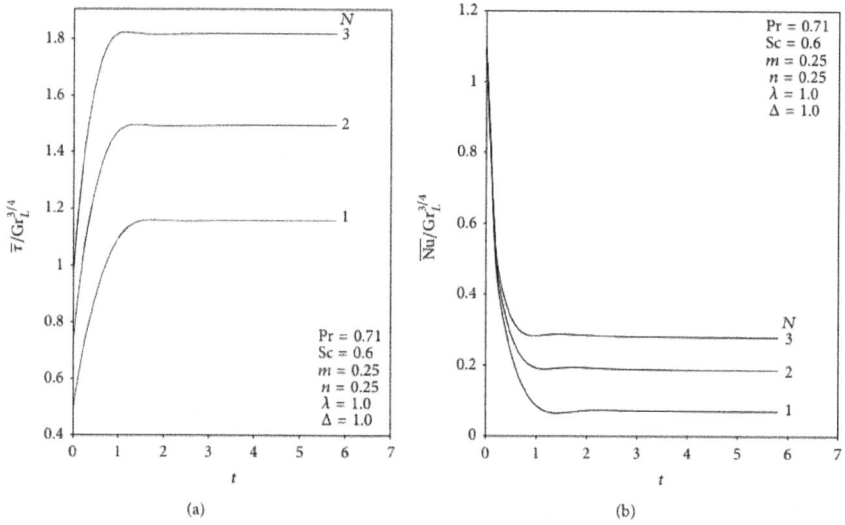

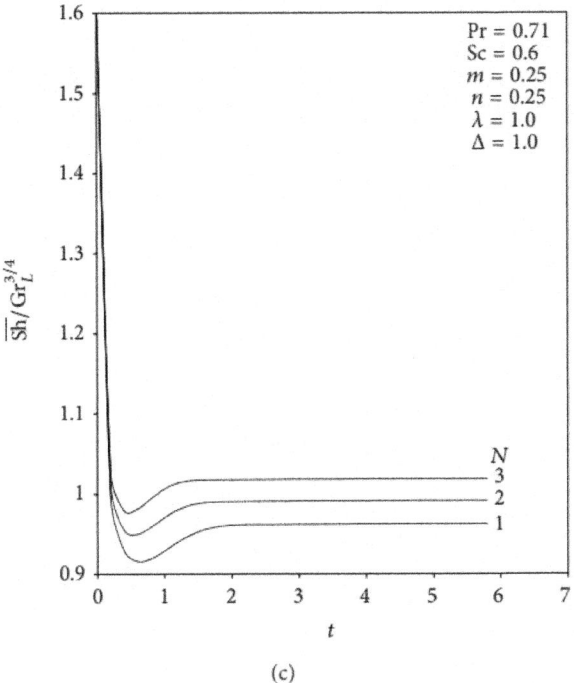

(c)

Figure 16: (a) Average skin friction profiles for different values of N in transient state. (b) Average Nusselt number profiles for different values of N in transient state. (c) Average Sherwood number profiles for different values of N in transient state

CONCLUSIONS

A mathematical model has been presented for the free convection flow from a vertical cone with heat generation/absorption and chemical reaction. The family of governing partial differential equations is solved by an implicit finite difference scheme of Crank-Nicholson type. A parametric study is performed to illustrate the influence of thermophysical parameters on the velocity, temperature, and concentration profiles. It has been observed that

(1)   the time taken to reach steady state increases with increasing Δ, Pr, λ, Sc, N, m, and n;

(2)   the fluid velocity increases for higher values of Δ, N and lower values of Pr, λ, Sc, and m. Temperature increases for larger values of Δ, Sc and smaller values of Pr, λ, m, and N, while the concentration of species decreases for smaller values of λ, n and larger values of Δ, Pr, Sc, m, and N;

(3)   the local skin friction increases for higher values of Pr, Δ, and N and for lower values of λ, Sc, and m. The local Nusselt number increases for higher values of Pr and N and lower values of λ, Sc, Δ, and m. The local Sherwood number increases for higher values of λ, Sc, Δ, m, n, and N and smaller values of Pr;

(4)   the average skin friction increases for larger values of Δ, Pr, and N and smaller values of λ, Sc, and m. The average Nusselt number increases for higher values of Pr and N and lower values of λ, Sc, Δ, m, and N. The average Sherwood number increases for higher values of λ, Sc, Δ, m, and N and lower values of Pr and n;

(5)   momentum boundary layers become thick for higher values of Sc, Δ, λ, and N and lower values Pr and m, the thermal boundary layer becomes thick for higher values of Sc, Δ, and m and lower values of Pr, λ, and N, and the concentration boundary layer becomes thick for larger values of λ and n and smaller values of Δ, Pr, Sc, N, m, and N.

Nomenclature

a :   Constant

b: Constant

C_p: Specific heat at constant pressure

D: Mass diffusivity $m^2 s^{-1}$

$f''(\infty, 0)$: Local skin friction in [8]

Gr_L: Thermal Grashof number

Gr^*: Mass Grashof number

g: Acceleration due to gravity ms^{-2}

k: Thermal conductivity $Wm^{-1}K^{-1}$

k_1:   Dimensional chemical reaction parameter J

L: Reference length m

m: Exponent in power law variation in surface concentration

n: Exponent in power law variation in surface temperature

N: Dimensionless buoyancy ratio

Nu_x: Local Nusselt number

$\overline{\mathrm{Nu}_L}$: Average Nusselt number

Nu_x: Nondimensional local Nusselt number

$\overline{\mathrm{Nu}}$: Nondimensional average Nusselt number

Pr: Prandtl number

Q_0: Dimensional heat generation/absorption coefficient $\mathrm{W\,m^{-3}}$

R: Dimensionless local radius

r: Local radius of the cone m

Sc: Schmidt number

T': Temperature $\mathrm{K^0}$

T: Dimensionless temperature

t': Time s

t: Dimensionless time

U: Dimensionless velocity in—X direction

u: Velocity component in x direction $\mathrm{ms^{-1}}$

V: Dimensionless velocity in—Y direction

v: Velocity component in—y direction $\mathrm{ms^{-1}}$

X: Dimensionless spatial coordinate along the cone generator

x: Spatial coordinate along the cone generator m

Y: Dimensionless spatial coordinate along the normal to the cone generator

y: Spatial coordinate along the normal to the cone generator m.

Greek Symbols

α: Thermal diffusivity $\mathrm{m^2 s^{-1}}$

β: Volumetric thermal expansion $\mathrm{^0 k^{-1}}$

β^*: Volumetric coefficient of expansion with concentration $\mathrm{^0 k^{-1}}$

Δ: Dimensionless heat generation and absorption parameter

λ: Nondimensional chemical reaction parameter

ρ: Density $\mathrm{kg\,m^{-3}}$

$-\theta'(\infty, 0)$: Local Nusselt number in [8]

Δt: Dimensionless time step

ΔX: Dimensionless finite difference grid size in X direction

ΔY: Dimensionless finite difference grid size in Y direction

μ: Dynamic viscosity $\mathrm{kg\,m^{-1} s^{-1}}$

ν: Kinematic viscosity $\mathrm{m^2 s^{-1}}$

τ_x: Local skin friction

τ_X: Dimensionless local skin friction

$\overline{\tau_L}$: Average skin friction

$\overline{\tau}$: Dimensionless average skin friction.

Subscripts

w: Condition on the wall

∞: Free stream condition.

CONFLICT OF INTERESTS

The authors declare that there is no conflict of interests regarding the publication of this paper.

ACKNOWLEDGMENTS

The authors thankfully acknowledge the financial support from UTM-Flagship Research Grant, Vote no. 01G40, MOHE, and Research Management Centre (RMC), Universiti Teknologi Malaysia for completion of this research work and thanks to the reviewers for their valuable suggestions and comments to improve this paper.

REFERENCES

1.   H. J. Merk and J. A. Prins, "Thermal convection in laminary boundary layers I," Applied Scientific Research, vol. 4, pp. 11–24, 1953.

2.   H. J. Merk and J. A. Prins, "Thermal convection laminar boundary layer II," Applied Scientific Research, vol. 4, pp. 195–206, 1954.

3.   R. G. Hering and R. J. Grosh, "Laminar free convection from a non-isothermal cone," International Journal of Heat and Mass Transfer, vol. 5, no. 11, pp. 1059–1068, 1962.

4.   R. G. Hering, "Laminar free convection from a non-isothermal cone at low Prandtl numbers,"International Journal of Heat and Mass Transfer, vol. 8, no. 10, pp. 1333–1337, 1965.

5.   N. G. Kafoussias, "Effects of mass transfer on free convective flow past a vertical isothermal cone surface," International Journal of Engineering Science, vol. 30, no. 3, pp. 273–281, 1992

6.   K. A. Yih, "Uniform transpiration effect on combined heat and mass transfer by natural convection over a cone in saturated porous media: Uniform wall temperature/concentration or heat/mass flux,"International Journal of Heat and Mass Transfer, vol. 42, no. 18, pp. 3533–3537, 1999.

7.   K. A. Yih, "Coupled heat and mass transfer by free convection over a truncated cone in porous media: VWT/VWC or VHF/VMF," Acta Mechanica, vol. 137, no. 1, pp. 83–97, 1999.

8.   A. J. Chamkha, "Coupled heat and mass transfer by natural convection about a truncated cone in the presence of magnetic field and radiation effects," Numerical Heat Transfer A: Applications, vol. 39, no. 5, pp. 511–530, 2001.

9.   A. J. Chamkha and M. M. A. Quadri, "Combined heat and mass transfer by hydromagnetic natural convection over a cone embedded in a non-

Darcian porous medium with heat generation/absorption effects," Heat and Mass Transfer/Waerme- und Stoffuebertragung, vol. 38, no. 6, pp. 487–495, 2002.

10. A. A. Afify, "The effect of radiation on free convective flow and mass transfer past a vertical isothermal cone surface with chemical reaction in the presence of a transverse magnetic field," Canadian Journal of Physics, vol. 82, no. 6, pp. 447–458, 2004.

11. A. J. Chamkha and A. Al-Mudhaf, "Unsteady heat and mass transfer from a rotating vertical cone with a magnetic field and heat generation or absorption effects," International Journal of Thermal Sciences, vol. 44, no. 3, pp. 267–276, 2005.

12. A. J. Chamkha, A. F. Al-Mudhaf, and I. Pop, "Effect of heat generation or absorption on thermophoretic free convection boundary layer from a vertical flat plate embedded in a porous medium," International Communications in Heat and Mass Transfer, vol. 33, no. 9, pp. 1096–1102, 2006.

13. S. M. M. El- Kabeir, M. Modather, and M. A. Mansour, "Effect of heat and mass transfer on free convection flow over a cone with uniform suction or injection in micro polar fluids," International Journal of Applied Mechanics and Enginering, vol. 11, no. 1, pp. 15–35, 2006.

14. S. M. M. El-Kabeir and M. M. M. Abdou, "Chemical reaction, heat and mass transfer on MHD flow over a vertical isothermal cone surface in micropolar fluids with heat generation/absorption," Applied Mathematical Sciences: Journal for Theory and Applications, vol. 1, no. 33–36, pp. 1663–1674, 2007.

15. S. M. M. El-Kabeir, M. A. El-Hakiem, and A. M. Rashad, "Group method analysis for the effect of radiation on MHD coupled heat and mass transfer natural convection flow water vapor over a vertical cone through porous medium," International Journal of Applied Mathematics and Mechanics, vol. 3, no. 2, pp. 35–53, 2007.

16. C. Cheng, "Natural convection heat and mass transfer from a vertical truncated cone in a porous medium saturated with a non-Newtonian fluid with variable wall temperature and concentration,"International Communications in Heat and Mass Transfer, vol. 36, no. 6, pp. 585–589, 2009.

17. C. Y. Cheng, "Soret and Dufour effects on natural convection heat and mass transfer from a vertical cone in a porous medium," International Communications in Heat and Mass Transfer, vol. 36, no. 10, pp. 1020–1024, 2009.

18. C. Cheng, "Soret and Dufour effects on heat and mass transfer by natural convection from a vertical truncated cone in a fluid-saturated porous medium with variable wall temperature and concentration,"International Communications in Heat and Mass Transfer, vol. 37, no. 8, pp. 1031–1035, 2010.

19. A. S. N. Murti, P. K. Kameswaran, and K. T. Poorna, "Radiation, chemical reaction, double dispersion effects on heat and mass transfer in non-Newtonian fluids," International Journal of Engineering, vol. 4, no. 1, pp. 13–25, 2010.

20. P. M. Kishore, V. Rajesh, and S. Vijayakumar Verma, "Viscoelastic buoyancy- driven MHD free convective heat and mass transfer past avertical cone with thermal radiation and viscous dissipation effects," International Journal of Mathematics and Mechanics, vol. 6, no. 15, pp. 67–87, 2010.

21. A. Mahdy, "Effect of chemical reaction and heat generation or absorption on double-diffusive convection from a vertical truncated cone in porous media with variable viscosity," International Communications in Heat and Mass Transfer, vol. 37, no. 5, pp. 548–554, 2010.

22. S. G. Mohiddin, S. Vijayakumar Verma, and N. Ch. S. N. Iyengar, "Radiation and mass transfer effects on MHD free convective flow past a vertical cone with variable surface conditions in the presence of viscous dissipation," International Electronic Engineering Mathematical Society, vol. 8, pp. 22–37, 2010.

23. S. Gouse Mohiddin, V. R. Prasad, S. V. K. Varma, and O. Anwar Bég, "Numerical study of unsteady free convective heat and mass transfer in a Walters-B visco elastic flow along a vertical cone," International Journal of Applied Mathematics and Mechanics, vol. 6, no. 15, pp. 88–114, 2010.

24. P. M. Patil and I. Pop, "Effects of surface mass transfer on unsteady mixed convection flow over a vertical cone with chemical reaction," International Journal of Heat Mass Transfer, vol. 47, no. 11, pp. 1453–1464, 2011.

25. S. M. M. EL-Kabeir and E. A. EL-Sayed, "Effects of thermal radiation and viscous dissipation on MHD viscoelastic free convection past a vertical isothermal cone surface with chemical reaction," International Journal of Energy & Technology, vol. 4, no. 10, pp. 1–7, 2012.

26. F. G. Awad, P. Sibanda, S. S. Motsa, and O. D. Makinde, "Convection from an inverted cone in a porous medium with cross-diffusion effects," Computers & Mathematics with Applications, vol. 61, no. 5, pp. 1431–1441, 2011.

27. M. Narayana, F. G. Awad, and P. Sibanda, "Free magnetohydrodynamic flow and convection from a vertical spinning cone with cross-diffusion effects," Applied Mathematical Modelling. Simulation and Computation for Engineering and Environmental Systems, vol. 37, no. 5, pp. 2662–2678, 2013.

28. A. Basiri Parsa, M. M. Rashidi, and T. Hayat, "MHD boundary-layer flow over a stretching surface with internal heat generation or absorption," Heat Transfer—Asian Research, vol. 42, no. 6, pp. 500–514, 2013.

29. V. M. Soundalgekar and P. Ganesan, "Finite-difference analysis of transient free convection with mass transfer on an isothermal vertical flat plate," International Journal of Engineering Science, vol. 19, no. 6, pp. 757–770, 1981.

30. B. Pullepu, K. Ekambavanan, and A. J. Chamkha, "Unsteady laminar natural convection from a non-isothermal vertical cone," Nonlinear Analysis: Modelling and Control, vol. 12, no. 4, pp. 525–540, 2007.

31. R. Muthukumaraswamy and P. Ganesan, "Unsteady flow past an impulsively started vertical plate with heat and mass transfer," International Journal of Heat Mass Transfer, vol. 34, no. 2-3, pp. 187–193, 1998.

32. R. Muthukumaraswamy and P. Ganesan, "Natural convection on a moving isothermal vertical plate with chemical reaction," Journal of Engineering Physics and Thermophysics, vol. 75, no. 1, pp. 113–119, 2002.

33. P. Bapuji, K. Ekambavanan, and I. Pop, "Finite difference analysis of laminar free convection flow past a non isothermal vertical cone," Heat and Mass Transfer, vol. 44, no. 5, pp. 517–526, 2008.

Chapter 9

EFFECTS OF THERMAL RADIATION AND RADIATION ABSORPTION ON FLOW PAST AN IMPULSIVELY STARTED INFINITE VERTICAL PLATE WITH NEWTONIAN HEATING AND CHEMICAL REACTION

Swetha Ravi[1], Jagdish Prakash[2], Viswanatha Reddy Gottam[3], Vijaya Kumar Varma Sibyala[3]

[1]Department of Mathematics, Gudlavalleru Engineering College, Gudlavalleru, India

[2]Department of Mathematics, University of Botswana, Gaborone, Botswana

[3]Department of Mathematics, S. V. University, Tirupati, India

ABSTRACT

A perfect solution to the present natural convective flow problem of a vertical transfinite plate owing to the impulsive motion in the ubiety of first ordered chemical reaction, radiation absorption, radiation, Newtonian heating and species concentration in its plane is evolved by applying the method of Laplace transforms in closed form at the plate. Exact results for velocity, temperature, concentration fields are prevailed and expressions for heat and mass transfer rates are also found. The effects are analyzed for the respective invariables for both ammonia and water vapor.

INTRODUCTION

On chemical reaction, the field of mass and heat transfer is of good pragmatic importance to applied scientists owing to its general occurrence in various fields of engineering and science. Especially, the subject of mass and heat transfer with heat radiation, chemical reaction has significant role in hydrometallurgical and chemical industries. For a moving plate, a chemical reaction takes place in legion chemical processes between a fluid and foreign mass. This sue is involved in many industrial usages such as glassware or ceramics manufacturing, food processing and production of polymers. The convection study with mass and heat transfer plays a major role in the dispersion and formation of fog, design of chemical processing equipment's, temperature distribution, and moisture over agricultural fields and in the paper drying process.

Ahmed et al. [1] have identified the analysis for MHD rotating heat or mass transport phenomenon bounded by a vertical oscillating surface in the Mein of Darcian porous regime by using Numerical/Laplace transform. Characteristics of the heat and mass transfer in the Mien of chemical reaction and thermal radiation for a Newtonian incompressible fluid across an extending vertical surface having temperature dependent viscosity was studied by Kandasamy et al. [2] . Makinde [3] examined the free transient convection interaction of an absorbing, emitting plate with thermal radiation. Mukhopadhyay [4] performed an investigation on the results of heat transfer and thermal radiation on a mixed unsteady convective flow across an extending porous surface in porous medium.

Heat transfer analysis of a forced convective flow of the fluid past an embedded plate in a porous medium for an incompressible fluid was examined by Mukhopadhyay and Layek [5]. Muthucumaraswamy and Ganesan [6] looked at the impulsively started transient radiation-convection flow with vertical temperature consequences. An analysis of the chemical reaction, theoretically a result, with variable temperature on a vertical oscillating plate was given by Muthucumaraswamy [7] . Reddy et al. [8] investigated the consequences of unsteady natural MHD convective flow in a porous medium with constant mass diffusion and Newtonian heating. The effects of MHD radiating and chemically reacting fluid past a non-isothermal impulsively started vertical surface adjacent to a porous regime by using numerical analysis was discussed by Sahin Ahmed [9] .

The importance of the present flow problem is to analyze the effects of thermal radiation and radiation absorption on the flow past an impulsively started infinite vertical plate with Newtonian heating and chemical reaction.

MATHEMATICAL ANALYSIS

Free convective unsteady flow of the fluid for a vertical transfinite plate with Newtonian heating, past an impulsively started incompressible viscous fluid in the Mien of radiation and radiation absorption is studied. Along the plate and in the vertical upward direction, axis $\overline{x}^*$ is chosen and normal to the plate, axis

$\bar{y}^*$ is considered. Initially the fluid and the plate are having same temperature $\bar{T}_\infty^*$ and the concentration $\bar{C}_\infty^*$ at all points in a stationary state for time $\bar{t}^* \leq 0$. The coordinate system and the flow model are shown in Figure 1. The plate is fixed with a velocity u_0 in the vertical direction into impulsive motion versus the gravitational field at time $\bar{t}^* > 0$. We assumed that i) heat transfer rate and the local surface temperature $\bar{T}^*$ are proportional to one another from the surface, and near the plate concentration rises to $\bar{C}_w^*$ and ii) the consequences of viscous dissipation are negligible in the energy equation. Among the fluid and diffusing species, there is a first order chemical reaction. Since all the physical quantities are expressed in terms of $\bar{y}^*$, $\bar{t}^*$ only and are free from $\bar{x}^*$ and in the direction of $\bar{x}^*$, the plate is considered transfinite.

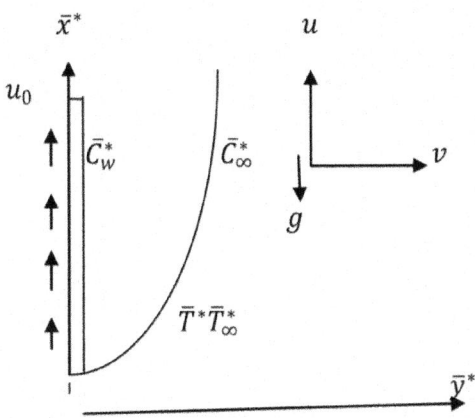

Figure 1: A sketch of flowmodel and coordinate system.

The equations for this present flow, by the Boussinesq estimation are as follows

$$\frac{\partial \bar{u}^*}{\partial \bar{t}^*} = g\beta\left(\bar{T}^* - \bar{T}_\infty^*\right) + g\bar{\beta}\left(\bar{C}^* - \bar{C}_\infty^*\right) + v\frac{\partial^2 \bar{u}^*}{\partial \bar{y}^{*2}}$$

(1)

$$\rho C_p \frac{\partial \bar{T}^*}{\partial \bar{t}^*} = \kappa \frac{\partial^2 \bar{T}^*}{\partial \bar{y}^{*2}} + \bar{Q}_1^*\left(\bar{C}^* - \bar{C}_\infty^*\right) - \frac{\partial \bar{q}_r^*}{\partial \bar{y}^*}$$

(2)

$$\frac{\partial \bar{C}^*}{\partial \bar{t}^*} = D\frac{\partial^2 \bar{C}^*}{\partial \bar{y}^{*2}} - K_r\left(\bar{C}^* - \bar{C}_\infty^*\right)$$

(3)

and the connected conditions for this flow are

$$\left.\begin{array}{l} \text{at } \bar{t}^* \leq 0 : \bar{u}^* = 0, \bar{T}^* = \bar{T}_\infty^*, \bar{C}^* = \bar{C}_\infty^* \ \forall \bar{y}^* \\[2mm] \text{at } \bar{t}^* > 0 : \bar{u}^* = u_0, \dfrac{\partial \bar{T}^*}{\partial \bar{y}^*} = -\dfrac{h}{\kappa} T^*, \bar{C}^* = \bar{C}_w^* \ \text{ for } \bar{y}^* = 0 \\[2mm] \bar{u}^* = 0, \bar{T}^* \to \bar{T}_\infty^*, \bar{C}^* \to \bar{C}_\infty^* \ \text{ as } \bar{y}^* \to \infty \end{array}\right\}$$

(4)

Here $\quad A = \dfrac{u_0^2}{\upsilon}$

The term of radiative heat flux by the Rosseland estimation is given by

$$\bar{q}_r^* = -\frac{4\sigma_s}{3k_c}\frac{\partial \bar{T}^{*4}}{\partial \bar{y}^*}$$

(5)

But here with in the flow, presuming that the deviation in temperatures can be showed as a linear combination of the temperatures and around $\bar{T}_\infty^*$, which is found by expanding $\bar{T}^{*4}$ in a Taylor's series as follows:

$$\bar{T}^{*4} = \bar{T}_\infty^{*4} + 4\bar{T}_\infty^{*3}\left(\bar{T}^* - \bar{T}_\infty^*\right) + 6\bar{T}_\infty^{*2}\left(\bar{T}^* - \bar{T}_\infty^*\right) + \cdots$$

(6)

and ignoring the higher ordered terms, beyond the first degree, we get

$$\bar{T}^{*4} \cong 4\bar{T}_\infty^{*3}\bar{T}^* - 3\bar{T}_\infty^{*4}$$

(7)

Differentiating Equation (5) with respect to $\bar{y}^*$ and applying Equation (6), we get

$$\frac{\partial \bar{q}_r^*}{\partial \bar{y}^*} = -\frac{16\sigma_s \bar{T}_\infty^{*3}}{3k_c}\frac{\partial^2 \bar{T}^*}{\partial \bar{y}^{*2}}$$

(8)

We have inserted the non-dimensional succeeding measures

$$U = \frac{\bar{u}^*}{u_0}, \; t = \frac{\bar{t}^* u_0^2}{v}, \; y = \frac{\bar{y}^* u_0}{v}, \; Pr = \frac{\mu c_p}{\kappa}, \; Gr = \frac{g\beta v \bar{T}_\infty^*}{u_0^3},$$

$$\theta = \frac{\bar{T}^* - \bar{T}_\infty^*}{\bar{T}_\infty^*}, \; Sc = \frac{v}{D}, \; Q_1 = \frac{v\bar{Q}_1^* \left(\bar{C}_w^* - \bar{C}_\infty^*\right)}{\rho C_p u_0^2 \bar{T}_\infty^*}, \; Gm = \frac{g\bar{\beta} v \left(\bar{C}_w^* - \bar{C}_\infty^*\right)}{u_0^3},$$

$$C = \frac{\bar{C}^* - \bar{C}_\infty^*}{\bar{C}_w^* - \bar{C}_\infty^*}, \; K = \frac{K_r v}{u_0^2}, \; N_r = \frac{16\sigma_s \bar{T}_\infty^{*3}}{3\kappa k_c}$$

$$(9)$$

The Equations (1)-(3) are reduced into the following forms by using the Equations (8) and (9) as follows

$$\frac{\partial U}{\partial t} = G_r \theta + G_m C + \frac{\partial^2 u}{\partial y^2}$$

$$(10)$$

$$\frac{\partial \theta}{\partial t} = \left(\frac{1 + N_r}{P_r}\right) \frac{\partial^2 \theta}{\partial y^2} + Q_1 C$$

$$(11)$$

$$\frac{\partial C}{\partial t} = \frac{1}{S_c} \frac{\partial^2 C}{\partial y^2} - KC$$

$$(12)$$

In non-dimensional form, the conditions reduce to as follows

$$\left. \begin{array}{l} \text{at } t \le 0 : U = 0, \theta = 0, C = 0 \quad \forall y \\[2mm] \text{at } t > 0 : U = 1, \dfrac{\partial \theta}{\partial y} = -(1 + \theta), C = 1 \text{ for } y = 0 \\[2mm] \quad\quad U = 0, \theta \to 0, C \to 0 \quad\quad \text{as } y \to \infty \end{array} \right\}$$

$$(13)$$

SOLUTION OF THE PROBLEM

The non-dimensional Equations (10)-(12) associated with the conditions given by Equations (13) are evolved by the method of Laplace transforms, and therefore the results of concentration, temperature and velocity are given by

$$C(y,t) = B_6$$

$$(14)$$

$$\theta(y,t) = -\frac{1}{\sqrt{a}} B_1 - B_2 + \frac{1}{\sqrt{a}} B_3 + N_1 \left(M_1 B_1 + M_2 B_2 + M_3 B_4 + M_4 B_5 + M_5 B_3\right) + N_2 \left(B_6 - B_7\right)$$

$$(15)$$

$$U(y,t) = A_1 + C_1\left(A_4 + A_3 - A_5 - A_{27} + A_2 - A_{28}\right)$$
$$+ C_2\left(A_{12} + A_{13} + A_6 + A_{11} + A_7 + A_{10} + A_{14} + A_8 + A_{15} + A_9\right)$$
$$+ C_3\left[M_6\left(m_1 - m_2\right) + M_7\left(A_1 - A_{12}\right) + M_8\left(A_{16} - A_{22}\right)\right.$$
$$+ M_9\left(A_{17} - A_{23}\right) + M_{10}\left(A_{18} - A_{24}\right) + M_{11}\left(A_{19} - A_{25}\right)$$
$$\left. + M_{12}\left(A_{20} - A_{26}\right)\right] + C_4\left[A_1 - A_{29} - A_{30} + A_{31}\right] \tag{16}$$

THE RATE OF HEAT TRANSFER

In dimensionless form, heat transfer rate from the temperature gradient is

$$Nu = -\left(\frac{\partial \theta}{\partial y}\right)_{y=0} \tag{17}$$

From Equations (15) and (17), we get

$$Nu = \exp(at)\operatorname{erfc}\left(-\sqrt{at}\right) + N_1\left[\frac{M_2}{\sqrt{\pi a t}} - M_3\frac{b}{\sqrt{a}}\exp(-bt)\operatorname{erfc}\left(i\sqrt{bt}\right)\right.$$
$$+ \frac{i\sqrt{b}}{\sqrt{\pi a t}}e^{-2bt}\left(M_4 - M_3\right) - M_4\frac{b}{\sqrt{a}}\exp(-bt)\operatorname{erfc}\left(-i\sqrt{bt}\right)$$
$$\left. + M_5\sqrt{a}\exp(at)\operatorname{erfc}\left(-\sqrt{at}\right) + M_5\frac{1}{\sqrt{\pi t}}\right] + N_2\sqrt{KS_c}\operatorname{erf}\left(\sqrt{Kt}\right)$$
$$- N_2\exp(-bt)\sqrt{S_c(K-b)}\operatorname{erf}\left(\sqrt{(K-b)t}\right) \tag{18}$$

THE RATE OF MASS TRANSFER

In dimensionless form, mass transfer rate from the concentration gradient is

$$Sh = -\left(\frac{\partial c}{\partial y}\right)_{y=0} \tag{19}$$

From Equations (14) and (19), we get

$$Sh = \sqrt{KS_c}\operatorname{erf}\left(\sqrt{Kt}\right) + \sqrt{\frac{S_c}{\pi t}}\exp(-Kt) \tag{20}$$

DEDUCTION

The effects of this analysis are in good agreement with the results given by Rajesh [8] in the absence of parameters radiation absorption (Q_1) and radiation (N_r), where

$$B_1 = \frac{1}{\sqrt{\pi t}} \exp\left(-\frac{y^2}{4at}\right), \quad B_2 = \mathrm{erfc}\left(\frac{y}{2\sqrt{at}}\right), \quad A_1 = \mathrm{erfc}\left(\frac{y}{2\sqrt{t}}\right)$$

$$B_3 = \frac{1}{\sqrt{\pi t}} \exp\left(-\frac{y^2}{4at}\right) + \sqrt{a}\exp\left(-y+at\right)\mathrm{erfc}\left(-\sqrt{at}+\frac{y}{2\sqrt{at}}\right)$$

$$B_4 = \frac{1}{\sqrt{\pi t}} \exp\left(-\frac{y^2}{4at}\right) - i\sqrt{b}\exp\left(\frac{iy\sqrt{b}}{\sqrt{a}}-bt\right)\mathrm{erfc}\left(i\sqrt{bt}+\frac{y}{2\sqrt{at}}\right)$$

$$B_5 = \frac{1}{\sqrt{\pi t}} \exp\left(-\frac{y^2}{4at}\right) + i\sqrt{b}\exp\left(-\frac{iy\sqrt{b}}{\sqrt{a}}-bt\right)\mathrm{erfc}\left(-i\sqrt{bt}+\frac{y}{2\sqrt{at}}\right)$$

$$B_6 = \frac{1}{2}\left[\mathrm{erfc}\left(\frac{y\sqrt{S_c}}{2\sqrt{t}}-\sqrt{Kt}\right)\exp\left(-y\sqrt{KS_c}\right)+\mathrm{erfc}\left(\frac{y\sqrt{S_c}}{2\sqrt{t}}+\sqrt{Kt}\right)\exp\left(y\sqrt{KS_c}\right)\right]$$

$$B_7 = \frac{\exp(-bt)}{2}\left\{\mathrm{erfc}\left(\frac{y\sqrt{S_c}}{2\sqrt{t}}-\sqrt{(K-b)t}\right)\exp\left(-y\sqrt{S_c(K-b)}\right)\right.$$

$$\left.+\mathrm{erfc}\left(\frac{y\sqrt{S_c}}{2\sqrt{t}}+\sqrt{(K-b)t}\right)\exp\left(y\sqrt{S_c(K-b)}\right)\right\}$$

$$A_2 = \frac{1}{b(b-d)}\left[\frac{\exp(-bt)}{2}\left\{\mathrm{erfc}\left(\frac{y\sqrt{S_c}}{2\sqrt{t}}-\sqrt{(K-b)t}\right)\exp\left(-y\sqrt{S_c(K-b)}\right)\right.\right.$$

$$\left.\left.+\mathrm{erfc}\left(\frac{y\sqrt{S_c}}{2\sqrt{t}}+\sqrt{(K-b)t}\right)\exp\left(y\sqrt{S_c(K-b)}\right)\right\}\right]$$

$$A_3 = \frac{1}{d(d-b)}\left[\frac{\exp(-dt)}{2}\left\{\mathrm{erfc}\left(\frac{y\sqrt{S_c}}{2\sqrt{t}}-\sqrt{(K-d)t}\right)\exp\left(-y\sqrt{S_c(K-d)}\right)\right.\right.$$

$$\left.\left.+\mathrm{erfc}\left(\frac{y\sqrt{S_c}}{2\sqrt{t}}+\sqrt{(K-d)t}\right)\exp\left(y\sqrt{S_c(K-d)}\right)\right\}\right]$$

$$A_4 = \frac{1}{bd}\left[\frac{1}{2}\left\{\mathrm{erfc}\left(\frac{y\sqrt{S_c}}{2\sqrt{t}}-\sqrt{Kt}\right)\exp\left(-y\sqrt{S_cK}\right)+\mathrm{erfc}\left(\frac{y\sqrt{S_c}}{2\sqrt{t}}+\sqrt{Kt}\right)\exp\left(y\sqrt{S_cK}\right)\right\}\right]$$

$$A_5 = \frac{1}{b(b-d)}\left[\frac{\exp(-bt)}{2}\operatorname{erfc}\left(\frac{y}{2\sqrt{t}}-\sqrt{-bt}\right)\exp\left(-y\sqrt{-b}\right)+\operatorname{erfc}\left(\frac{y}{2\sqrt{t}}+\sqrt{-bt}\right)\exp\left(y\sqrt{-b}\right)\right]$$

$$A_6 = -\frac{1}{\sqrt{\pi a t}}\exp\left(-\frac{y^2}{4t}\right), \qquad A_7 = -A_1, \qquad A_8 = -\sqrt{a}A_{16},$$

$$A_9 = -aA_{17},$$

$$A_{10} = \frac{1}{\sqrt{a}}\left[\frac{1}{\sqrt{\pi t}}\exp\left(-\frac{y^2}{4t}\right)+\sqrt{a}\exp\left(-y\sqrt{a}+at\right)\operatorname{erfc}\left(-\sqrt{a}t+\frac{y}{2\sqrt{t}}\right)\right]$$

$$A_{11} = \frac{1}{\sqrt{a}}\left[\frac{1}{\sqrt{\pi t}}\exp\left(-\frac{y^2}{4at}\right)\right], \qquad A_{12} = \operatorname{erfc}\left(\frac{y}{2\sqrt{at}}\right), \qquad A_{13} = \sqrt{a}A_{22},$$

$$A_{14} = aA_{23},$$

$$A_{15} = \frac{-1}{\sqrt{a}}\left[\frac{1}{\sqrt{\pi t}}\exp\left(-\frac{y^2}{4at}\right)+\sqrt{a}\exp\left(-y+at\right)\operatorname{erfc}\left(-\sqrt{a}t+\frac{y}{2\sqrt{at}}\right)\right]$$

$$A_{16} = 2\sqrt{\frac{t}{\pi}}\exp\left(-\frac{y^2}{4t}\right)-y\operatorname{erfc}\left(\frac{y}{2\sqrt{t}}\right)$$

$$A_{17} = \left(t+\frac{y^2}{2}\right)\operatorname{erfc}\left(\frac{y}{2\sqrt{t}}\right)-y\sqrt{\frac{t}{\pi}}\exp\left(-\frac{y^2}{4t}\right)$$

$$A_{18} = \frac{1}{\sqrt{\pi t}}\exp\left(-\frac{y^2}{4t}\right)-i\sqrt{b}\exp\left(i\sqrt{b}y-bt\right)\operatorname{erfc}\left(i\sqrt{b}t+\frac{y}{2\sqrt{t}}\right)$$

$$A_{19} = \frac{1}{\sqrt{\pi t}}\exp\left(-\frac{y^2}{4t}\right)+i\sqrt{b}\exp\left(-i\sqrt{b}y-bt\right)\operatorname{erfc}\left(-i\sqrt{b}t+\frac{y}{2\sqrt{t}}\right)$$

$$A_{20} = \frac{1}{\sqrt{\pi t}} \exp\left(-\frac{y^2}{4t}\right) + \sqrt{a}\exp\left(-\sqrt{a}y + at\right)\mathrm{erfc}\left(-\sqrt{a}t + \frac{y}{2\sqrt{t}}\right),$$

$$A_{22} = 2\sqrt{\frac{t}{\pi}}\exp\left(-\frac{y^2}{4at}\right) - \frac{y}{\sqrt{a}}\mathrm{erfc}\left(\frac{y}{2\sqrt{at}}\right) \qquad A_{28} = \frac{1}{bd}A_1$$

$$A_{23} = \left(t + \frac{y^2}{2a}\right)\mathrm{erfc}\left(\frac{y}{2\sqrt{at}}\right) - y\sqrt{\frac{t}{a\pi}}\exp\left(-\frac{y^2}{4at}\right)$$

$$A_{24} = \frac{1}{\sqrt{\pi t}}\exp\left(-\frac{y^2}{4at}\right) - i\sqrt{b}\exp\left(i\sqrt{b}\frac{y}{\sqrt{a}} - bt\right)\mathrm{erfc}\left(i\sqrt{b}t + \frac{y}{2\sqrt{at}}\right)$$

$$A_{25} = \frac{1}{\sqrt{\pi t}}\exp\left(-\frac{y^2}{4at}\right) + i\sqrt{b}\exp\left(-i\sqrt{b}\frac{y}{\sqrt{a}} - bt\right)\mathrm{erfc}\left(-i\sqrt{b}t + \frac{y}{2\sqrt{at}}\right)$$

$$A_{26} = \frac{1}{\sqrt{\pi t}}\exp\left(-\frac{y^2}{4at}\right) + \sqrt{a}\exp\left(-y + at\right)\mathrm{erfc}\left(-\sqrt{a}t + \frac{y}{2\sqrt{at}}\right)$$

$$A_{27} = \frac{1}{d(d-b)}\left[\frac{\exp(-dt)}{2}\mathrm{erfc}\left(\frac{y}{2\sqrt{t}} - \sqrt{-d}t\right)\exp\left(-y\sqrt{-d}\right) + \mathrm{erfc}\left(\frac{y}{2\sqrt{t}} + \sqrt{-d}t\right)\exp\left(y\sqrt{-d}\right)\right]$$

$$A_{29} = \frac{\exp(-dt)}{2}\left\{\mathrm{erfc}\left(\frac{y}{2\sqrt{t}} - \sqrt{-d}t\right)\exp\left(-y\sqrt{-d}\right) + \mathrm{erfc}\left(\frac{y}{2\sqrt{t}} + \sqrt{-d}t\right)\exp\left(y\sqrt{-d}\right)\right\}$$

$$A_{30} = \frac{1}{2}\left[\mathrm{erfc}\left(\frac{y\sqrt{S_c}}{2\sqrt{t}} - \sqrt{Kt}\right)\exp\left(-y\sqrt{KS_c}\right) + \mathrm{erfc}\left(\frac{y\sqrt{S_c}}{2\sqrt{t}} + \sqrt{Kt}\right)\exp\left(y\sqrt{KS_c}\right)\right]$$

$$A_{31} = \frac{\exp(-dt)}{2}\left\{\mathrm{erfc}\left(\frac{y\sqrt{S_c}}{2\sqrt{t}} - \sqrt{(K-d)t}\right)\exp\left(-y\sqrt{S_c(K-d)}\right)\right.$$

$$\left. + \mathrm{erfc}\left(\frac{y\sqrt{S_c}}{2\sqrt{t}} + \sqrt{(K-d)t}\right)\exp\left(y\sqrt{S_c(K-d)}\right)\right\}$$

RESULTS AND DISCUSSION

In this field of study, in order to examine the consequences of velocity field, temperature profile and concentration profiles by allotting numerical values for several arguments for both water vapor and ammonia, Prandtl number (Pr) and Schmidt number (Sc) values are considered.

For several values of different arguments the velocities are analyzed and are presented in Figures 2-7 at time t = 0.4 respectively, for both the types of heating (Gr < 0, Gm < 0) plate and cooling (Gr > 0, Gm > 0) plate. Figure 2 depicts the result of Schmidt number (Sc) at time t = 0.4 on the flow. With an increase in Sc, it is noticed that the velocity increases for heating of the plate and decreases for cooling of the plate and as Sc increases at t = 0.4 from 0.22 - 0.30 and to 0.60, the maximum velocity of the fluid decreases by 5.7% - 10% for ammonia and by 5.6% - 10.9% for water vapor in the case of cooling plate and the minimum velocity of the fluid increases by 10% and 16.58% for ammonia and by 10.26% - 17% for water vapor in the case of heating plate. Prandtl number (Pr) effects on the flow are expressed in Figure 4 at time t = 0.4. It is identified that there is an increase in velocity near the plate and then decreases with a point of separation moving far away from the plate in the type of

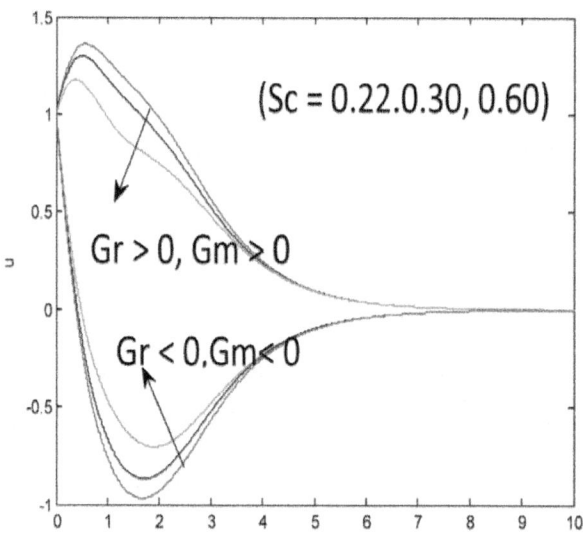

Figure 2: Velocity profile shows the effect of Sc and Pr = 0.71, K = 0.2, Nr = 0.1, Q_1 = 0.1, t = 0.4.

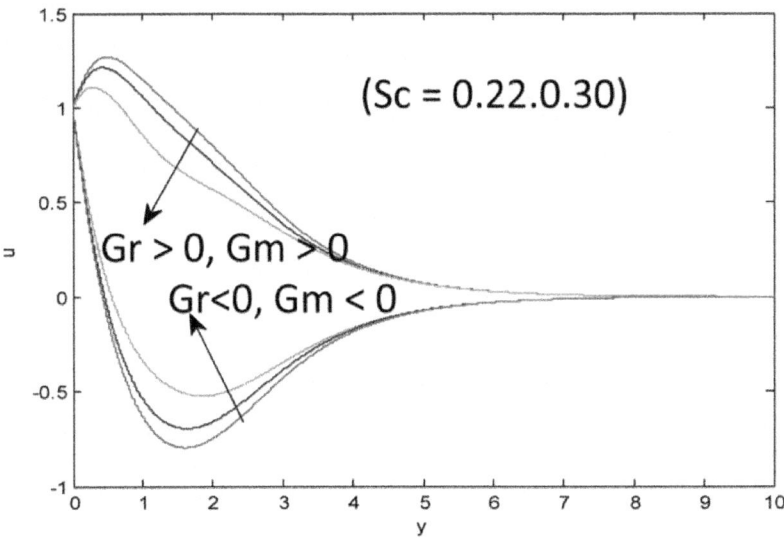

Figure 3: Velocity profile displays the result of Sc and Pr = 0.71, K = 0.2, Nr = 0, Q_1 = 0, t = 0.4.

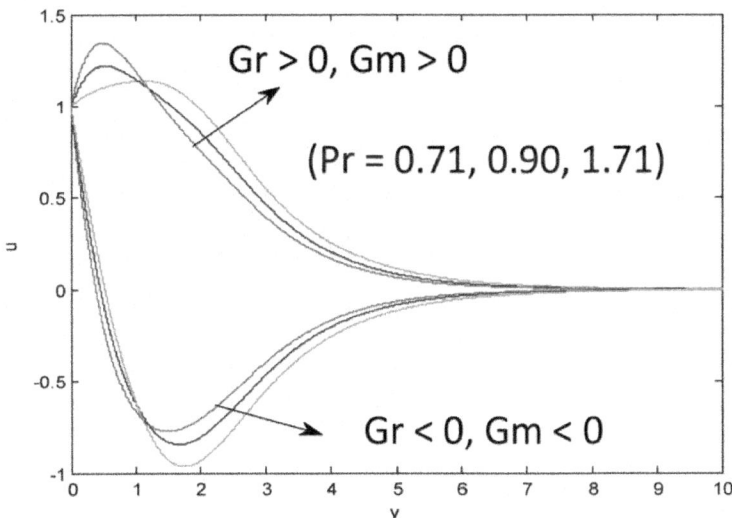

Figure 4: Velocity profile shows the effect of Pr and Sc = 0.22, K = 0.2, Nr = 0.1, Q_1 = 0.1, t = 0.4.

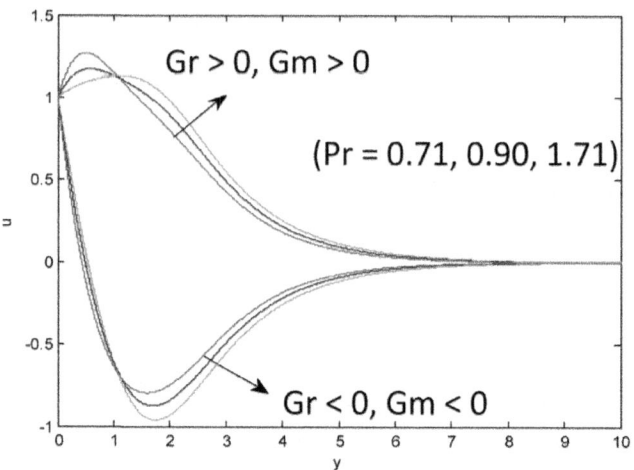

Figure 5: Velocity profile displays the result of Pr and Sc = 0.22, K = 0.2, Nr = 0, Q_1 = 0, t = 0.4.

heating and in the instance of cooling plate, the reverse effect is found with the increase of Pr. Owing to the variation in parameter of chemical reaction (K), Figure 6 reveals the consequence of velocity profiles at time t = 0.4. As K increases, it is found that, there is an increase velocity in the plate of heating type and the velocity decreases in the plate of cooling type and as K increases at t = 0.4 from 0.2 - 2 and to 5, the maximum velocity of

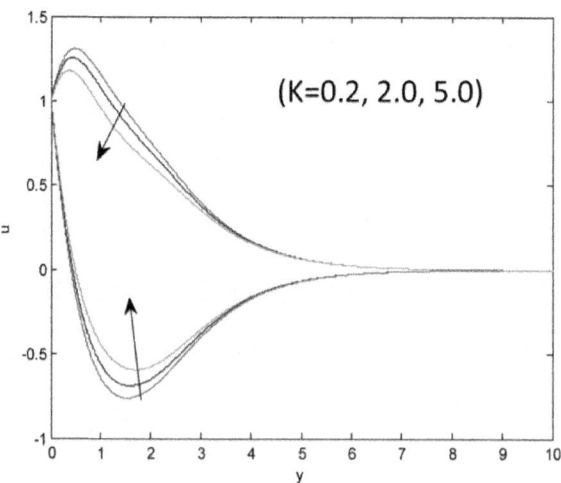

Figure 6: Velocity profile shows the effect of K and Sc = 0.22, Pr = 0.71, Nr = 0.1, Q_1 = 0.1, t = 0.4.

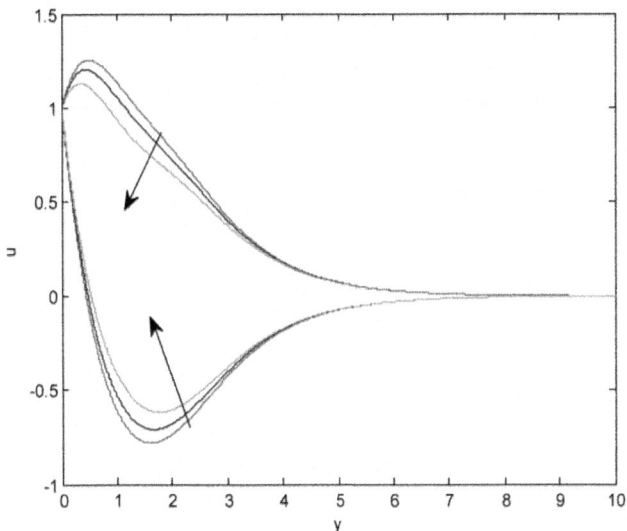

Figure 7: Velocity profile displays the result of K and Sc = 0.22, Pr = 0.71, Nr = 0, $Q_1 = 0$, t = 0.4.

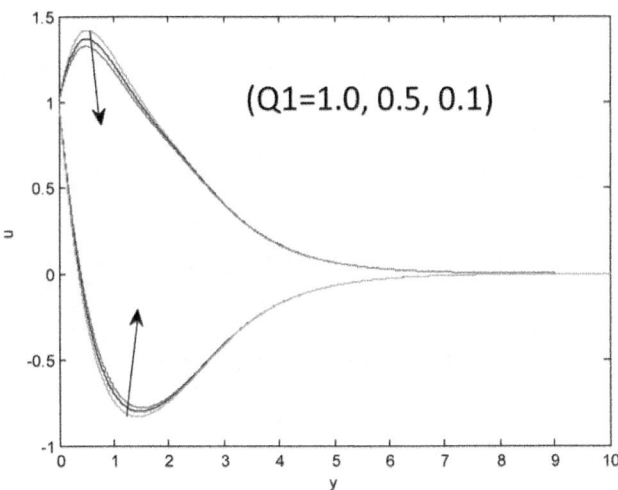

Figure 8: Velocity profile shows the effect of K and Sc = 0.22, Pr = 0.71, Nr = 0.1, $Q_1 = 0.1$, t = 0.4.

the fluid decreases by 7.4% - 6.8% for ammonia and by 7.1% - 7.4% for water vapor in the case of cooling plate and the minimum velocity of the fluid increases by 8.4% - 9.04% for ammonia and by 5.56% - 9.61% for water vapor in the case of heating plate.

For various values of Nr (radiation parameter), the velocity profile is shown in Figure 8 at t = 0.4. It is seen that there is a decrease in velocity near the plate and then increases with a point of separation moving away from the plate in the cooling case and the phenomenon is reversed in the case of heating type with the decrease of Nr. Figure 9 describes the effects of Q_1 (radiation absorption parameter) at t = 0.4. As Q_1 decreases, it is observed that the velocity increases for heating plate and decreases for cooling plate. Figure 10 reveals the result of velocity profile at several times (t = 0.4, 0.6, 0.8). It is identified that there is a considerable decrease in velocity in the heating type and increase when the plate is cooled as time (t) increases. The same results are noticed in the absence of thermal radiation and radiation absorption for different values of Sc, Pr, K, t which are shown in Figure 3, Figure 5, Figure 7, and Figure 11. Hence, these results are in good agreement with the results of Rajesh [10] .

The effect of temperature profile for several values of various parameters are studied and shown in Figures 12-17 at time t = 0.2. It is observed from Figures 12-14 that the temperature rises with the fall in Sc and K. In Figures 15-17, it is observed that the temperature rises with the increase of Nr, Q_1, and t. Moreover Figure 13 shows the results of velocity for different values of Pr. It is noticed from the values that, velocity increases near the plate and falls far away from the plate with a point of separation.

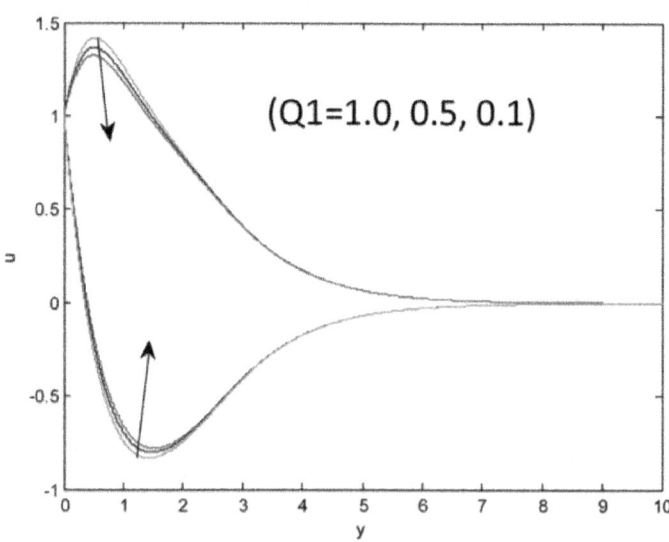

Figure 9: Velocity profile displays the result of K and Sc = 0.22, Pr = 0.71, Nr = 0, Q1 = 0, t = 0.4.

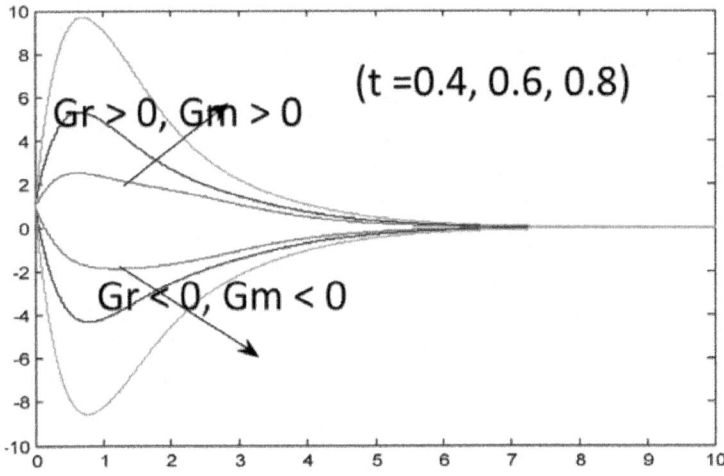

Figure 10: The effect of velocity profile fortand Pr = 0.71, K = 0.2, Nr = 0.1, Sc = 0.22, Q_1 = 0.1.

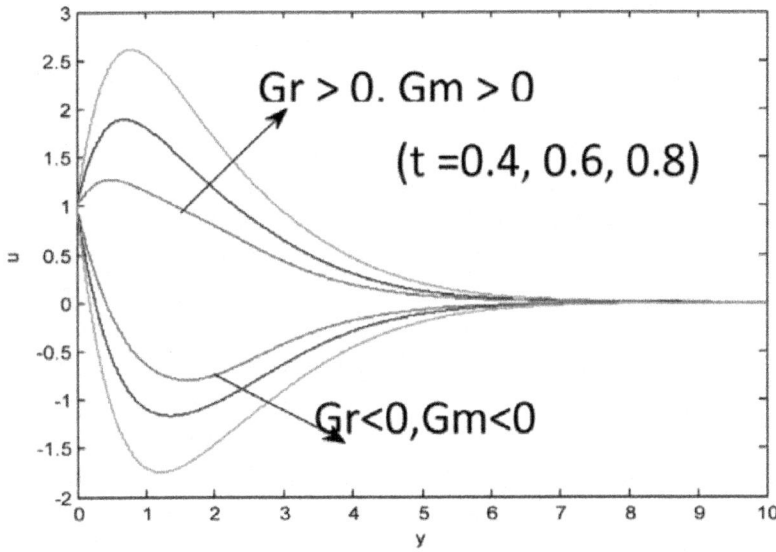

Figure 11: The result of velocity profile for tand Sc = 0.22, K = 0.2, Nr = 0.1, Q_1 = 0.1, Pr = 0.71.

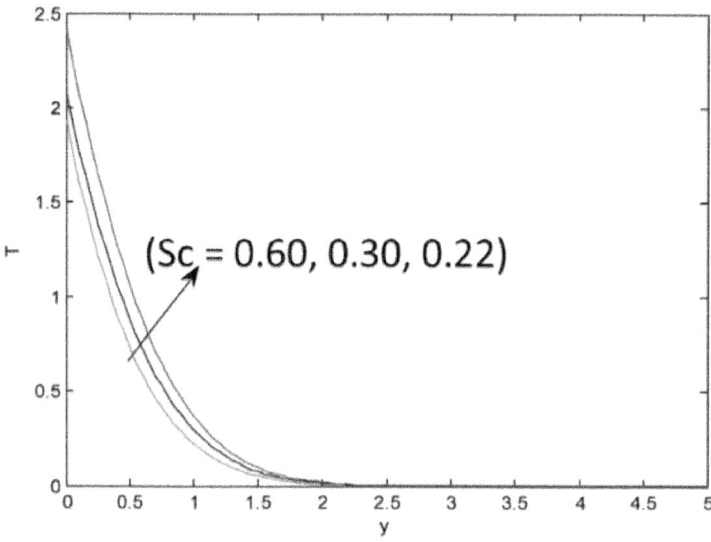

Figure 12: Temperature profile presents the effect of Sc and Pr = 0.71, K = 0.2, Nr = 0.1, Q_1 = 0.1, t = 0.2.

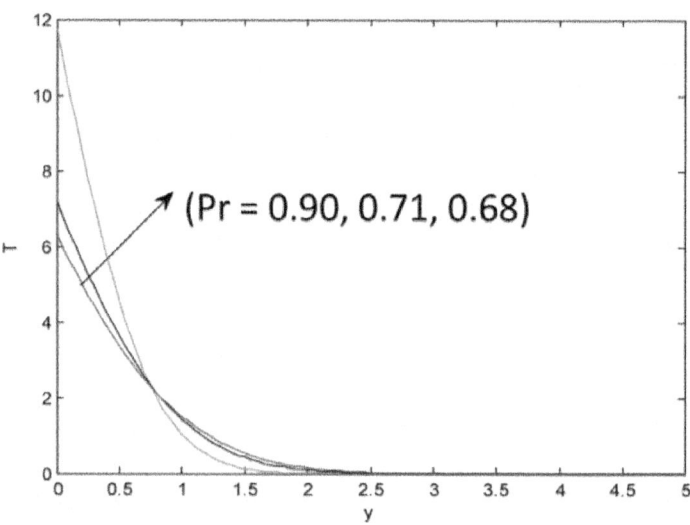

Figure 13: Temperature profile shows the result of Pr and Sc = 0.22, K = 0.2, Nr = 0.1, Q_1 = 0.1, t = 0.2.

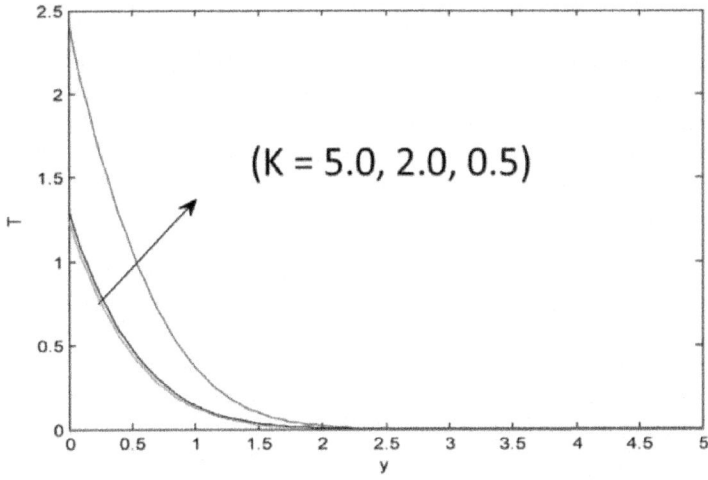

Figure 14: Temperature profile displays ensue of K and Sc = 0.22, Pr = 0.71, Nr = 0.1, Q_1 = 0.1, t = 0.2.

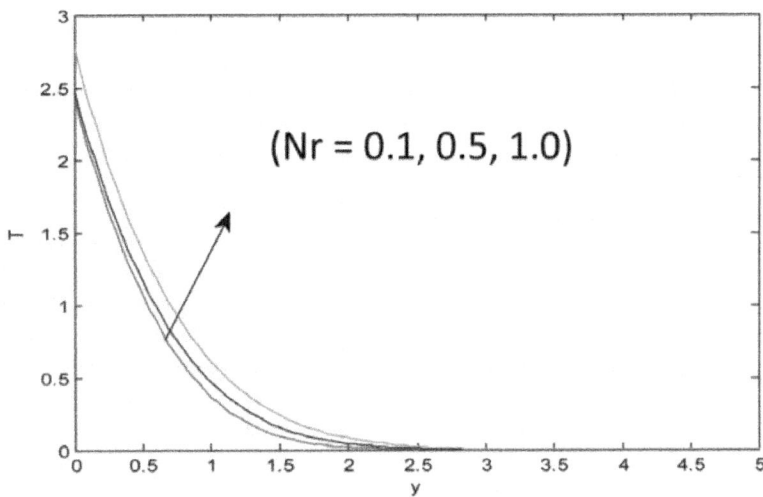

Figure 15: Temperature profile indicates the effect of Nr and 0.22, K = 0.2, Q_1 = 0.1, Pr = 0.71, t = 0.2.

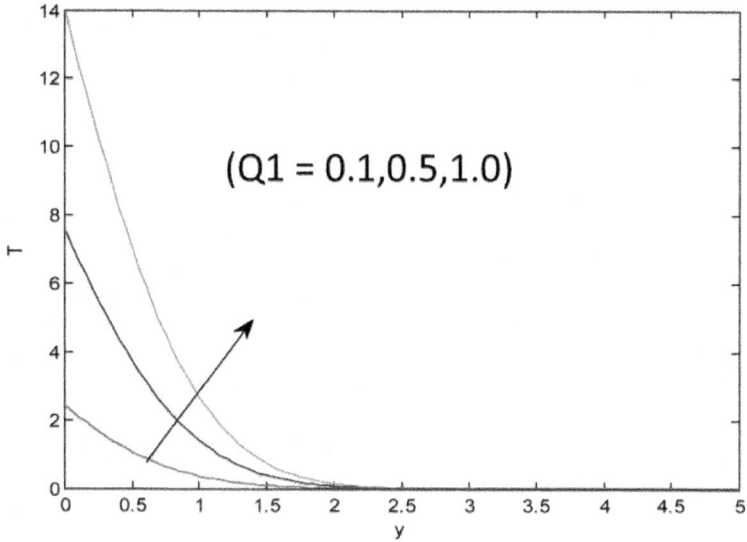

Figure 16: The effect of temperature profile for Q_1 and Pr = 0.71, Sc = 0.22, Nr = 0.1, K = 0.2, t = 0.2.

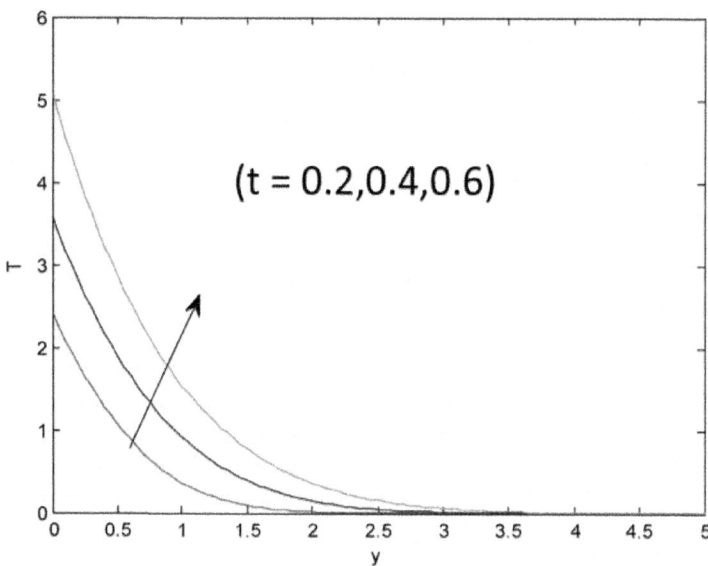

Figure 17: Theresult of temperature profile for tand Sc = 0.22, K = 0.2, Q_1 = 0.1, Pr = 0.1, Nr = 0.1.

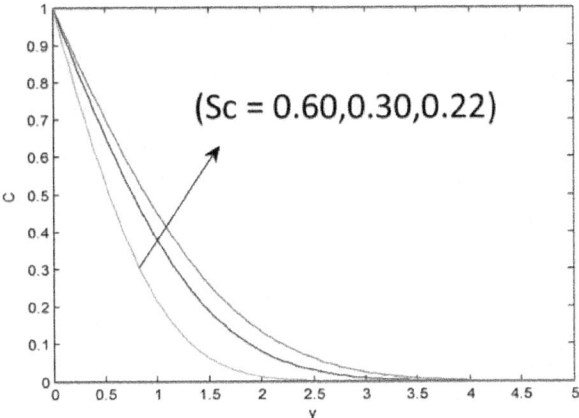

Figure 18: Concentration profile displays the effect of Sc and K = 0.2, t = 0.2.

The effect of concentration profile for several values of various parameters is studied and is presented in Figures 18-20 at times 0.4. From Figure 18 and Figure 19, it is identified that with the decrease in Sc and K, the concentration increases. And from Figure 20, it is found that there is a rise in concentration with the rise in time t. For several values of various arguments, the Sherwood number versus time is presented in Figure 21 and Figure 22. It is identified that, there is an increase in Sherwood number for both hydrogen and water vapor with the increase of Sc and K. For different values of various arguments for both hydrogen and water vapor, the Nusselt number versus time is shown in Figure 23. From this figure, it is observed that, there is a rise in Nusselt number with the fall in Pr.

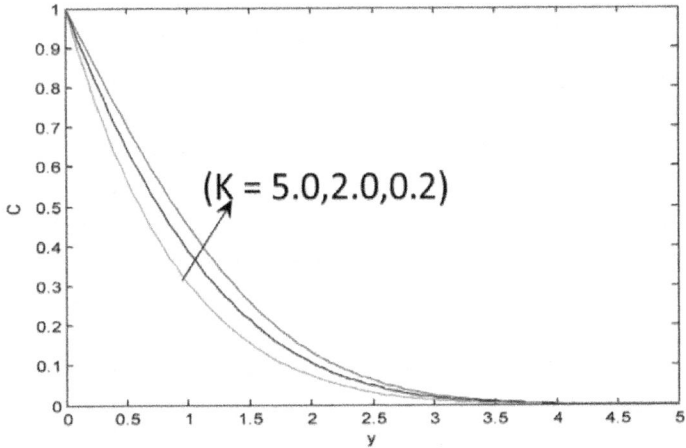

Figure 19: Concentration profile shows theresult of K and Sc = 0.22, t = 0.2.

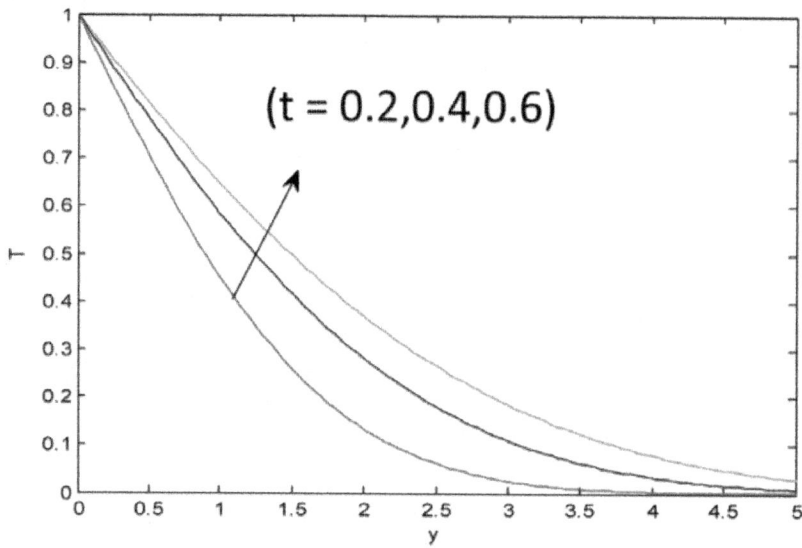

Figure 20: Concentration profile presents theensue of t and Sc = 0.22, K = 0.2.

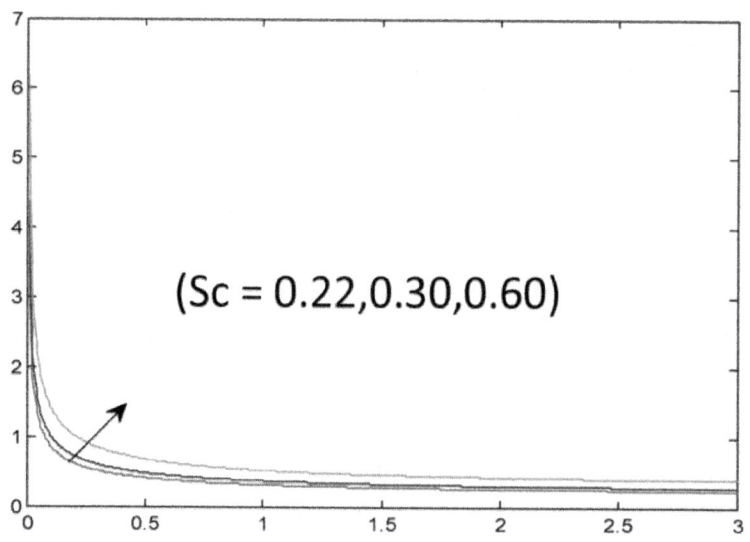

Figure 21: Effect of Sc on the real part of Sherwood number and K = 0.2.

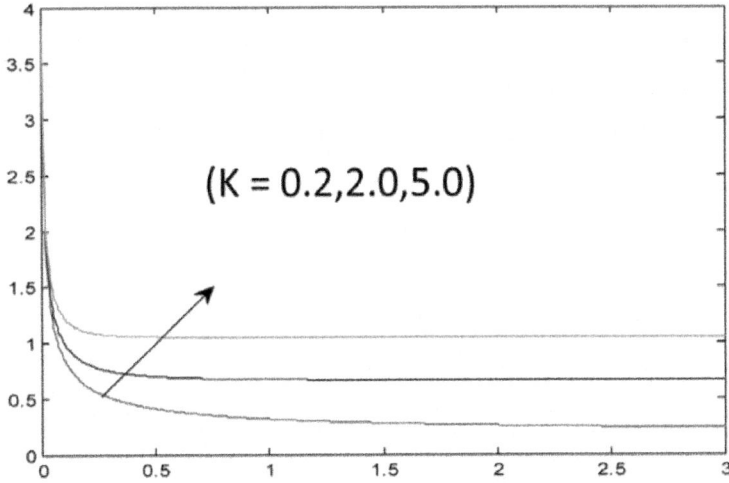

Figure 22: Result of K on the real part of Sherwood number and Sc = 0.22.

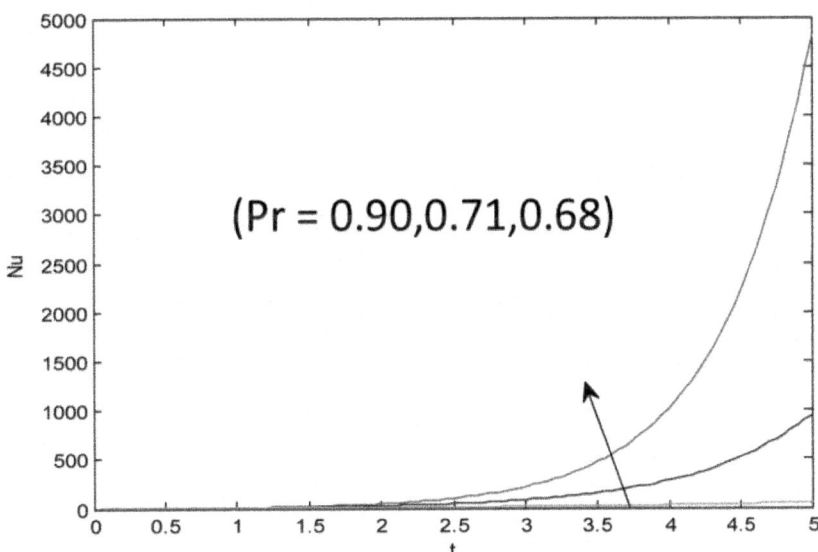

Figure 23: Effect of Pr on the real part of nusselt number and K = 0.2, Sc = 0.22, Nr = 0.1, Q_1 = 0.1.

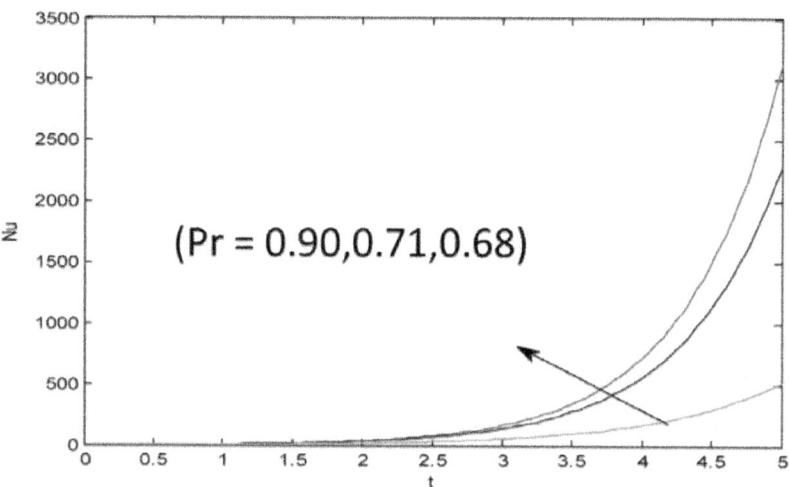

Figure 24: Result of Pr on the real part of nusselt number and K = 0.2, Sc = 0.22, Nr = 0, Q_1 = 0.

And the same results are noticed in the absence of thermal radiation and radiation absorption for different values of Pr which is shown in Figure 24. Hence, these results are in good agreement with the results of Rajesh [10] .

CITE THIS PAPER

SwethaRavi,JagdishPrakash,Viswanatha   ReddyGottam,Vijaya   Kumar VarmaSibyala, (2015) Effects of Thermal Radiation and Radiation Absorption on Flow Past an Impulsively Started Infinite Vertical Plate with Newtonian Heating and Chemical Reaction. *Open Journal of Fluid Dynamics*,**05**,364-379. doi: 10.4236/ojfd.2015.54036

REFERENCES

1.   Ahmed, S., Batin, A. and Chamka, A.J. (2015) Numerical/Laplace Transform Analysis for MHD Rotating Heat/ Mass Transport in a Darcian Porous Regime Bounded by an Oscillating Vertical Surface. Alexandria Engineering Journal, 54, 45-54. http://dx.doi.org/10.1016/j. aej.2014.11.006

2.   Kandasamy, R., Muhaimin, I. and Saim, H.B. (2010) Group Analysis for the Effects of the Temperature Dependent Fluid Viscosity and Chemical Reaction on Free Convective Heat and Mass Transfer. Journal

of Applied Mechanics and Technical Physics, 51, 887-897. http://dx.doi. org/10.1007/s10808-010-0110-2

3. Makinde, O.D. (2005) Free Convection Flow with Thermal Radiation and Mass Transfer Past a Moving Vertical Porous Plate. International Communications in Heat and Mass Transfer, 32, 1411-1419. http:// dx.doi.org/10.1016/j.icheatmasstransfer.2005.07.005

4. Mukhopadhyay, S. (2009) Effect of Thermal Radiation on Unsteady Mixed Convection Flow and Heat Transfer over a Porous Stretching Surface in Porous Medium. International Journal of Heat and Mass Transfer, 52, 3261-3265. http://dx.doi.org/10.1016/j.ijheatmasstransfer.2008.12.029

5. Mukhopadhyay, S. and Layek, G.C. (2009) Radiation Effect on Forced Convective Flow and Heat Transfer over a Porous Plate in a Porous Medium. Meccanica, 44, 587-597. http://dx.doi.org/10.1007/s11012-009-9211-5

6. Muthucumaraswamy, R. and Ganesan, P. (2003) Radiation Effects on Flow Past an Impulsively Started Infinite Vertical Plate with Variable Temperature. International Journal of Applied Mechanics and Engineering, 8, 125-129.

7. Muthucumaraswamy, R. (2010) Chemical Reaction Effects on Vertical Oscillating Plate with Variable Temperature. Chemical Industry & Chemical Engineering Quarterly, 16, 167-173.

8. Hussanan, A., Ismail, Z., Khan, I., Hussein, A.G. and Shafie, S. (2014) Unsteady Boundary Layer MHD Free Convection Flow in a Porous Medium with Constant Mass Diffusion and Newtonian Heating. The European Physical Journal Plus, 129, 46.

9. Ahmed, S. (2014) Numerical Analysis for MHD Chemically Reacting and Radiating Fluid Past a Non Isothermal Impulsively Started Vertical Surface Adjacent to a Porous Regime. Ain Shams Engineering Journal, 5, 923-933. http://dx.doi.org/10.1016/j.asej.2014.02.005

10. Rajesh, V. (2012) Effects of Mass Transfer on Flow Past an Impulsively Started Infinite Vertical Plate with Newtonian Heating and Chemical Reaction. Journal of Engineering Physics and Thermophysics, 85, 221-228. http://dx.doi.org/10.1007/s10891-012-0642-9

Chapter 10

A DAG SCHEDULING SCHEME ON HETEROGENEOUS COMPUTING SYSTEMS USING TUPLE-BASED CHEMICAL REACTION OPTIMIZATION

Yuyi Jiang, Zhiqing Shao, and Yi Guo

College of Information Science and Engineering, East China University of Science and Technology, Shanghai 200237, China

ABSTRACT

A complex computing problem can be solved efficiently on a system with multiple computing nodes by dividing its implementation code into several parallel processing modules or tasks that can be formulated as directed acyclic graph (DAG) problems. The DAG jobs may be mapped to and scheduled on the computing nodes to minimize the total execution time. Searching an optimal DAG scheduling solution is considered to be NP-complete. This paper proposed a tuple molecular structure-based chemical reaction optimization (TMSCRO) method for DAG scheduling on heterogeneous computing systems, based on a very recently proposed metaheuristic method, chemical reaction optimization (CRO). Comparing with other CRO-based algorithms for DAG scheduling, the design of tuple reaction molecular structure and four elementary reaction operators of TMSCRO is more reasonable. TMSCRO also applies the concept of constrained critical paths (CCPs), constrained-critical-path directed acyclic graph (CCPDAG) and super molecule for accelerating convergence. In this paper, we have also conducted simulation experiments to verify the effectiveness and efficiency of TMSCRO upon a large set of randomly generated graphs and the graphs for real world problems.

INTRODUCTION

Modern computer systems with multiple processors working in parallel may enhance the processing capacity for an application. The effective scheduling of parallel modules of the application may fully exploit the parallelism. The application modules may communicate and synchronize several times during the processing. The limitation of the overall application performance may be incurred by a large communication cost on heterogeneous systems with a combination of GPUs, multicore processors and CELL processors, or distributed memory systems. And an effective scheduling may greatly improve the performance of the application.

Scheduling generally defines not only the processing order of application modules but also the processor assignment of these modules. The concept of makespan (i.e., the schedule length) is used to evaluate the scheduling solution quality including the entire execution and communication cost of all the modules. On the heterogeneous systems [1–4], searching optimal schedules minimizing the makespan is considered as a NP-complete problem. Therefore, two classes of scheduling strategies have been proposed to solve this problem by finding the suboptimal solution with lower time overhead, such as heuristic scheduling and metaheuristic scheduling.

Heuristic scheduling strategies try to identify a good solution by exploiting the heuristics. An important subclass of heuristic scheduling is list scheduling with an ordered task list for a DAG job on the basis of some greedy heuristics. Moreover, the ordered tasks are selected to be allocated to the processors which minimize the start times in list scheduling algorithms. In heuristic scheduling, the attempted solutions are narrowed down by greedy heuristics to a very small portion of the entire solution space. And this limitation of the solution searching leads to the low time complexity. However, the higher complexity DAG scheduling problems have, the harder greedy heuristics produce consistent results on a wide range of problems, because the quality of the found solutions relies on the effectiveness of the heuristics, heavily.

Metaheuristic scheduling strategies such as ant colony optimization (ACO), genetic algorithms (GA), Tabu search (TS), simulated annealing (SA), and so forth take more time cost than heuristic scheduling strategies, but they can produce consistent results with high quality on the problems with a wide range by directed searching solution spaces.

Chemical reaction optimization (CRO) is a new metaheuristic method proposed very recently and has shown its power to deal with NP-complete problem. There is only one CRO-based algorithm called double molecular structure-based CRO (DMSCRO) for DAG scheduling on heterogeneous system as far as we know. DMSCRO has a better performance on makespan and convergence rate than genetic algorithm (GA) for DAG scheduling on heterogeneous systems. However, the rate of convergence of DMSCRO as a metaheuristic method is still defective. This paper proposes a new CRO-based algorithm, tuple molecular structure-based CRO (TMSCRO), for the mentioned problem, encoding the two basic components of DAG scheduling, module execution order and module-to-processor mapping, into an array of tuples. Combining this kind of molecular structure with the elementary reaction operator designed in TMSCRO has a better capability of intensification and diversification than DMSCRO. Moreover, in TMSCRO, the concept of constrained critical paths (CCPs) [5] and constrained-critical-path directed acyclic graph (CCPDAG) are applied to creating initial population in order to speed up the convergence of TMSCRO. In addition, the first initial molecule, InitS, is also considered to be a super molecule [6] for accelerating convergence, which is converted from the scheduling result of the algorithm constrained earliest finish time (CEFT). In theory, a metaheuristic method will gradually approach the optimal result if it runs for long enough, based on No-Free-Lunch Theorem, which means the performances of the search for optimal solution of each metaheuristic algorithm are alike when averaged over all possible fitness functions. We have conducted the simulation experiments over the graphs abstracted from two well-known real applications: Gaussian elimination and molecular dynamics application and also a large set of randomly generated graphs. The experiment results show that the proposed TMSCRO can achieve similar performance as DMSCRO in the literature in terms of makespan and outperforms the heuristic algorithms.

There are three major contributions of this work.(1)Developing TMSCRO based on CRO framework by designing a more reasonable molecule encoding method and elementary chemical reaction operators on intensification and diversification search than DMSCRO.(2)For accelerating convergence, applying CEFT and CCPDAG to the data pretreatment, utilizing the concept of CCPs in the initialization, and using the first initial molecule, InitS, to be a super molecule in TMSCRO.(3)Verifying the effectiveness and efficiency of the proposed TMSCRO by simulation experiments. The simulation results of this paper show that TMSCRO is able to approach similar makespan as DMSCRO, but it finds good solutions faster than DMSCRO by 12.89% on average (by 26.29% in the best case).

RELATED WORK

Most of the scheduling algorithms can be categorized into heuristic scheduling (including list scheduling, duplication-based scheduling, and cluster scheduling) and metaheuristic (i.e., guided-random-search-based) scheduling. These strategies are to generate the scheduling solution before the execution of the application. The approaches adopted by these different scheduling strategies are summarized in this section.

Heuristic Scheduling

Heuristic methods usually provide near-optimal solutions for a task scheduling problem in less than polynomial time. The approaches adopted by heuristic method search only one path in the solution space, ignoring other possible ones [7]. Three typical kinds of algorithms based on heuristic scheduling for the DAG scheduling problem are discussed as below, such as list scheduling [7, 8], cluster scheduling [9, 10], and duplication-based scheduling [11, 12].

The list scheduling [7, 13–21] generates a schedule solution in two primary phases. In phase 1, all the tasks are processed in a sequence order by their assigned priorities, which are normally based on the task execution and communication costs. There are two attributes used in most list scheduling algorithms, such as b-level and t-level, to assign task priorities. In a DAG, b-level of a node (task) is the length of the longest path from the end node to the node; however, t-level of a node is the length of the longest path from the entry node to the node. In phase 2, the processors are assigned to each task in the sequence.

The heterogeneous earliest finish time (HEFT) scheduling algorithm [16] assigns the scheduling task priorities based on the earliest start time of each task. HEFT allocates a task to the processor which minimizes the task's start time.

The modified critical path (MCP) scheduling [22] considers only one CP (critical path) of the DAG and assigns the scheduling priority to tasks based on their latest start time. The latest start times of the CP tasks are equal to their t-levels. MCP allocates a task to the processor which minimizes the task's start time.

Dynamic-level scheduling (DLS) [23] uses the concept of the dynamic level, which is the difference between the b-level and earliest start time of a task on a processor. Each time the (task, processor) pair with the largest dynamic-level value is chosen by DLS during the task scheduling.

Mapping heuristic (MH) [24] assigns the task scheduling priorities based on the static b-level of each task, which is the b-level without the communication costs between tasks. Then, a task is allocated to the processor which gives the earliest start time.

Levelized-min time (LMT) [17] assigns the task scheduling priority in two steps. Firstly, it groups the tasks into different levels based on the topology of the DAG, and then in each level, the task with the highest priority is the one with the largest execution cost. A task is allocated to the processor which minimizes the sum of the total communication costs with the tasks in the previous level and the task's execution cost.

There are two heuristic algorithms for DAG scheduling on heterogeneous systems proposed in [8]. One algorithm named HEFT_T uses the sum of t-level and b-level to assign the priority to each task. In HEFT_T, the critical tasks are attempted to be on the same processor, and the other tasks are allocated to the processor that gives earliest start time. The other algorithm named HEFT_B applies the concept of b-level to assign the priority (i.e., scheduling order) to each task. After the priority assignment, a task is allocated to the processor that minimizes the start time. The extensive experiment results in [8] demonstrate that HEFT_B and HEFT_T outperform (in terms of makespan) other representative heuristic algorithms in heterogeneous systems, such as DLS, MH, and LMT.

Comparing with the list scheduling algorithms, the duplication-based algorithms [23, 25–29] attempt to duplicate the tasks to the same processor on heterogeneous systems, because the duplication may eliminate the communication cost of these tasks and it may effectively reduce the total schedule length.

The clustering algorithms [8, 11, 30–32] regard task collections as clusters to be mapped to appropriate processors. These algorithms are mostly used in the homogeneous systems with unbounded number of processors and they will use as many processors as possible to reduce the schedule length. Then, if the number of the processors used for scheduling is more than that of the available processors, the task collections (clusters) are processed further to fit in with a limited number of processors.

Metaheuristic Scheduling

In comparison with the algorithms based on heuristic scheduling, the metaheuristic (guided-random-search-based) algorithms use a combinatorial process for solution searching. In general, with robust performance on many kinds of scheduling problems, the metaheuristic algorithms need sampling candidate solutions in the search space, sufficiently. Many metaheuristic algorithms have been applied to solve the task scheduling problem successfully, such as GA, chemical reaction optimization (CRO), energy-efficient stochastic [33], and so forth.

GA [15, 31, 34–36] is the mostly used metaheuristic method for DAG scheduling. In [15], a solution for scheduling is encoded as one-dimensional string representing an ordered list of tasks to be allocated to a processor. In each string of two parent solutions, the crossover operator selects a crossover point randomly and then merges the head portion of one parent with the tail portion of the other. Mutation operator exchanges two tasks in two solutions, randomly. The concept of makespan is used to evaluate the scheduling solution quality by fitness function.

Chemical reaction optimization (CRO) was proposed very recently [20, 30, 37–39]. It mimics the interactions of molecules in chemical reactions. CRO has good performance already in solving many problems, such as quadratic assignment problem (QAP), resource-constrained project scheduling problem (RCPSP), channel assignment problem (CAP) [39], task scheduling in grid computing (TSGC) [40], and 0-1 knapsack problem (KP01) [41]. So far as we know, double molecular structure-based chemical reaction optimization (DMSCRO) recently proposed in [37] is the only one CRO-based algorithm with two molecular structures for DAG scheduling on heterogeneous systems. CRO-based algorithm (just DMSCRO) mimics the chemical reaction process in a closed container and accords with energy conservation. In DMSCRO, one solution for DAG scheduling including two essential components, task execution order and task-to-processor mapping, corresponds to a double-structured molecule with two kinds of energy, potential energy (PE) and kinetic energy (KE). The value of PE of a molecule is just the fitness value (objective value), makespan, of the corresponding solution, which can be calculated by the fitness function designed in DMSCRO, and KE with a nonnegative value is to help the molecule escape from local optimums. There are four kinds of elementary reactions used to do the intensification and diversification search in the solution space to find the solution with the minimal makespan, and the principle of the reaction selection is in detail presented in Section 3.2. Moreover, a central buffer is also applied in DMSCRO for energy interchange and conservation during the searching progress. However, as a metaheuristic method for DAG scheduling, DMSCRO still has very large time expenditure and the rate of convergence of this algorithm needs to be improved. Comparing with GA, DMSCRO is similar in model and workload to TMSCRO proposed in this paper.

Our work is concerned with the DAG scheduling problems and the flaw of CRO-based method for DAG scheduling, proposing a tuple molecular structure-based chemical reaction optimization (TMSCRO). Comparing with DMSCRO, TMSCRO applies CEFT [5] to data pretreatment to take the advantage of CCPs as heuristic information for accelerating convergence.

Moreover, the molecule structure and elementary reaction operators design in TMSCRO are more reasonable than those in DMSCRO on intensification and diversification of searching the solution space.

BACKGROUND

Ceft

Constrained earliest finish time (CEFT) based on the constrained critical paths (CCPs) was proposed for heterogeneous system scheduling in [5]. In contrast to other approaches, the CEFT strategy takes account of a broader view of the input DAG. Moreover, the CCPs can be scheduled efficiently because of their static generation.

The constrained critical path (CCP) is a collection with the tasks ready for scheduling only. A task is ready when all its predecessors were processed. In CEFT, a critical path (CP) is generally the longest path from the start node to the end node for scheduling in the DAG. The DAG is initially traversed and critical paths are found. Then it is pruned off the nodes that constitute a critical path. The subsequent traversals of the pruned graph produce the remaining critical paths. While the nodes are being removed from the task graph, a pseudo-edge to the start or end node is added if a node has no predecessors or no successors, respectively. The CCPs are subsequently formed by selecting ready nodes in the critical paths in a round-robin fashion. Each CCP may be assigned a single processor which has the minimum finish time of processing all the tasks in the CCP. All the tasks in a CCP not only reduce the communication cost, but also benefit from a broader view of the task graph.

Consider the CEFT algorithm generates schedules for n tasks with $|P|$ heterogeneous processors. Some specific terms and their usage are indicated in Table 1

Table 1: Specific terms and their usage for the CEFT algorithm.

$EC_{P_r}(w)$	Execution cost of a node w using processor P_r
$CM(w, P_r, v, P_x)$	Communication cost from node v to w, if P_x has been assigned to node v and P_r is assigned to node w
$ST_{P_r}(w, v)$	Possible start time of node w which is assigned the processor P_r, with the v node being any predecessor of w which has already been scheduled
$EFT_{P_r}(w)$	Finish time of node w using processor P_r
$AEFT_w$	Actual finish time of node w
$CEFT_{P_r}(CCP_j)$	Finish time of the constrained critical path Q_j when processor P_r is assigned to it
AT_{P_r}	Availability time of P_r
$Pred(w)$	Set of predecessors of node w
$Succ(w)$	Set of successors of node w
$AEC(w)$	Average execution cost of node w

The CEFT scheduling approach (Algorithm 1) works in two phases. (1) The critical paths are generated according to the description in the second paragraph of Section 3.1. The critical paths are traversed and the ready nodes are inserted into the constrained critical paths (CCPs) CCP_j, $\forall_j = 1, 2, \ldots,$ $|Q|$. If no more ready nodes are in a critical path, the constrained critical path takes nodes from the next critical path following round-robin traversal of the critical paths. (2) All the CCPs are traversed in order (line 12).Then, STP_r (w, k), the maximum of ATP_r and the start time of the predecessors of each node w, is calculated (1). $EFTP_r$ (w) is computed as the sum of STP_r (w, k) and ECP_r (w) (2). EP_r (Q_j) is the maximum of the finish times of all the CCP nodes on the same processor Pr (3). The processor is then assigned to constrained-criticalpath CCP_j which minimizes the $CEFTP_r$ ($CCP j$) value (line 20). After the actual finish time $AEFTw$ of each task w in CCP_j is updated, the processor assignment continues iteratively.

```
(1)  //PHASE 1: Find the constrained critical paths (CCPs)
(2)  Find set of critical paths CP according to the description in the second paragraph of Section 3.1.
(3)  j = 1
(4)  for i = 1 to |CP| do
(5)      while there exist ready nodes in CP_i do
(6)          Insert ready node v_k into constrained critical path Queue (Q_j).
(7)      end while
(8)      j ← j + 1
(9)      i ← i% |CP|
(10) end for
(11) //PHASE 2: Assign and schedule tasks
(12) for j = {1, 2, ..., |Q|} do
(13)     for each processor P_r ∈ P do
(14)         for each node w ∈ Q_j do
(15)             Find the start time of node k, which is the predecessor of w
                 ST_{P_r}(w, k) = max (((AEFT_k) + CM (w, P_r, k, P_x)), AT_{P_r})
(16)             Find the finish time of the node
                 EFT_{P_r}(w) = max (ST_{P_r}(w, k))_{∀k∈Pred(w)} + EC_{P_r}(w)
(17)         end for
(18)         Find the finish time of the CCP Q_j
                 CEFT_{P_r}(Q_j) = max ((EFT_{P_r}(w))_{∀w∈Q_j})
(19)     end for
(20)     Assign the processor to CCP Q_j which minimizes CEFT_{P_r}(Q_j).
(21)     Let P_x be assigned, update AEFT_w of each task w in Q_j
                 (AEFT_w)_{∀w∈Q_j} = (EFT_{P_x}(w))_{∀w∈Q_j}
(22) end for
```

Algorithm 1: CEFT.

CRO

Chemical reaction optimization (CRO) mimics the process of a chemical reaction where molecules undergo a series of reactions between each other or with the environment in a closed container. The molecules are manipulated agents with a profile of three necessary properties of the molecule, including the following. (1) The molecular structure S: S actually structure represents the positions of atoms in a molecule. Molecular structure can be in the form of a number, a vector, a matrix, or even a graph which is independent of the problem, (2) (Current) potential energy (PE): PE is the objective function value of the current molecular structure ω, that is, $PE_\omega = (\omega)$. (3) (Current) kinetic energy (KE): KE is a nonnegative number and it helps the molecule escape from local optimums. There is a central energy buffer implemented in CRO. The energy in CRO may accord with energy conservation and can be exchanged between molecules and the buffer

Four kinds of elementary reactions may happen in CRO, which are defined as below.

(1)   On-wall ineffective collision: on-wall ineffective collision is a unimolecule reaction with only one molecule. In this reaction, a molecule ω is allowed to change to another one ω', if their energy values accord with the following inequality:

$$PE_\omega + KE_\omega \geq PE_{\omega'};$$

(1)

after this reaction, KE will be redistributed in CRO. The redundant energy with the value $KE_{\omega'} = (PE_\omega + KE_\omega - PE_{\omega'}) \times t$ will be stored in the central energy buffer. Parameter t is a random number from KELossRate to 1 and KELossRate, a system parameter set during the CRO initialization, is the KE loss rate less than 1.

(2)   Decomposition: decomposition is the other unimolecule reaction in CRO. A molecule ω may decompose into two new molecules, ω'_1 and ω'_2, if their energy values accord with inequality (2), in which buf denotes the energy in the buffer, representing the energy interactions between molecules and the central energy buffer:

$$PE_\omega + KE_\omega + buf \geq PE_{\omega'_1} + PE_{\omega'_2};$$

(2)

after this reaction, buf is updated by (3) and the KEs of ω_1' and ω_2' are, respectively, computed as (4) and (5), where Edecomp $= (PE_\omega + KE_\omega) - (PE_{\omega_1'} + PE_{\omega_2'})$ and $\mu1$, $\mu2$, $\mu3$, $\mu4$ is a number randomly selected from the range of [0, 1]. Consider

$$buf = \text{Edecomp} + buf - \left(PE_{\omega_1'} + PE_{\omega_2'}\right),$$
(3)

$$KE_{\omega_1'} = (\text{Edecomp} + buf) \times \mu1 \times \mu2,$$
(4)

$$KE_{\omega_2'} = \left(\text{Edecomp} + buf - KE_{\omega_1'}\right) \times \mu3 \times \mu4.$$
(5)

(3) Intermolecular ineffective collision: intermolecular ineffective collision is an intermolecule reaction with two molecules. Two molecules, ω_1 and ω_2, may change to two new molecules, ω_1' and ω_2', if their energy values accord with the following inequality:

$$PE_{\omega_1} + PE_{\omega_2} + KE_{\omega_1} + KE_{\omega_2} \geq PE_{\omega_1'} + PE_{\omega_2'};$$
(6)

after this reaction, the KEs of ω_1' and ω_2', $KE_{\omega_1'}$ and $KE_{\omega_2'}$, will share the spare energy Eintermole calculated by (7). $KE_{\omega_1'}$ and $KE_{\omega_2'}$ are computed as (8) and (9), respectively, where $\mu1$ is a number randomly selected from the range of [0, 1]. Consider

$$\text{Eintermole} = \left(PE_{\omega_1} + PE_{\omega_2} + KE_{\omega_1} + KE_{\omega_2}\right)$$
$$- \left(PE_{\omega_1'} + PE_{\omega_2'}\right),$$
(7)

$$KE_{\omega_1'} = \text{Eintermole} \times \mu1,$$
(8)

$$KE_{\omega_2'} = \text{Eintermoler} \times \left(1 - \mu1\right).$$
(9)

(4) Synthesis: synthesis is also an intermolecule reaction. Two molecules, ω_1 and ω_2, may be combined to a new molecule, ω', if their energy values accord with inequality (10). The KE of ω' is computed as (11):

The canonical CRO works as follows. Firstly, the initialization of CRO is to set system parameters, such as PopSize (the size of the molecules),

KELossRate, InitialKE (the initial energy of molecules), buf (initial energy in the buffer), and MoleColl (MoleColl is a threshold value to determine whether to perform a unimolecule reaction or an intermolecule reaction). Then the CRO processes a loop. In each iteration, whether to perform a unimolecule reaction or an intermolecule reaction is first decided in the following way. A number ε is randomly selected from the range of [0, 1]. If ε is bigger than MoleColl, a unimolecule reaction will be chosen, or an intermolecular reaction is to occur. If it is a unimolecular reaction, a parameter θ as a threshold value is used to guide the further choice of on-wall collision or decomposition. NumHit is the parameter used to record the total collision number of a molecule. It will be updated after a molecule undergoes a collision. If the NumHit of a molecule is larger than θ, a decomposition will then be selected. Similarly, a parameter ϑ is used to further decide selection of an intermolecule collision reaction or a synthesis reaction. ϑ specifies the least KE of a molecule. Synthesis reaction will be chosen when both KEs of the molecules ω_1 and ω_2 are less than ϑ, or intermolecular ineffective collision reaction will take place. When the stopping criterion satisfies (e.g., a better solution cannot be found after a certain number of consecutive iterations), the loop will be stopped and the best solution is just the molecule that possesses the lowest PE.

MODELS

This section discusses the system, application, and task scheduling model assumed in this work. The definition of the notations can be found in the Notations section.

System Model

In this paper, there are multiple heterogeneous processors in the target system, which are presented by $P = \{pi \mid i = 1, 2, 3, \ldots, |P|\}$. They are fully interconnected with high speed network. Each task in a DAG can only be executed on one processor on heterogeneous system. The edges of the graph are labeled with communication cost that should be taken into account if its start and end tasks are executed on different processors. The communication cost is zero when the same processor is assigned to two communicating modules.

We assume a static computing system model in which the constrained relations and the execution costs of tasks are known a priori and the execution and communication can be performed simultaneously by the processors. In this paper, the heterogeneity is represented by $ECP_r(w)$, which means the execution cost of a node w using processor Pr. As the assumption of the MHM model, the heterogeneity in the simulations is set as follows to make a processor have different speed for different tasks. The value of each $ECPr(w)$

is randomly chosen within the scope of $[1 - g\%, 1+g\%]$ by using a parameter g $(g \in (0, 1))$. Therefore, the heterogeneity level can be formulated as $(1 + \%)/(1 - g\%)$. g is set as the value that makes the heterogeneity level 2 in this paper unless otherwise specified

Application Model

In DAG scheduling, finding optimal schedules is to find the scheduling solution with the minimum schedule length. The schedule length encompasses the entire execution and communication cost of all the modules and is also termed as makespan. In this paper, the task scheduling problem is to map a set of tasks to a set of processors, aiming at minimizing the makespan. It takes as input a directed acyclic graph DAG = (, E), with $|V|$ nodes representing tasks, and $|E|$ edges representing constrained relations among the tasks. $V= (v_1, v_2, \ldots, v_i, \ldots, v_{|V|})$ is a node sequence in which the hypothetical entry node (with no predecessors) V_1 and end node (with no successors) $V|V|$, respectively, represent the beginning and the end of execution.

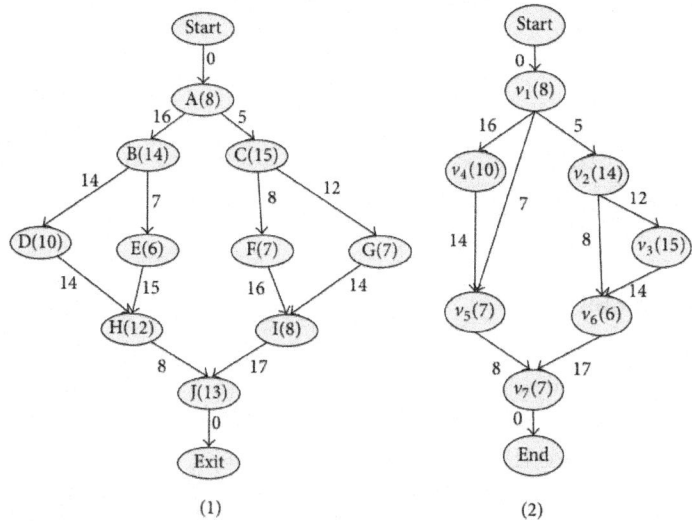

Figure 1: Two simple DAG models with 7 and 10 tasks

The execution cost value of Vi on processor pk is denoted as $EC_{p_k}(v_i)$, , and the average computation cost of V_i, denoted as $\overline{W(v_i)}$, can be calculated by (12). The parameter for the amounts of computing power available at each node in a heterogeneous system and its heterogeneous level value is given in the 5th paragraph of Section 6 and Table 1. $E = \{E_i \mid i = 1, 2, 3, \ldots, |E|\}$ is an edge set in which $E_i = (ev_s, ev_e, ew_{s,e})$, with ev_s & $ev_e \in \{v_1, v_2, \ldots, v_{|V|}\}$ representing its start and end nodes, and the value of communication cost between ev_s and ev_e is denoted as $ews_{s,e}$. The DAG topology of an exemplar application model and system model is shown in Figures 1 and 2, respectively.

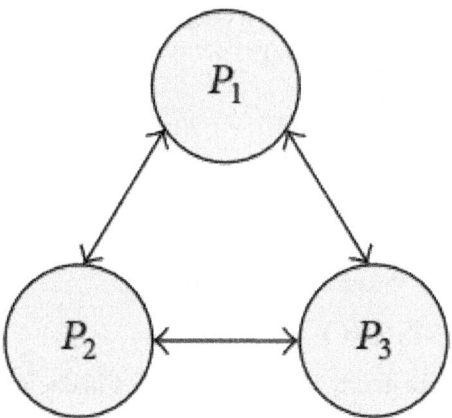

Figure 2: A fully connected parallel system with 3 heterogeneous processors.

Consider

$$\overline{W(v_i)} = \sum_{k=1}^{|P|} \frac{EC_{p_k}(v_i)}{|P|}.$$

(12)

The constrained-critical-path sequence of DAG $= (V, E)$ is denoted as CCP $= (CCP_1, CCP_2, \ldots, CCP_{|CCP|})$ with $CCP_i = (cv_{i,1}, cv_{i,2}, \ldots, cv_{i,|CCP_i|})$ in which the set $\{cv_{i,1}, cv_{i,2}, \ldots, cv_{i,|CCP_i|}\} \subseteq \{v_1, v_2, \ldots, v_{|V|}\}$.

The start time of the task V_i on processor p_k is denoted as $STp_k(V_i)$, which can be calculated using (13), where $Pred(V_i)$ is the set of the predecessors of the task V_i. And the earliest finish time of the task V_i on processor pk is denoted as $EFT_{pk}(V_i)$, which can be calculated using (14):

$$ST_{p_k}(v_i) = \begin{cases} 0, & v_i = v_1 \\ \max_{v_j \in Pred(v_i)} EFT_{p_k}(v_i), & p_k = p_m \\ \max_{v_j \in Pred(v_i)} \left(EFT_{p_m}(v_j) + ew_{j,i} \right), & p_k \neq p_m \end{cases}$$

(13)

$$EFT_{p_k}(v_i) = ST_{p_k}(v_i) + EC_{p_k}(v_i).$$

(14)

The communication to computation ratio (CCR) can be used to indicate whether a DAG is communication intensive or computation intensive. For a given DAG, it is computed by the average communication cost divided by the average computation cost on a target computing system. The computation can be formulated as follows:

$$CCR = \frac{\sum_{(v_i,v_j,ew_{i,j})\in E} ew_{i,j}}{\overline{W(v_i)}}.$$

(15)

DESIGN OF TMSCRO

TMSCRO mimics the interactions of molecules in chemical reactions with the concepts of molecule, atoms, molecular structure, and energy of a molecule. The structure of a molecule is unique, which represents the atom positions in a molecule. The interactions of molecules in four kinds of basic chemical reactions, on-wall ineffective collision, decomposition, intermolecular ineffective collision, and synthesis, aim to transform to the molecule with more stable states which has lower energy. In DAG scheduling, a scheduling solution including a task and processor allocation corresponds to a molecule in TMSCRO. This paper also designs the operators on the encoded scheduling solutions (tuple arrays). These designed operators correspond to the chemical reactions and change the molecular structures. The arrays with different tuples represent different scheduling solutions, and we can calculate the corresponding makespan of the scheduling solution. A scheduling solution makespan corresponds to the energy of a molecule.

In this section, we first present the data pretreatment of the TMSCRO. After the presentation of the encoding of scheduling solutions and the fitness function used in the TMSCRO, we present the design of four elementary chemical reaction operators in each part of the TMSCRO. Finally, we outline the framework of the TMSCRO scheme and discuss a few important properties in TMSCRO.

Molecular Structure, Data Pretreatment, and Fitness Function

This subsection first presents the encoding of scheduling solutions (i.e., the molecular structure) and data pretreatment, respectively. Then we give the statement of the fitness function for optimization designed in TMSCRO.

Molecular Structure and Data Pretreatment

A reasonable initial population in CRO-based methods may increase the scope of searching over the fitness function [20] to support faster convergence and to result in a better solution. Constrained critical paths (CCPs) can be seen as the classification of task sequences constructed by constrained earliest finish time (CEFT) algorithm, which takes into account all factors in DAG (i.e., the average of each task execution cost, the communication costs, and the graph topology). Therefore, TMSCRO utilizes the CCPs to create a reasonable initial population based on a broad view of DAG.

The data pretreatment is to generate the CCPDAG from DAG and to construct CCPS for the initialization of TMSCRO. The CCPDAG is a directed acyclic graph with |CCP| nodes representing constrained critical paths, (CCP_s), two virtual nodes (i.e., start and end) representing the beginning and exit of execution, respectively, and |CE| edges representing dependencies among the nodes. The edges of CCPDAG are not labeled with communication overhead which is different from DAG. The data pretreatment includes two steps.

(1)   The CCP and the processor allocation of each element of CCP in DAG can be obtained by executing CEFT and the first initial CCP solution,

$$\text{Init} \quad ((CCP_1, sp_1), (CCP_2, sp_2), \ldots, (CCP_{|CCP|}, sp_{|CCP|})),$$

can also be got, in which $((CCP_i, sp_i))$ is sorted as the generated order of CCP_i and sp_i is processor assignment of CCP_i after executing CEFT. Consider the graph as shown in Figure 1; the resulting CCPs are indicated in Table 2.

(2)   After the execution of CEFT for DAG, the CCPDAG is generated with the input of CCP and DAG. A detailed description is given in Algorithm 2.

Table 2: CCP corresponding to the DAG as shown in Figure 1(1).

i	CCP_i
1	A-B-D
2	C-G
3	F
4	E
5	H
6	I
7	J

(1) **for** each $E_i = (ev_s, ev_e, w_{s,e})$ in E
(2) CCP_s = BelongCCP (ev_s);
(3) CCP_e = BelongCCP (ev_e);
(4)   **if** $(CCP_s \neq CCP_e)$ & $(CCPE(CCP_s, CCP_e))$ does not exist
(5)   create CCPE (CCP_s, CCP_e)
(6)   **end if**
(7)   add Start and End
(8)   add edges among Start and CCP nodes
(9)   add edges among End and CCP nodes
(10) **end for**

Algorithm 2: Gen_CCPDAG(DAG, CCP) generating CCPDAG.

As shown in Algorithm 1, the edge E_i of DAG with the start node CCPs and the end node CCP_e is obtained in each loop (line 1). BelongCCP(V_i) represents which CCP_j in CCPV$_i$ belongs to (line 2 and line 3). If CCP_s and CCP_e are different CCP_s and there is no edge between them (line 4), then the edge between CCP_s and CCP_e is generated (line 5). Finally, the nodes, start and end, and the edges among them and CCP nodes are added (line 7, line 8, and line 9). Consider the DAG as shown in Figure 1 and the CCP as indicated in Table 1. The resulting CCPDAG is shown in Figure 3.

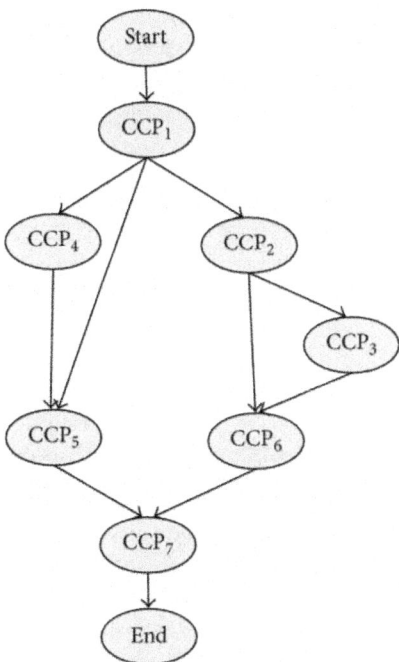

Figure 3: CCPDAG corresponding to the DAG as shown in Figure 1 and the CCP as indicated in Table 1.

In this paper, there are two kinds of molecular structures of TMSCRO, CCPS, and S. CCP molecular structure CCPS is just used in the initialization of TMSCRO, which can be formulated as in (16). Whereas the reaction molecular structure S converted from CCPS is used to participate in the elementary reaction of TMSCRO. In CCPS,$((CCP_i,sp_i))$s are sorted as the topology of CCPDAG in which CCPi is constrained critical path (CCP), and sp$_i$ is the processor assigned to CCP$_i$. $|CCP| \leq |V|$ because the number of elements in each SCCP$_i$ is greater than or equal to one. A reaction molecule S can be formulated as in (17), which consists of an array of atoms (i.e., tuples) representing a solution of DAG scheduling problem. A tuple includes three integers V_i, f_i, and p_i. The reaction molecular structure S is encoded with each integer in the permutation representing a task in DAG, the constraint relationship between a tuple and the one before it, and the processor pi. In each reaction molecular structure S, V_i represents a task in DAG and $(V_1, V_2,..., V|V|)$ is a topological sequence of DAG. In S, if V_A of the tuple A, which is before tuple B, is the predecessor of VB of tuple B in DAG, the second integer of tuple B, fB, will be 1, or it will be 0. p_i represents the processor allocation of each V_i in the tuple. The sequence of the tuples in a reaction molecular structure S represents the scheduling order of each task in DAG:

CCPS

$$= \left((CCP_1, sp_1), (CCP_2, sp_2), \ldots, \left(CCP_{|CCP|}, sp_{|CCP|}\right)\right), \tag{16}$$

$$S = \left((v_1, f_1, p_1), (v_2, f_2, p_2), \ldots, \left(v_{|V|}, f_{|V|}, p_{|V|}\right)\right). \tag{17}$$

Fitness Function

The initial molecule generator is used to generate the initial solutions for TMSCRO to manipulate. The first molecule InitS is converted from InitCCPS. Part three ^{sp_i} of each tuple is generated by a random perturbation in the first InitCCPS. A detailed description is given in Algorithms 3 and 4 and presents how to convert a CCPS to an S.

```
(1)  InitS = ConvertMole(InitCCPS);
(2)  update each f_i in molecule InitS as defined in the last paragraph of Section 5.1.1
(3)  MoleN = 1;
(4)  while MoleN ≤ PopSize − 1 do
(5)      for each CCP_i in CCP molecule CCPS
(6)          find the first successor Succ(i) in CCPDAG from i to the end;
(7)          for each CCP_j, j∈ (i, Succ(i))
(8)              find the first predecessor Pred(j) from Succ(i) to the begin in CCP molecule CCPS;
(9)              if Pred(j) < i
(10)                 interchanged position of (CCP_i, sp_i) and (CCP_j, sp_j) in CCP molecule CCPS;
(11)             end if
(12)         end for
(13)     end for
(14)     Generate a new CCP molecule CCPS';
(15)     S = ConvertMole(CCPS')
(16)     update each f_i in reaction molecule S as defined in the last paragraph of Section 5.1.1
(17)     MoleN ← MoleN + 1;
(18) end while
```

Algorithm 3: InitTMolecule(InitCCPS) generating the initial population.

```
(1)  for i = 1; i ≤ |V|; i++
(2)      for each CCP_j in molecule CCPS
(3)          for each cv_k in CCP_j
(4)              v_i = cv_k;
(5)              f_i = 0;
(6)              p_i = sp_j;
(7)              Generate a new tuple (v_i, f_i, p_i)
(8)          end for
(9)      end for
(10) end for
(11) Generate a new reaction molecule S = ((v_1, f_1, p_1), (v_2, f_2, p_2), ..., (v_{|V|}, f_{|V|}, p_{|V|}));
(12) for each (v_i, f_i, p_i) in reaction molecule S
(13)     find the first successor Succ(v_i) in DAG from i to the end;
(14)     for each v_j ∈ (v_i, Succ(v_i))
(15)         find the first predecessor v_k = Pred(v_j) from Succ(v_i) to the begin in reaction molecule S;
(16)         if k < i
(17)             interchanged position of (v_i, f_i, p_i) and (v_j, f_j, p_j) in reaction molecule S;
(18)         end if
(19)     end for
(20) end for
(21) for each p_i in reaction molecule S to randomly change;
(22)     change p_i randomly
(23) end for
(24) return S;
```

Algorithm 4: ConvertMole(CCPS) converting a CCPS to an .

Potential energy (PE) is defined as the objective function (fitness function) value of the corresponding solution represented by S. The overall schedule length of the entire DAG, namely, makespan, is the largest finish time among all tasks, which is equivalent to the actual finish time of the end node in DAG. For the DAG scheduling problem by TMSCRO, the goal is to obtain the scheduling that minimizes makespan and ensure that the precedence of the tasks is not violated. Hence, each fitness function value is defined as

$$PE_S = makespan = Fit\,(S).$$

(18)

Algorithm 5 presents how to calculate the value of the optimization fitness function Fit(S).

```
(1) slength = 0;
(2) for each node v in S = ((v₁, f₁, p₁), (v₂, f₂, p₂), ... , (v_{|V|}, f_{|V|}, p_{|V|})) do
(3)    Calculate the start time of predecessor node pv of v
       ST_{p_v}(v, pv) = max ((EFT_{pv} + CM (v, p_v, pv, p_{pv})), AT_{P_v});
(4)    Find the finish time of v
       EFT_{p_v}(v) = max ((ST_{p_v}(v, pv))_{∀pv∈Pred(v)} + EC_{p_v}(v));
(5)    if slength < EFT_{p_v}(v)
(6)       update scheduling length
          slength = EFT_{p_v}(v);
(7)    end if
(8) end for
(9) return slength;
```

Algorithm 5: Fit(S) calculating the fitness value of a molecule and the processor allocation optimization.

Elementary Chemical Reaction Operators

This subsection presents four elementary chemical reaction operators for sequence optimization and processor allocation optimization designed in TMSCRO, including on-wall collision, decomposition, intermolecular collision, and synthesis.

On-Wall Ineffective Collision

In this paper, the operator, OnWallT, is used to generate a new molecule S' from a given reaction molecule S for optimization. OnWallT works as follows. (1) The operator randomly chooses a tuple (V_i, f_i, p_i) with $f_i = 0$ in S and then exchanges the positions of (V_i, f_i, p_i) and $(V_{i-1}, f_{i-1}, p_{i-1})$. (2) f_{i-1}, fi and f_{i+1} in S are modified as defined in the last paragraph of Section 5.1.1. (3) The operator changes pi randomly. In the end, the operator generates a new molecule S' from S as an intensification search. Figures 4 and 5 show the example which is the molecule corresponding to the DAG as shown in Figure 1(2).

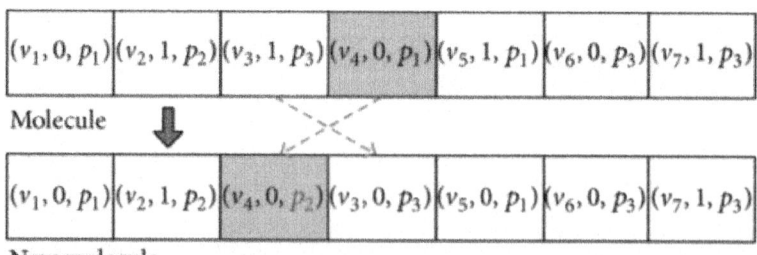

Figure 4: Illustration of molecular structure change for on-wall ineffective collision.

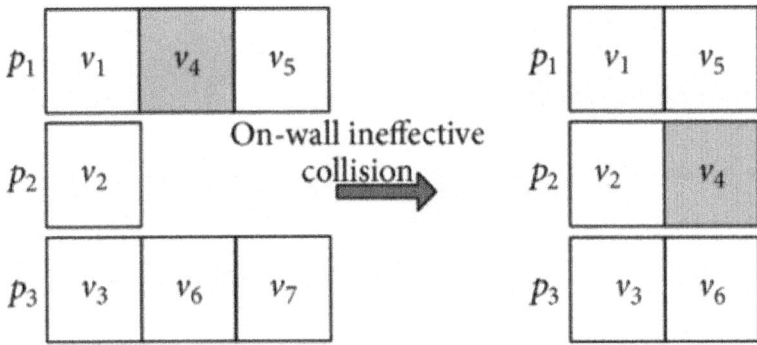

Figure 5: Illustration of the task-to-computing-node mapping for on-wall ineffective collision

Decomposition

In this paper, the operator, DecompT, is used to generate new molecules S_1' and S_2' from a given reaction molecule S. DecompT works as follows. (1) The operator randomly chooses two tuples (tuples)(V_i, f_i, p_i) with $f_i = 0$ and (V_t, f_t, p_t) with $ft = 0$ in S and then finds the tuple with the first predecessor of (Vi, f_i, p_i), such as (V_j, f_j, p_j), from the selection position to the beginning of reaction molecule S. (2) A random number $k \in [j+1, i-1]$ is generated, and the tuple (V_i, f_i, p_i) is stored in a temporary variable temp, and then from the position $i-1$, the operator shifts each tuple by one place to the right position until a position k. (3) The operator moves the tuple temp to the position k. The rest of the tuples in $S_{1'}$ are the same as those in S. (4) $fi, fi+1$ and f_k in S are modified as defined in the last paragraph of Section 5.1.1. (5) The operator generates the other new molecule S_2' as the former steps. The only difference is that, in step 2, we use (V_t, f_t, p_t) instead of (V_i, f_i, p_i). (6) The operator keeps the tuples in $S_{1'}$, which is at the odd position in S, and retains the tuples in S_2', which is at the even position in S, and then changes the remaining $p_x s$ of tuples in S_1' and S_2', randomly. In the end, the operator generates two new molecules S_1' and S_2' from S as a diversification search. Figures 6 and 7 show the example which is the molecule corresponding to the DAG as shown in Figure 1(2).

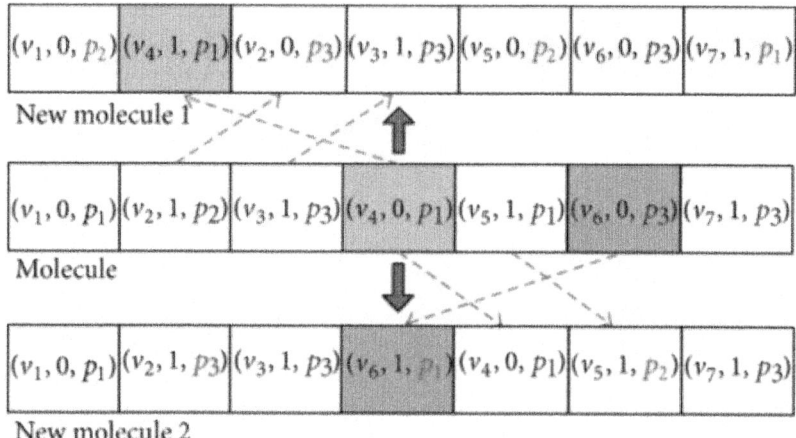

Figure 6: Illustration of molecular structure change for decomposition.

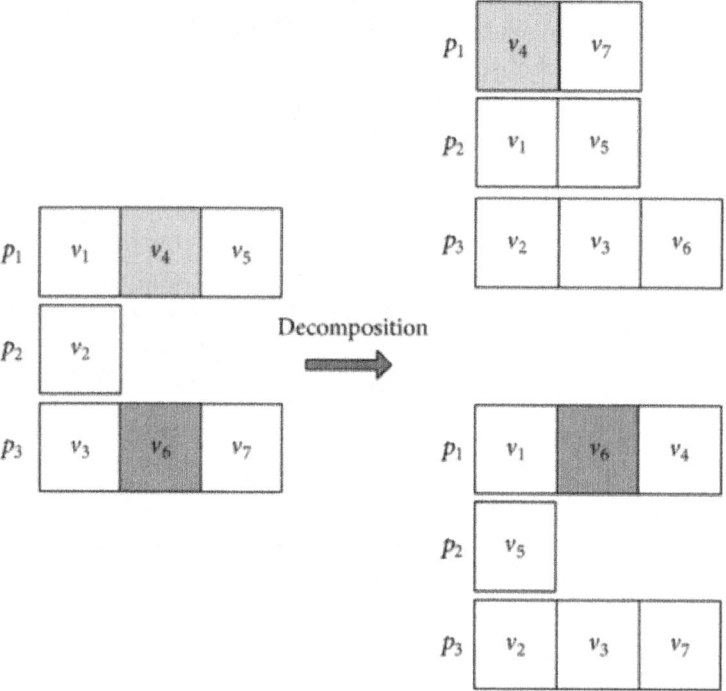

Figure 7: Illustration of the task-to-computing-node mapping for decomposition.

Intermolecular Ineffective Collision

In this paper, the operator, IntermoleT, is used to generate new molecules S_1' and S_2' from given molecules S_1 and S_2. This operator first uses the steps in OnWallT to generate S_1' from S_1, and then the operator generates the other new molecule S_1' from S_1 in similar fashion. In the end, the operator generates two new molecules S_1' and S_2' from S_1 and S_2 as an intensification search. Figures 8 and 9 show the example which is the molecule corresponding to the DAG as shown in Figure 1(2).

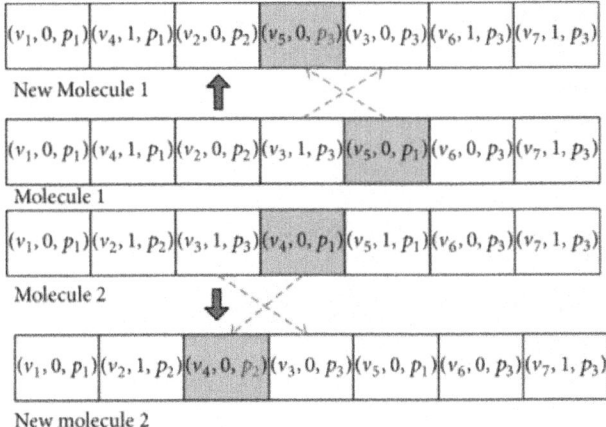

Figure 8: Illustration of molecular structure change for intermolecular ineffective collision.

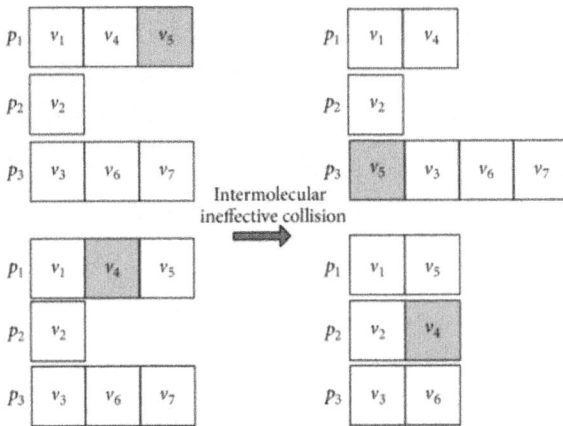

Figure 9: Illustration of the task-to-computing-node mapping for intermolecular ineffective collision.

Synthesis

In this paper, the operator, SynthT, is used to generate a new molecule $\bar{S}'$ from given molecules S_1 and S_2 for optimization. SynthT works as follows. (1) If $|V|$ is plural, then the integer $i = |V|/2$; else $i = (|V|+1)/2$. (2) S_1 and S_2 are cut off at the position i to become the left and right segments. (3) The left segments of $\bar{S}'$ are inherited from the left segments of S_1, randomly. (4) Each tuple in the right segments of $\bar{S}'$ comes from the tuples in S_2 that do not appear in the left segment of $\bar{S}'$, with their fx modified as defined in the last paragraph of Section 5.1.1 as well. (5) The operator keeps the tuples in $\bar{S}'$, which are at the same position in S_1 and S_2 with the same pxs, and then changes the remaining pys in $\bar{S}'$, randomly. As a result, the operator generates $\bar{S}'$ from S_1 and S_2 as a diversification search. Figures 10 and 11 show the example which is the molecule corresponding to the DAG as shown in Figure 1(2)

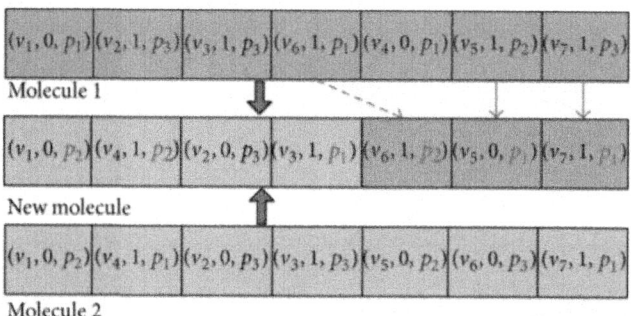

Figure 10: Illustration of molecular structure change for synthesis.

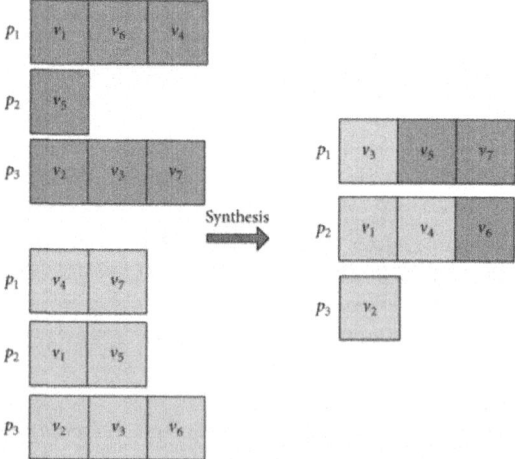

Figure 11: Illustration of the task-to-computing-node mapping for synthesis.

The Framework and Analysis of TMSCRO

The framework of TMSCRO is shown as an outline to schedule a DAG job in Algorithm 6 and the output of Algorithm 6 is just the resultant near-optimal solution for the corresponding DAG scheduling problem. In this framework, TMSCRO first initializes the process. Then, the process enters a loop. In each iteration, one of the elementary chemical reaction operators for optimization is performed to generate new molecules and PE of newly generated molecules will be calculated. The whole working of TMSCRO for DAG scheduling on heterogeneous problem is as presented in the last paragraph in Section 3.2. However, InitS is considered to be a super molecule [6], so it will be tracked and only participates in on-wall ineffective collision and intermolecular ineffective collision to explore as much as possible the solution space in its neighborhoods and the main purpose is to prevent InitS from changing dramatically. The iteration repeats until the stopping criteria are met. The stopping criteria may be set based on different parameters, such as the maximum amount of CPU time used, the maximum number of iterations performed, an objective function value less than a predefined threshold obtained, and the maximum number of iterations performed without further performance improvement. The stopping criterion of TMSCRO in the experiments of this paper is that the makespan is not changed after 5000 consecutive iterations in each loop. The time complexity of TMSCRO is $O(\text{iters} \times [2 \times (|V|^2 + |E| \times |P|)]),$, where iters is the number of iterations in TMSCRO, respectively.

It is very difficult to theoretically prove the optimality of the CRO (as well as DMSCRO and TMSCRO) scheme [37]. However, by analyzing the molecular structure, chemical reaction operators, and the operational environment in TMSCRO, it can be shown to some extent that TMSCRO scheme has the advantage of three points in comparison with GA, SA, and DMSCRO

First, just like DMSCRO, TMSCRO enjoys the advantages of GA and SA to some extent by analyzing the chemical reaction operators designed in TMSCRO and the operator environment of TMSCRO: (1) the OnWallT and IntermoleT in TMSCRO exchange the partial structure of two different molecules like the crossover operator in GA. (2) The energy conservation requirement in TMSCRO is able to guide the searching of the optimal solution in a similar way as the Metropolis Algorithm of SA guides the evolution of the solutions in SA.

(1) Initialize PopSize, KELossRate, MoleColl and InitialKE, θ and ϑ;
(2) Call Algorithm 2 to generate the initial population of TMSCRO, CROPop;
(3) Call Algorithm 3 to calculate PE of each molecule in CROPop;
(4) **while** the stopping criteria is not met **do**
(5)   Generate $\varepsilon \in [0, 1]$;
(6)   **if** $\varepsilon >$ MoleColl
(7)   Select a reaction molecule S from CROPop randomly;
(8)   **if** $((\text{NumHit}_S - \text{MinHit}_S) > \theta)$ & $(S \neq \text{InitS})$
(9)   Call DecompT to generate new molecules S_1' and S_2';
(10)   Call Algorithm 3 to calculate $PE_{S_1'}$ and $PE_{S_2'}$;
(11)   **if** Inequality (2) holds
(12)   Remove S from CROPop;
(13)   Add S_1' and S_2' to CROPop;
(14)   **end if**
(15)   **else**
(16)   Call OnWallT to generate a new molecules S';
(17)   Call Algorithm 3 to calculate $PE_{S'}$;
(18)   **If** $(S = \text{InitS})$
(19)   InitS $= S'$;
(20)   **end if**
(21)   Remove S from CROPop;
(22)   Add S' to CROPop;
(23)   **end if**
(24)   **else**
(25)   Select two molecules S_1 and S_2 from CROPop randomly;
(26)   **if** $(\text{KE}_{S_1} < \vartheta)$ & $(\text{KE}_{S_2} < \vartheta)$ & $(S_1 \neq \text{InitS})$ & $(S_2 \neq \text{InitS})$
(27)   Call SynthT to generate a new molecule S';
(28)   Call Algorithm 3 to calculate $PE_{S'}$;
(29)   **if** Inequality (10) holds
(30)   Remove S_1 and S_2 from CROPop;
(31)   Add S' to CROPop;
(32)   **end if**
(33)   **else**
(34)   Call IntermoleT to generate two new molecules S_1' and S_2';
(35)   Call Algorithm 3 to calculate $PE_{S_1'}$ and $PE_{S_2'}$;
(36)   **if** $(S_1 = \text{InitS})$
(37)   InitS $= S_1'$;
(38)   **else if** $(S_2 = \text{InitS})$
(39)   InitS $= S_2'$;
(40)   **end if**
(41)   Remove S_1 and S_2 from CROPop;
(42)   Add S_1' and S_2' to CROPop;
(43)   **end if**
(44)   **end if**
(45) **end while**
(46) **return** the molecule with the lowest PE in CROPop;

Algorithm 6: TMSCRO (DAG) The TMSCRO outline (framework).

Second, constrained earliest finish time (CEFT) algorithm constructs constrained critical paths (CCPs) by taking into account a broader view of the input DAG [5]. TMSCRO applies CEFT and CCPDAG to the data pretreatment and utilizes CCPs in the initialization of TMSCRO to create a more reasonable initial population than DMSCRO for accelerating convergence, because a wide distributed initial population in CRO-based methods may increase the scope of searching over the fitness function [20] to support faster convergence and to result in a better solution. Moreover, to some degree, InitS is also similar to the super molecule in super molecule-based CRO or the "elite" in GA [6]. However, the "elite" in GA is usually generated from two chromosomes, while InitS is based on the whole input DAG by executing CEFT. Third, the operators with the molecular structure in TMSCRO are designed more reasonably than DMSCRO. In CRO-based algorithm, the operators of on-wall collision and intermolecular collision are used for intensifications, while the operators of decomposition and synthesis are for diversifications. The better the operator can get the better the search results of intensification and diversification are. This feature of CRO is very important, which gives CRO more opportunities to jump out of the local optimum and explore the wider areas in the solution space. In TMSCRO, the operators of OnWallT and IntermoleT every time only exchange the positions of one tuple and its former neighbor in the molecule with better capability of intensification on sequence optimization than DMSCRO, of which the reaction operators, OnWall (ω_1) and Intermole (ω_1, ω_2) [37] (ω_1 and ω_2 are big molecules in DMSCRO), may change the task sequence(s) dramatically. Moreover, under the consideration that the optimization includes not only sequence but also processor assignment optimization, all reaction operators in TMSCRO can change the processor assignment, but DMSCRO has only two reactions, on-wall and synthesis [37], for processor assignment optimization. On the one hand, TMSCRO has 100% probability of searching the processor assignment solution space by four elementary reactions, with better capability of diversification and intensification on processor assignment optimization than DMSCRO, of which the chance to search this kind of solution space is only 50%. On the other hand, the division of diversification and intensification of four reactions in TMSCRO is very clear; however, this is not in DMSCRO. In each iteration, the diversification and intensification search in TMSCRO have the same probability to be conducted, whereas the possibility of diversification or intensification search in DMSCRO is uncertainty. This design enhances the ability to get better rapidity of convergence and search result in the whole solution space, which is demonstrated by the experimental results in Section 6.3.

SIMULATION AND RESULTS

The simulations have been performed to test TMSCRO scheduling algorithm in comparison with heuristic (HEFT_B and HEFT_T) [8] for DAG scheduling and with two metaheuristic algorithms, double molecular structure-based chemical reaction optimization (DMSCRO) [37], by using two sets of graph topology such as the real world application (Gaussian elimination and molecular dynamics code) and randomly generated application. The task graph for Gaussian elimination for input matrix of size 7 is shown in Figure 12, whereas a molecular dynamics code graph is shown in Figure 13. Figure 14 shows a random graph with 10 nodes. The baseline performance is the makespan obtained by DMSCRO.

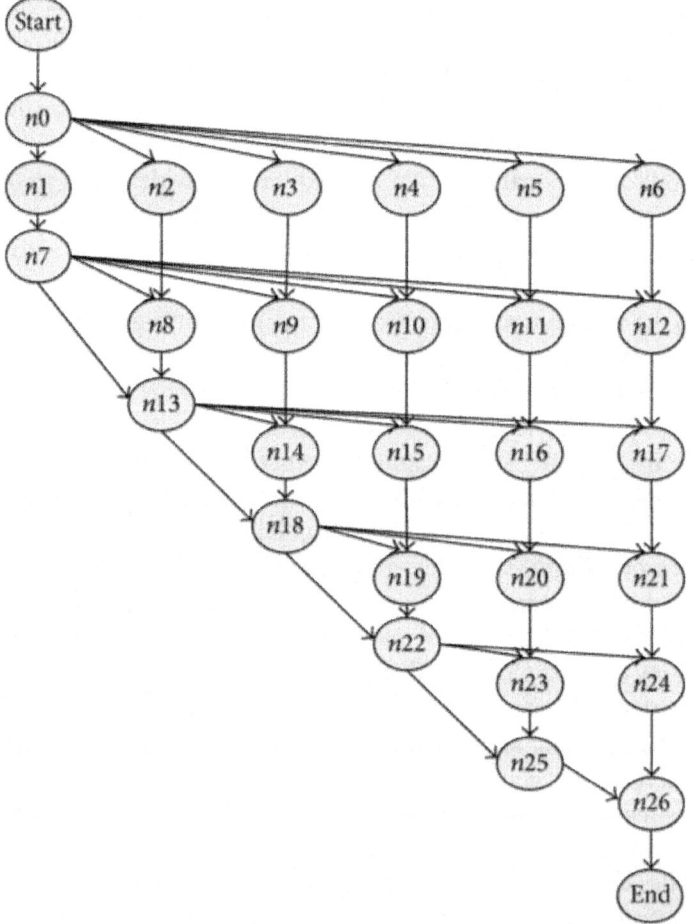

Figure 12: Gaussian elimination for a matrix of size 7.

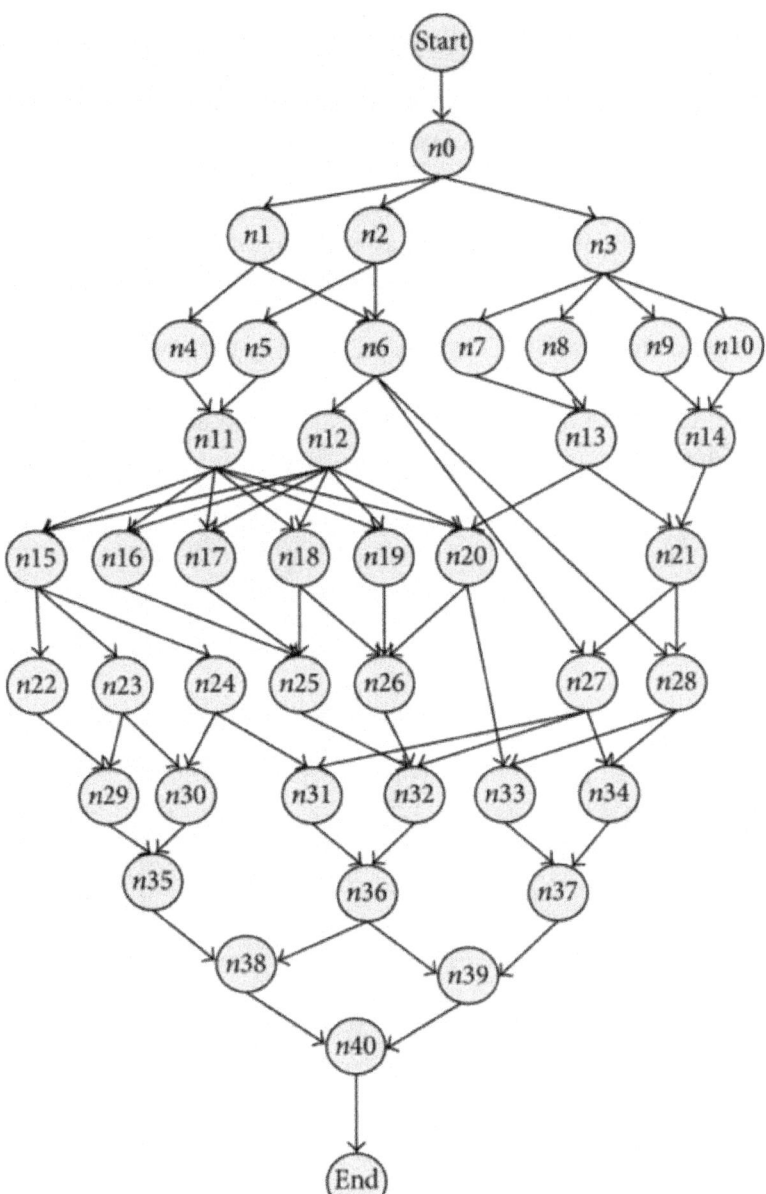

Figure 13: A molecular dynamics code.

Considering that HEFT_B and HEFT_T have better performance than other heuristics algorithms for DAG scheduling on heterogeneous computing systems, as proposed in the 8th paragraph in Section 2.1, these two algorithms are used to be the representatives of heuristics in the simulation. There are three reasons why we regard the makespan performance of DMSCRO [37] scheduling as the baseline performance. (1) So far as we know, DMSCRO is the only one CRO-based algorithm for DAG scheduling which takes into account the searching of the task order and processor assignment. (2) As discussed in the 3rd paragraph of Section 2.2, DMSCRO [37] has the closest system model and workload to that of TMSCRO. (3) In [37], CRO-based scheduling algorithm is considered as absorbing the strengths of SA and GA. However, the underlying principles and philosophies of SA are very different from DMSCRO, and because the DMSCRO is also proved to be more effective than genetic algorithm (GA) [15] as presented in [37], we just use DMSCRO to represent the metaheuristic algorithms. We propose to make a comparison between TMSCRO and DMSCRO to validate the advantages of TMSCRO over DMSCRO.

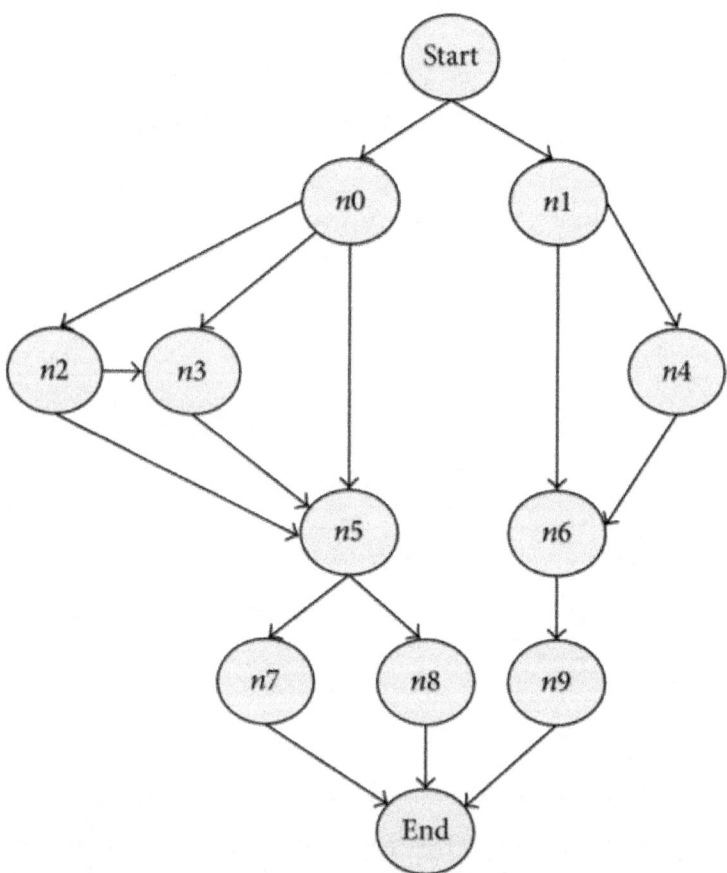

Figure 14: A random graph with 10 nodes.

The performance has been evaluated by the parameter makespan. The makespan values plotted in the bar graph of makespan and the chart of converge trace are, respectively, the average result of 50 and 25 independent runs to validate the robustness of TMSCRO. The communication cost is calculated by using computation costs and the computation cost ratio (CCR) values. The computation can be formulated as in (17):

$$Communication\ Cost = CCR * Computation\ Cost.$$

(19)

All the suggested values for the other parameters of the simulation of TMSCRO and their values are listed in Table 3. These values are proposed in [20].

Table 3: Configuration parameters for the simulation of TMSCRO.

Parameter	Value
InitialKE	1000
θ	500
ϑ	10
Buffer	200
KELossRate	0.2
MoleColl	0.2
PopSize	10
g	0.33
Number of runs	50

Real World Application Graphs

The real world application set is used to evaluate the performance of TMSCRO, which consists of two real world problem graph topologies, Gaussian elimination [22] and molecular dynamics code [19].

Gaussian Elimination

Gaussian elimination is a well-known method to solve a system of linear equations. Gaussian elimination converts a set of linear equations to the upper triangular form by applying elementary row operators on them systematically. As shown in Figure 12, the matrix size of the task graph of Gaussian elimination algorithm is 7, with 27 tasks in total. In [37], this DAG has been used for the simulation of DMSCRO, and we also apply it to the evaluation of TMSCRO in this paper. Under the consideration that graph structure is fixed, the variable parameters are only 22 the communication to computation ratio (CCR) value and the heterogeneous processor number. In the simulation, CCR values were set as 0.1, 0.2, 1, 2, and 5, respectively. Considering the identical operator is executed on each processor and the information communicated between heterogeneous processors is the same in Gaussian elimination, the execution cost of each task is supposed to be the same and all communication links have the same communication cost.

The parameters and their values of the Gaussian elimination graphs performed in the simulation are given in Table 4.

Table 4: Configuration parameters for the Gaussian elimination graphs.

Parameter	Possible values
CCR	$\{0.1, 0.2, 1, 2, 5\}$
Number of processors	$\{4, 8, 16, 32\}$
Number of tasks	27

The makespan of TMSCRO, DMSCRO, HEFT_B, and HEFT_T under the increasing processor number is shown in Figure 15. As shown in Figure 15, it can also been seen that as the processor number increases, the average makespan declines, and the advantage of TMSCRO and DMSCRO over HEFT_B and HEFT_T also decreases, because when more computing nodes are contributed to run the same scale of tasks, less intelligent scheduling algorithms are needed in order to achieve good performance.

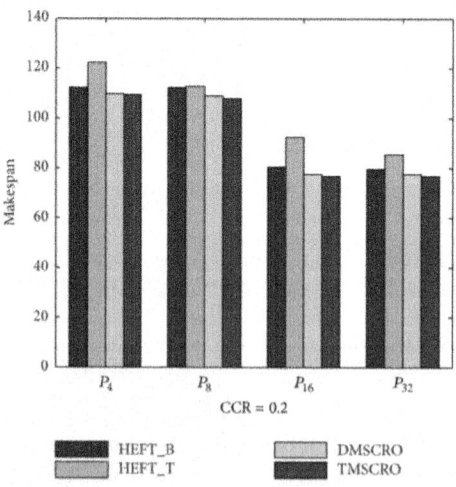

Figure 15: Average makespan for Gaussian elimination.

As the intelligent random search algorithms, TMSCRO and DMSCRO search a wider area of the solution space than HEFT_B, HEFT_T, or other heuristic algorithms, which narrow the search down to a very small portion of the solution space. This is the reason why TMSCRO and DMSCRO are more likely to obtain better solutions and outperform HEFT_B and HEFT_T.

The simulation results show that the performance of TMSCRO and DMSCRO is very similar to the fundamental reason that these algorithms are metaheuristic algorithms. Based on No-Free-Lunch Theorem in the field of metaheuristics, the performances of all well-designed metaheuristic search algorithms for optimal solution are the same, when averaged over all possible objective functions. The optimal solution will be gradually approached by a well-designed metaheuristic algorithm in theory, if it runs for long enough. The DMSCRO developed in [37] is well-designed, and we use it in the simulations of this paper. Therefore similar simulation results of the performances of TMSCRO and DMSCRO indicate that TMSCRO we developed is also well-designed. The detailed experiment result is shown in Table 5.

Table 5: The experiment results for the Gaussian elimination graph under different processors, CCR = 0.2.

The number of processors	HEFT_B (the average makespan)	HEFT_T (the average makespan)	DMSCRO (the average makespan)	TMSCRO (the average makespan)	TMSCRO (the best makespan)	TMSCRO (the worst makespan)	TMSCRO (the variance of resultant makespans)
4	112.2	122.227	109.9	109.31	109.2	109.9	0.2473
8	112.2	112.648	108.9	107.83	107.1	108.9	0.9613
16	80.4	92.354	77.5	76.62	76.3	78.9	1.6696
32	79.64	85.454	77.5	76.62	76.1	78.9	1.7201

In Figure 15, the figure shows that TMSCRO is superior to DMSCRO slightly. There will be only one reason for it: the stopping criteria set in this simulation are that the makespan stays unchanged for 5000 consecutive iterations in the search loop. As discussed in the last paragraph of Section 5, all metaheuristic methods that search for optimal solutions are the same in performance when averaged over all possible objective functions. And these experimental stopping criteria make TMSCRO and DMSCRO run for long enough to gradually approach the optimal solution. Moreover, better convergence of TMSCRO makes it more efficient in searching good solutions than DMSCRO by running much less iteration times. More detailed experiment results in this regard will be presented in Section 6.3.

Figure 16 shows that the average makespan of these four algorithms increases rapidly under the CCR increasing. The reason for it is because as CCR increases, the application becomes more communication intensive, making the heterogeneous processors in the idle state for longer. As shown in

Figure 16, TMSCRO and DMSCRO outperform HEFT_B and HEFT_T with the advantage being more obvious as CCR becomes larger. These experimental results suggest that, for communication-intensive applications, TMSCRO and DMSCRO can deliver more consistent performance and perform more effectively than heuristic algorithms, HEFT_B and HEFT_T, in a wide range of scenarios for DAG scheduling. The detailed experiment result is shown in Table 6.

Table 6: The experiment results for the Gaussian elimination graph under different CCRs; the number of processors is 8.

CCR	HEFT_B (the average makespan)	HEFT_T (the average makespan)	DMSCRO (the average makespan)	TMSCRO (the average makespan)	TMSCRO (the best makespan)	TMSCRO (the worst makespan)	TMSCRO (the variance of resultant makespans)
0.1	108.2	110.312	106.78	105.04	104.76	106.6	1.7271
0.2	112.2	112.648	108.9	107.83	107.1	108.9	0.9613
1	120.752	124.536	115.63	114.717	114.3	115.4	0.3787
2	207.055	197.504	189.4	188.303	188.1	188.75	0.1522
5	263.8	263.8	252.39	250.671	250.3	251.79	0.9178

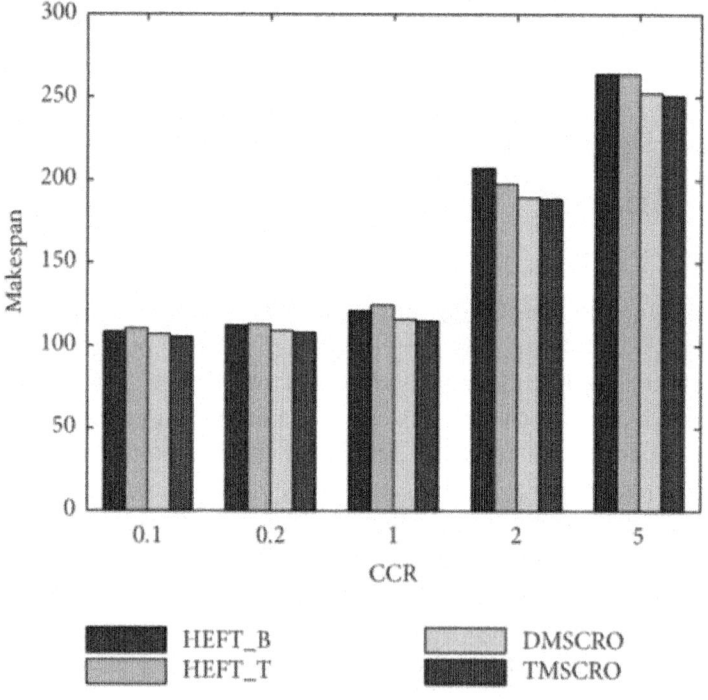

Figure 16: Average makespan for Gaussian elimination; the number of processors is 8.

Molecular Dynamics Code

Figure 13 shows the DAG of a molecular dynamics code as presented in [19]. As the experiment of Gaussian elimination, the structure of graph and the number of processors are fixed. The varied parameters are the number of heterogeneous processors and the CCR values which are used in our simulation are 0.1, 0.2, 1, 2, and 5.

The parameters and their values of the molecular dynamics code graphs performed in the simulation are given in Table 7.

Table 7: Configuration parameters for the molecular dynamics code graphs.

Parameter	Possible values
CCR	$\{0.1, 0.2, 1, 2, 5\}$
Number of processors	$\{4, 8, 16, 32\}$
Number of tasks	41

As shown in Figures 18 and 19, under different heterogeneous processor number and different CCR values, the average makespans of TMSCRO and DMSCRO are over HEFT_B and HEFT_T, respectively. In Figure 17, it can be observed that, with the number of heterogeneous processors increasing, the average makespan decreases. The average makespan with respect to different CCR values is shown in Figure 18. The average makespan increases with the value of CCR increasing. The detailed experiment results are shown in Tables 8and 9, respectively.

Table 8: The experiment results for the molecular dynamics code graph under different processors, CCR= 1.0.tab9

The number of processors	HEFT_B (the average makespan)	HEFT_T (the average makespan)	DMSCRO (the average makespan)	TMSCRO (the average makespan)	TMSCRO (the best makespan)	TMSCRO (the worst makespan)	TMSCRO (the variance of resultant makespans)
4	149.205	142.763	139.51	138.13	137.87	138.6	0.1749
8	131.031	122.265	118.8	116.9	116.2	117.33	0.2764
16	124.868	115.584	113.52	113.36	113.1	113.43	0.0237
32	120.047	103.784	102.617	101.29	101.023	101.47	0.0442

Table 9: The experiment results for the molecular dynamics code graph under different CCRs; the number of processors is 16.

CCR	HEFT_B (the average makespan)	HEFT_T (the average makespan)	DMSCRO (the average makespan)	TMSCRO (the average makespan)	TMSCRO (the best makespan)	TMSCRO (the worst makespan)	TMSCRO (the variance of resultant makespans)
0.1	82.336	90.136	80.53	77.781	77.3	78.9	0.9459
0.2	82.356	87.504	80.53	78.704	78.21	79.13	0.2002
1	124.868	115.584	113.52	113.36	113.1	113.43	0.0237
2	216.735	174.501	167.612	164.7	164.32	164.91	0.0742
5	274.7	274.7	265.8	262.173	262.022	262.6	0.1344

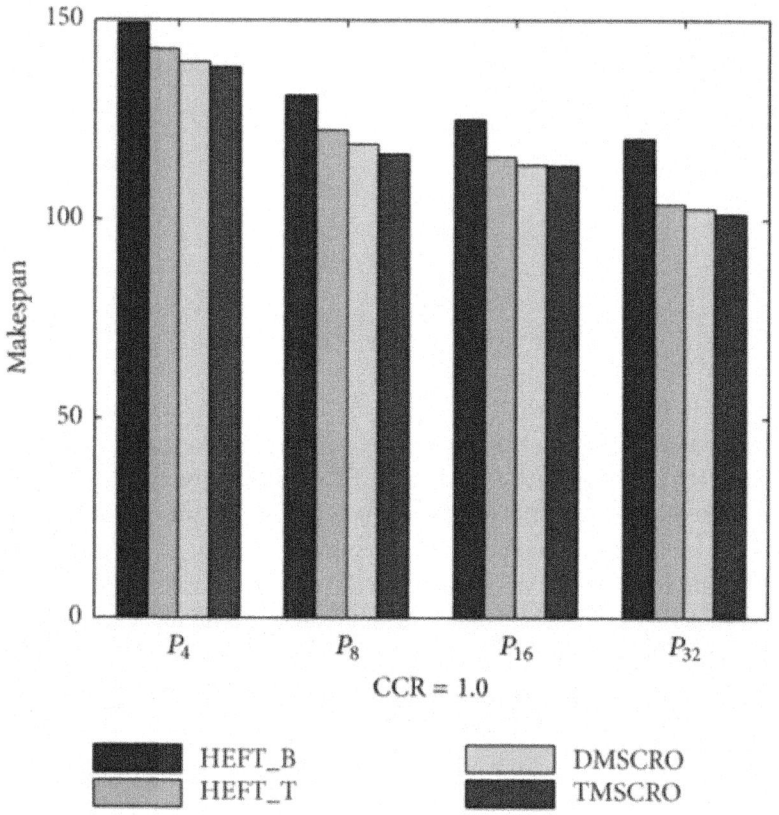

Figure 17: Average makespan for the molecular dynamics code.

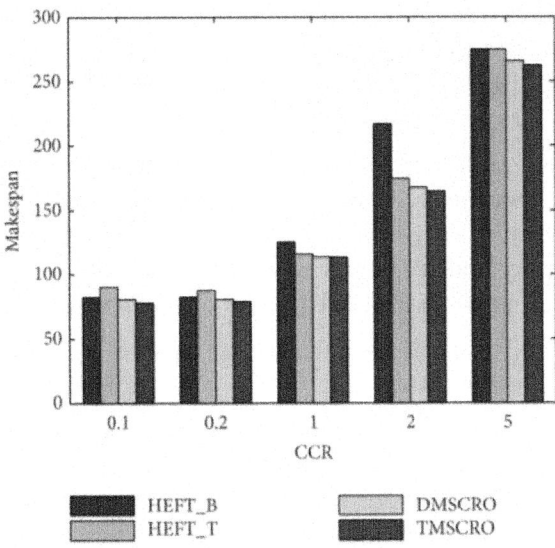

Figure 18: Average makespan for the molecular dynamics code; the number of processors is 16.

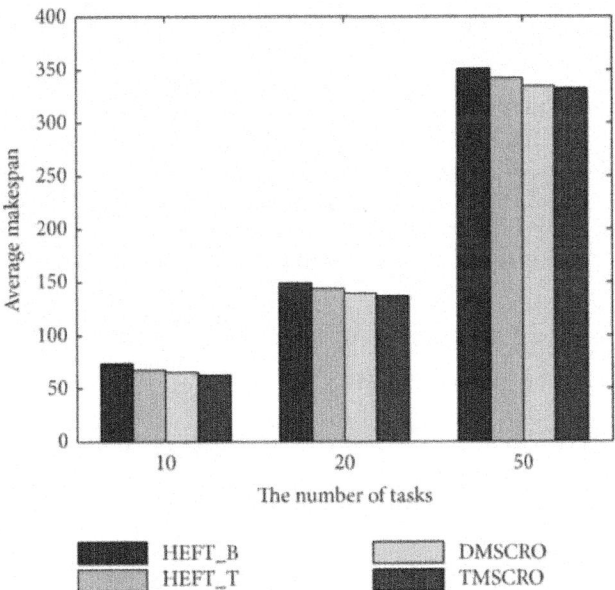

Figure 19: Average makespan of different task numbers, CCR= 10; the number of processors is 32.

Random Generated Application Graphs

An effective mechanism to generate random graph for various applications is proposed in [42]. By using the probability for an edge between any two nodes, it can generate a random graph without incline towards a specific topology

In the random graph generation of this mechanism, the topological order is used to guarantee the precedence constraints; that is, an edge exists between two nodes V_1 and V_2 only if $V_1 < V_2$. For probability pb, $\lfloor |V| * pb \rfloor$ edges are created from every node m to another node $(N_1 + (1/pb) * i) \bmod |V|$, where $1 \leq i \leq \lfloor |V| * pb \rfloor$, and $\lfloor V \rfloor$ is the total account of task nodes in DAG.

The parameters and their values of the random graphs performed in the simulation are given in Table 10.

Figure 19 shows that TMSCRO always outperforms HEFT_B, HEFT_T, and DMSCRO with the number of tasks in a DAG increasing. The comparison of the average makespan of four algorithms under the increase of heterogeneous processor number is shown in Figures 20 and 21. As can be seen from these figures, the performance of TMSCRO is better than the other three algorithms in all cases. The reasons for these two figures are the same as those explained in Figure 15. The detailed experiment results are shown in Tables 11, 12, and 13, respectively

Table 10: Configuration parameters for random graphs.

Parameter	Possible values
CCR	$\{0.1, 0.2, 1, 2, 5, 10\}$
Number of processors	$\{4, 8, 16, 32\}$
Number of tasks	$\{10, 20, 50\}$

Table 11: The experiment results for the random graph under different task numbers, CCR = 10; the number of processors is 32.

The number of tasks	TMSCRO (the average makespan)	TMSCRO (the best makespan)	TMSCRO (the worst makespan)	TMSCRO (the variance of resultant makespans)
10	73	67	65.1	62.2
20	148.9	143.9	139.421	136.8
50	350.7	341.7	334.17	331.9

Table 12: The experiment results for the random graph under different processors, CCR = 0.2; the number of tasks is 50.

The number of processors	HEFT_B (the average makespan)	HEFT_T (the average makespan)	DMSCRO (the average makespan)	TMSCRO (the average makespan)	TMSCRO (the best makespan)	TMSCRO (the worst makespan)	TMSCRO (the variance of resultant makespans)
4	167.12	178.023	159.234	157.63	157.12	158.3	0.3923
8	136.088	145.649	128.17	127.178	127.06	127.7	0.1949
16	119.292	125.986	115.9	114.33	114.1	115.2	0.4753
32	111.866	120.065	108.7	108.71	108.31	108.9	0.0733

Table 13: The experiment results for the random graph under different processors, CCR = 1.0; the number of tasks is 50.

The number of processors	HEFT_B (the average makespan)	HEFT_T (the average makespan)	DMSCRO (the average makespan)	TMSCRO (the average makespan)	TMSCRO (the best makespan)	TMSCRO (the worst makespan)	TMSCRO (the variance of resultant makespans)
4	178.662	175.52	168.12	167.703	167.42	168	0.0857
8	138.572	136.47	131.8	131.451	131.1	131.9	0.178
16	125.772	124.31	122.91	122.32	122.1	122.432	0.0233
32	117.11	116.4	114.124	113.127	112.9	113.54	0.1348

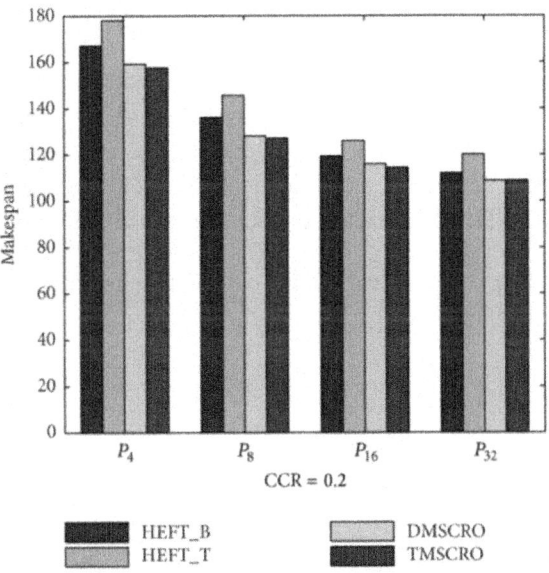

Figure 20: Average makespan of four algorithms under different processor numbers and the low communication costs; the number of tasks is 50.

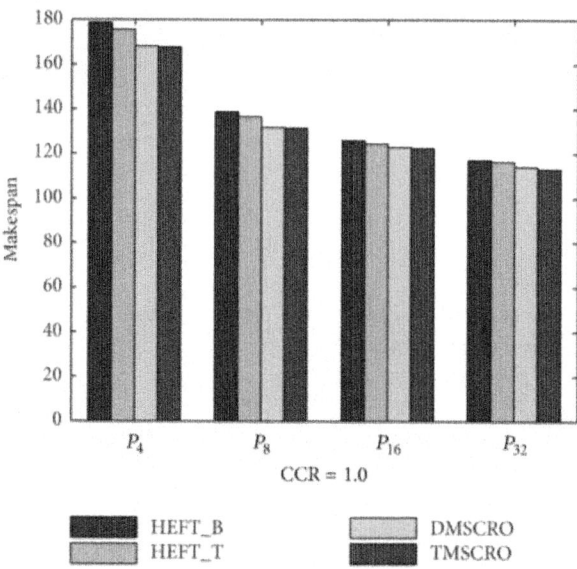

Figure 21: Average makespan of four algorithms under different processor numbers and the low communication costs; the number of tasks is 50.

As shown in Figure 22, it can be observed that the average makespan approached by TMSCRO increases rapidly with CCR values increasing. This may be because as CCR increases, the application becomes more communication intensive, making the heterogeneous processors in the idle state for longer. The detailed experiment results are shown in Table 14.

Table 14: The experiment results for the random graph under different task CCRs, the number of tasks is 50.

CCR	The number of processors is 4	The number of processors is 8	The number of processors is 16	The number of processors is 32
0.1	156.97	115.724	110.3	101.87
0.2	157.63	127.178	114.33	108.71
1	167.703	131.451	122.32	113.127
2	294.042	289.878	273.375	269.514
5	473.5	467.61	429.13	428.13

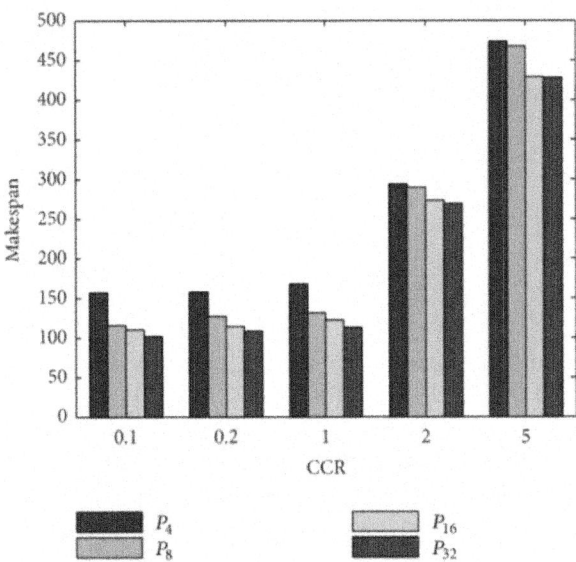

Figure 22: Average makespan of TMSCRO under different values of CCR; the number of tasks is 50.

Convergence Trace of TMSCRO

The result of the experiments in the previous subsections is the final makespan obtained by TMSCRO and DMSCRO, showing that TMSCRO can obtain similar makespan performance as DMSCRO. Moreover, in some cases the final makespan achieved by TMSCRO is even better than that by DMSCRO after the stop criteria are satisfied. In this section, the change of makespan in the experiments as TMSCRO and DMSCRO progress during the search is demonstrated by comparing the convergence trace of these two algorithms. These experiments help further reveal the better performance of TMSCRO on convergence and can also help explain why the TMSCRO sometimes outperforms DMSCRO in some cases.

The parameters and their values of the Gaussian elimination, molecular dynamics code, and random graphs performed in the simulation are given in Tables 15, 16, and 17, respectively.

Table 15: Configuration parameters of convergence experiment for the Gaussian elimination graph.

Parameter	Value
CCR	0.2
Number of processors	8
Number of tasks	27

Table 16: Configuration parameters of convergence experiment for the molecular dynamics graph.

Parameter	Value
CCR	1
Number of processors	16
Number of tasks	41

Table 17: Configuration parameters of convergence experiment for the random graphs.

Parameter	Values
CCR	{0.2, 1}
Number of processors	{8, 16}
Number of tasks	{10, 20, 50}

Figures 23 and 24, respectively, plot the convergence traces for processing Gaussian elimination and the molecular dynamics code. Figures 25, 26, and 27 show the convergence traces when processing the sets of randomly generated DAGs and each set contains the DAGs of 10, 20, and 50 tasks, respectively. These figures demonstrated that the makespan performance decreases quickly as both TMSCRO and DMSCRO progress and that the decreasing trends tail off when the algorithms run for long enough. These figures also show that, in most cases, the convergence traces of both algorithms are rather different even though the final makespans obtained by them are almost the same.

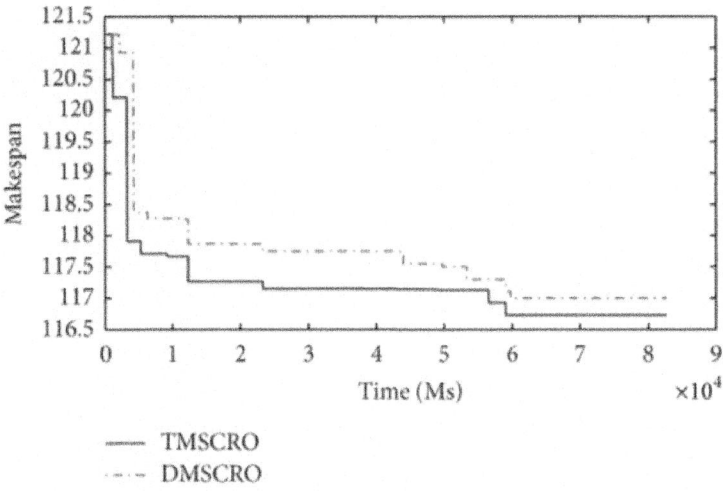

Figure 23: The convergence trace for Gaussian elimination; $^{ccr\,=\,0.2}$; the number of processors is 8.

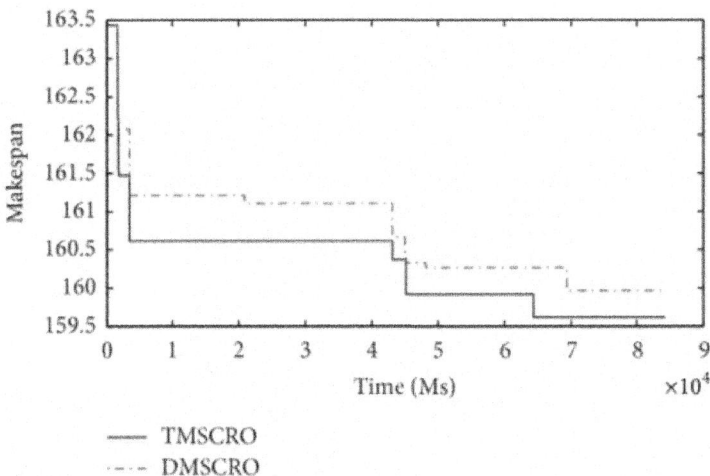

Figure 24: The convergence trace for the molecular dynamics code; $^{ccr\,=\,1}$; the number of processors is 16.

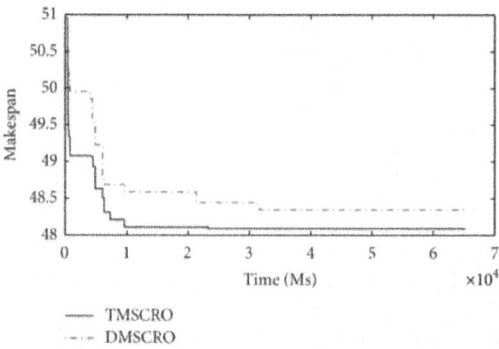

Figure 25: The convergence trace for the randomly generated DAGs with each containing 10 tasks.

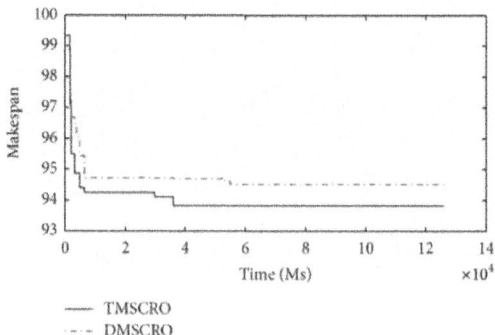

Figure 26: The convergence trace for the randomly generated DAGs with each containing 20 tasks.

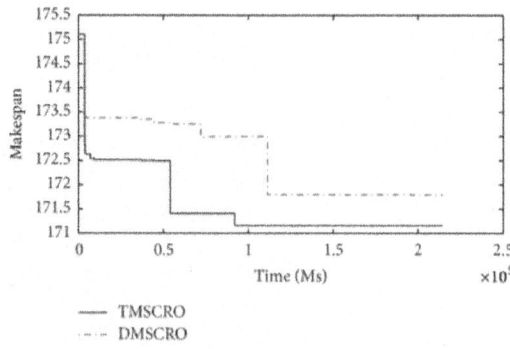

Figure 27: The convergence trace for the randomly generated DAGs with each containing 50 tasks.

The statistical analysis results over the average coverage rate at 5000 ascending sampling points from start time to end time of all the experiments are shown in Table 18 (the threshold of P is set as 0.05), which are obtained by Friedman test, and each experiment is carried out 25 times. We can find that the differences between two algorithms in performance are significant from a statistical point of view. The reason of it is because the super molecule makes TMSOCRO have a stronger convergence capability, especially early in each run. Moreover, the performance of TMSCRO on convergence is better than DMSCRO. Quantitatively, our records show that TMSCRO converges faster than DMSCRO by 12.89% on average in all the cases (by 23.27% on average in the best case).

Table 18: The results of the statistical analysis over the average coverage rate at different sampling times of all the experiments (the threshold of P is set as 0.05).

DAG	The value of P after Friedman test	Average convergence acceleration ratio
Gaussian elimination	7.10×10^{-8}	4.23%
Molecular dynamics code	2.54×10^{-8}	7.21%
Random graph with 10 tasks	4.26×10^{-8}	23.27%
Random graph with 20 tasks	3.48×10^{-8}	16.41%
Random graph with 50 tasks	2.58×10^{-8}	13.32%

In these experiments, the stopping criteria of the algorithms are that the algorithm stops when the makespan performance remains unchanged for a preset number of consecutive iterations in the search loop (in the experiments, it is 5000 iterations). In reality, the algorithms can also stop when the total processing time of it reaches a preset value (e.g., 180s). Moreover, both of TMSCRO and DMSCRO have the same initial population. In this case, the fact that TMSCRO outperforms DMSCRO on convergence means that the makespan achieved by TMSCRO could be much better than that by DMSCRO when the stopping criteria of the algorithm are satisfied. The reason for this can be explained by the analysis presented in the last paragraph of Section 5.3.

CONCLUSION

In this paper, we developed a TMSCRO for DAG scheduling on heterogeneous systems based on chemical reaction optimization (CRO) method. With a more reasonable reaction molecular structure and four designed elementary chemical reaction operators, TMSCRO has a better ability on intensification and diversification search than DMSCRO, which is the only one CRO-based algorithm for DAG scheduling on heterogeneous systems as far as we know. Moreover, in TMSCRO, the algorithm constrained earliest finish time (CEFT) and constrained-critical-path directed acyclic graph (CCPDAG) are applied to the data pretreatment, and the concept of constrained paths (CCPs) is also utilized in the initialization. We also use the first initial molecule, InitS, to be a super molecule for accelerating convergence. As a metaheuristic method, the TMSCRO algorithm can cover a much larger search space than heuristic scheduling approaches. The experiments show that TMSCRO outperforms HEFT_B and HEFT_T and can achieve a higher speedup of task executions than DMSCRO.

In future work, we plan to extend TMSCRO by applying synchronous communication strategy to parallelize the processing of TMSCRO. This kind of design will divide the molecules into groups and each group of molecules is handled by a CPU or GPU. So, multiple groups can be manipulated simultaneously in parallel and molecules can also be exchanged among the CPUs or GPUs from time to time in order to reduce the time cost.

NOTATIONS

$DAG = (V, E)$:   Input directed acyclic graph with $|V|$ nodes representing tasks, and $|E|$ edges representing constrained relations among the tasks

$V = (v_1, v_2, \ldots, v_{|V|})$:   Node sequence in which the hypothetical entry node (with no predecessors) v_1 and end node (wit no successors) $v_{|V|}$, respectively, represent the beginning and end of execution

$E = \{E_i \mid i = 1, 2, 3, \ldots, |E|\}$:   Edge set in which $E_i = (ev_s, ev_e, ew_{s,e})$, with ev_s & $ev_e \in \{v_1, v_2, \ldots, v_{|V|}\}$ representing its start and end node and the value of communication cost between ev_s and ev_e denoted a $ew_{s,e}$

$P = \{p_i \mid i = 1, 2, 3, \ldots, |P|\}$:   Set of multiple heterogeneous processors in target system

$CCP = (CCP_1, CCP_2, \ldots, CCP_{\lvert CCP \rvert})$:	Constrained-critical-path sequence of DAG $= (V, E)$
$CCP_i = (cv_{i,1}, cv_{i,2}, \ldots, cv_{i,\lvert CCP_i \rvert})$:	Constrained critical path in which the set $\{cv_{i,1}, cv_{i,2}, \ldots, cv_{i,\lvert CCP_i \rvert}\} \subseteq \{v_1, v_2, \ldots, v_{\lvert V \rvert}\}$
CCPDAG:	Directed acyclic graph with $\lvert CCP \rvert$ nodes representing CCPs, two virtual nodes (i.e., start and end) representing the beginning and exit of execution, respectively, and $\lvert CE \rvert$ edges representing dependencies among all nodes
$CCPS = ((CCP_1, sp_1), (CCP_2, sp_2), \ldots, (CCP_{\lvert CCP \rvert}, sp_{\lvert CCP \rvert}))$:	A CCP molecule used in the initialization of TMSCRO, in which sp_i is the processor assigned to the constrained-critical-path CCP_i
$S = ((v_1, f_1, p_1), (v_2, f_2, p_2), \ldots, (v_{\lvert V \rvert}, f_{\lvert V \rvert}, p_{\lvert V \rvert}))$:	A reaction molecule (i.e., solution) in TMSCRO
(v_i, f_i, p_i):	Atom (i.e., tuple) in S
InitCCPS:	The first CCP molecule for the initialization of TMSCRO
InitS:	The first molecule in TMSCRO
BelongCCP(w):	CCP_i that node w belongs to
CCPE($CCPs, CCP_e$):	Edge between CCPs and CCPe
$\overline{W(v)}$:	Average computation cost of node v
$EC_{P_r}(w)$:	Execution cost of a node w using processor P_r
$CM(w, P_r, v, P_x)$:	Communication cost from node v to w, if P_x has been assigned to node v and P_r is assigned to node w
$ST_{P_r}(w, v)$:	Possible start time of node w which is assigned the processor P_r with the v node being any predecessor of w which has already been scheduled

$EFT_{P_r}(w)$:	Finish time of node w using processor P_r
AT_{P_r}:	Availability time of P_r
Pred(w) :	Set of predecessors of node w
Succ(w) :	Set of successors of node w
CCR :	Communication to computation ratio
g :	The parameter to adjust the heterogeneity level in a heterogeneous system
PE:	Current potential energy of a molecule
KE:	Current kinetic energy of a molecule
InitialKE:	Initial kinetic energy of a molecule
θ:	Threshold value guiding the choice of on-wall collision or decomposition
ϑ:	Threshold value guiding the choice of intermolecule collision or synthesis
Buffer:	Initial energy in the central energy buffer

KELossRate:	Loss rate of kinetic energy
MoleColl:	Threshold value to determine whether to perform a unimolecule reaction or an intermolecule reaction
PopSize:	Size of the molecules
NumHit:	Total collision number of a molecule.

CONFLICT OF INTERESTS

The authors declare that there is no conflict of interests regarding the publication of this paper.

REFERENCES

1. J. L. R. L. Graham, E. L. Lawler, and A. R. Kan, "Optimization and approximation in deterministic sequencing and scheduling: a survey," Annals of Discrete Mathematics, vol. 5, pp. 287–326, 1979.

2. C. Papadimitriou and M. Yannakakis, "Towards an architecture-independent analysis of parallel algorithms," in Proceedings of the 20th Annual ACM Symposium on Theory of Computing (STOC '88), pp. 510–513, 1988.

3. V. Sarkar, Partitioning and Scheduling Parallel Programs for Multiprocessors, The MIT Press, Cambridge, Mass, USA, 1989.

4. P. Chrétienne, "Task scheduling with interprocessor communication delays," European Journal of Operational Research, vol. 57, no. 3, pp. 348–354, 1992.

5. M. A. Khan, "Scheduling for heterogeneous systems using constrained critical paths," Parallel Computing, vol. 38, no. 4-5, pp. 175–193, 2012.

6. J. Xu, Y. S. Albert Lam, and O. K. Victor Li, "Stock portfolio selection using chemical reaction optimization," in Proceedings of the International Conference on Operations Research and Financial Engineering (ICORFE '11), pp. 458–463, 2011.

7. Y.-K. Kwok and I. Ahmad, "Static scheduling algorithms for allocating directed task graphs to multiprocessors," ACM Computing Surveys, vol. 31, no. 4, pp. 406–471, 1999.

8. H. Topcuoglu, S. Hariri, and M.-Y. Wu, "Performance-effective and low-complexity task scheduling for heterogeneous computing," IEEE Transactions on Parallel and Distributed Systems, vol. 13, no. 3, pp. 260–274, 2002.

9. A. Amini, T. Y. Wah, M. R. Saybani, and S. R. A. S. Yazdi, "A study of density-grid based clustering algorithms on data streams," in Proceedings of the 8th International Conference on Fuzzy Systems and Knowledge Discovery (FSKD '11), pp. 1652–1656, Shanghai, China, July 2011.

10. H. Cheng, "A high efficient task scheduling algorithm based on heterogeneous multi-core processor," inProceedings of the 2nd International Workshop on Database Technology and Applications (DBTA '10), pp. 1–14, Wuhan, China, November 2010.

11. T. Tsuchiya, T. Osada, and T. Kikuno, "A new heuristic algorithm based on gas for multiprocessor scheduling with task duplication," in Proceedings of the 3rd International Conference on Algorithms and Architectures for Parallel Processing (ICAPP '97), pp. 295–308, Melbourne, Australia, December 1997.

12. R. Bajaj and D. P. Agrawal, "Improving scheduling of tasks in a heterogeneous environment," IEEE Transactions on Parallel and Distributed Systems, vol. 15, no. 2, pp. 107–118, 2004.

13. H.-W. Ge, L. Sun, Y.-C. Liang, and F. Qian, "An effective PSO and AIS-based hybrid intelligent algorithm for job-shop scheduling," IEEE Transactions on Systems, Man, and Cybernetics A: Systems and Humans, vol. 38, no. 2, pp. 358–368, 2008.

14. N. B. Ho and J. C. Tay, "Solving multiple-objective flexible job shop problems by evolution and local search," IEEE Transactions on Systems, Man and Cybernetics C: Applications and Reviews, vol. 38, no. 5, pp. 674–685, 2008.

15. E. S. H. Hou, N. Ansari, and H. Ren, "Genetic algorithm for multiprocessor scheduling," IEEE Transactions on Parallel and Distributed Systems, vol. 5, no. 2, pp. 113–120, 1994.

16. J.-J. Hwang, Y.-C. Chow, F. D. Anger, and C.-Y. Lee, "Scheduling precedence graphs in systems with interprocessor communication times," SIAM Journal on Computing, vol. 18, no. 2, pp. 244–257, 1989.

17. M. Iverson, F. Özgüner, and G. Follen, "Parallelizing existing applications in a distributed heterogeneous environment," in Proceedings of the IEEE International Conference on Heterogeneous Computing Workshop (HCW '95), pp. 93–100, 1995.

18. M. H. Kashani and M. Jahanshahi, "Using simulated annealing for task scheduling in distributed systems," in Proceedings of the International Conference on Computational Intelligence, Modelling, and Simulation (CSSim '09), pp. 265–269, Brno, Czech Republic, September 2009.

19. S. Kim and J. Browne, "A general approach to mapping of parallel computation upon multiprocessor architectures," in Proceedings of the International Conference on Parallel Processing, vol. 3, pp. 1–8, 1988.

20. A. Y. S. Lam and V. O. K. Li, "Chemical-reaction-inspired metaheuristic for optimization," IEEE Transactions on Evolutionary Computation, vol. 14, no. 3, pp. 381–399, 2010.

21. H. Li, L. Wang, and J. Liu, "Task scheduling of computational grid based on particle swarm algorithm," in Proceedings of the 3rd International Joint Conference on Computational Sciences and Optimization (CSO '10), vol. 2, pp. 332–336, Huangshan, China, May 2010.

22. M.-Y. Wu and D. D. Gajski, "Hypertool: a programming aid for message-passing systems," IEEE Transactions on Parallel and Distributed Systems, vol. 1, no. 3, pp. 330–343, 1990.

23. G. C. Sih and E. A. Lee, "Compile-time scheduling heuristic for interconnection-constrained heterogeneous processor architectures," IEEE Transactions on Parallel and Distributed Systems, vol. 4, no. 2, pp. 175–187, 1993.

24. H. El-Rewini and T. G. Lewis, "Scheduling parallel program tasks onto arbitrary target machines,"Journal of Parallel and Distributed Computing, vol. 9, no. 2, pp. 138–153, 1990.

25. F.-T. Lin, "Fuzzy job-shop scheduling based on ranking level (lambda, 1) interval-valued fuzzy numbers," IEEE Transactions on Fuzzy Systems, vol. 10, no. 4, pp. 510–522, 2002.

26. B. Liu, L. Wang, and Y.-H. Jin, "An effective PSO-based memetic algorithm for flow shop scheduling,"IEEE Transactions on Systems, Man, and Cybernetics B: Cybernetics, vol. 37, no. 1, pp. 18–27, 2007.

27. F. Pop, C. Dobre, and V. Cristea, "Genetic algorithm for DAG scheduling in Grid environments," inProceedings of the IEEE 5th International Conference on Intelligent Computer Communication and Processing (ICCP '09), pp. 299–305, Cluj-Napoca, Romania, August 2009.

28. R. Shanmugapriya, S. Padmavathi, and S. M. Shalinie, "Contention awareness in task scheduling using tabu search," in Proceedings of the IEEE International Advance Computing Conference (IACC '09), pp. 272–277, Patiala, India, March 2009.

29. L. Shi and Y. Pan, "An efficient search method for job-shop scheduling problems," IEEE Transactions on Automation Science and Engineering, vol. 2, no. 1, pp. 73–77, 2005.

30. P. Choudhury, R. Kumar, and P. P. Chakrabarti, "Hybrid scheduling of dynamic task graphs with selective duplication for multiprocessors under memory and time constraints," IEEE Transactions on Parallel and Distributed Systems, vol. 19, no. 7, pp. 967–980, 2008.

31. S. Song, K. Hwang, and Y.-K. Kwok, "Risk-resilient heuristics and genetic algorithms for security-assured grid job scheduling," IEEE Transactions on Computers, vol. 55, no. 6, pp. 703–719, 2006.

32. D. P. Spooner, J. Cao, S. A. Jarvis, L. He, and G. R. Nudd, "Performance-aware workflow management for grid computing," The Computer Journal, vol. 48, no. 3, pp. 347–357, 2005.

33. K. Li, X. Tang, and K. Li, "Energy-efficient stochastic task scheduling on heterogeneous computing systems," IEEE Transactions on Parallel and Distributed Systems, 2014.

34. J. Wang, Q. Duan, Y. Jiang, and X. Zhu, "A new algorithm for grid independent task schedule: genetic simulated annealing," in Proceedings of the World Automation Congress (WAC '10), pp. 165–171, Kobe, Japan, September 2010.

35. L. He, D. Zou, Z. Zhang, C. Chen, H. Jin, and S. Jarvis, "Developing

resource consolidation frameworks for moldable virtual machines in clouds," Future Generation Computer Systems, vol. 32, pp. 69–81, 2012.

36. Y. Xu, K. Li, J. Hu, and K. Li, "A genetic algorithm for task scheduling on heterogeneous computing systems using multiple priority queues," Information Sciences, vol. 270, pp. 255–287, 2014.

37. Y. Xu, K. Li, L. He, and T. K. Truonga, "A DAG scheduling scheme on heterogeneous computing systems using double molecular structure-based chemical reaction optimization," Journal of Parallel and Distributed Computing, vol. 73, no. 9, pp. 1306–1322, 2013.

38. J. Xu, A. Lam, and V. Li, "Chemical reaction optimization for the grid scheduling problem," inProceedings of the IEEE International Conference on Communications (ICC '10), pp. 1–5, Cape Town, South Africa, May 2010.

39. B. Varghese, G. Mckee, and V. Alexandrov, "Can agent intelligence be used to achieve fault tolerant parallel computing systems?" Parallel Processing Letters, vol. 21, no. 4, pp. 379–396, 2011.

40. J. Xu, A. Lam, and V. Li, "Chemical reaction optimization for task scheduling in grid computing," IEEE Transactions on Parallel and Distributed Systems, vol. 22, no. 10, pp. 1624–1631, 2011.

41. T. K. Truong, K. Li, and Y. Xu, "Chemical reaction optimization with greedy strategy for the 0-1 knapsack problem," Applied Soft Computing Journal, vol. 13, no. 4, pp. 1774–1780, 2013.

42. V. A. F. Almeida, I. M. M. Vasconcelos, J. N. C. Arabe, and D. A. Menasce, "Using random task graphs to investigate the potential benefits of heterogeneity in parallel systems," in Proceedings of the ACM/IEEE Conference on Supercomputing (Supercomputing '92), pp. 683–691, IEEE Computer Society Press, Los Alamitos, Calif, USA, 1992.

Chapter 11

THE SMALLEST CHEMICAL REACTION SYSTEM WITH BISTABILITY

Thomas Wilhelm

Theoretical Systems Biology, Institute of Food Research, Norwich Research Park, Colney Lane

ABSTRACT

Background

Bistability underlies basic biological phenomena, such as cell division, differentiation, cancer onset, and apoptosis. So far biologists identified two necessary conditions for bistability: positive feedback and ultrasensitivity.

Results

Biological systems are based upon elementary mono- and bimolecular chemical reactions. In order to definitely clarify all necessary conditions for bistability we here present the corresponding minimal system. According to our definition, it contains the minimal number of (i) reactants, (ii) reactions, and (iii) terms in the corresponding ordinary differential equations (decreasing importance from i-iii). The minimal bistable system contains two reactants and four irreversible reactions (three bimolecular, one monomolecular).

We discuss the roles of the reactions with respect to the necessary conditions for bistability: two reactions comprise the positive feedback loop, a third reaction filters out small stimuli thus enabling a stable 'off' state, and the fourth reaction prevents explosions. We argue that prevention of explosion is a third general necessary condition for bistability, which is so far lacking discussion in the literature.

Moreover, in addition to proving that in two-component systems three steady states are necessary for bistability (five for tristability, etc.), we also present a simple general method to design such systems: one just needs one production and three different degradation mechanisms (one production, five degradations for tristability, etc.). This helps modelling multistable systems and it is important for corresponding synthetic biology projects.

Conclusion

The presented minimal bistable system finally clarifies the often discussed question for the necessary conditions for bistability. The three necessary conditions are: positive feedback, a mechanism to filter out small stimuli and a mechanism to prevent explosions. This is important for modelling bistability with simple systems and for synthetically designing new bistable systems. Our simple model system is also well suited for corresponding teaching purposes.

BACKGROUND

Bistability is key for understanding basic phenomena of cellular functioning, such as decision-making processes in cell cycle progression, cell differentiation, and apoptosis [1]. It is also involved in loss of cellular homeostasis associated with early events in cancer onset [2] and in prion diseases [3]. A recent review discussed different bistability phenomena in bacteria, such as different phenotypes in clonal populations being important for the origin of new species [4].

Bistable switches are typically enabled by positive feedback loops in signal transduction networks. Here a sufficiently strong (external) signal switches on a self-amplifying process leading to expression of the corresponding target genes. Due to the corresponding hysteresis effect this process can retain its activity without a persistent signal. Such switches are therefore called 'decision-making'. One often discussed example is the restriction point control for the regulation of G1-S transition of the mammalian cell cycle, where recently a detailed small ordinary differential equation (ODE) model was presented [5]. The G2-M transition was also described as a toggle-switch [6]. Oocyte maturation is an example for the involvement of a bistable system in cell differentiation [7, 8]. Biochemical switches have also been found in nutrient utilization in bacteria [9], mating response in yeast [10], and synaptic memory processing [11]. Interestingly, just quite small parts of the large signal transduction systems (for instance a single layer of a MAPK cascade) can already induce bistability. This was demonstrated for the epidermal growth factor receptor system [12] and other kinase phosphatase systems [13, 14].

Given its outstanding biological importance, it is clear that bistable switches also attracted the attention of theoretical biologists. A frequently discussed problem is the necessary (and/or sufficient) condition for bistability. The central result goes back to the work of Clarke [15] and Thomas [16]: autonomous differential systems can possess multiple steady states only under the presence of positive feedback loops [17, 18]. It was also argued that the feedback needs 'some type of non-linearity' or 'ultrasensitivity' for inducing bistability [19]. Different types of ultrasensitivity have been discussed [20], but formulated in the most general manner that means the system needs some mechanism for filtering out small stimuli to enable a stable 'off' state [20–22]. Feinberg's chemical reaction network theory (CRNT) even gives necessary and sufficient conditions for bistability, by restricting to special mass-action kinetic (MAK) systems [23]. In a recent application example a single layer of a MAPK cascade was studied and the region in parameter space being relevant for bistability was analytically described [24]. However, for larger systems CRNT leads to cumbersome calculations, but the applicability could recently be improved by just studying important subnetworks [25] which are based on the concept of elementary flux modes [26].

Another approach for identifying necessary structural conditions for any dynamic behaviour is the identification of the corresponding minimal systems. Such systems have the advantage of being "simple enough to understand at an intuitive level" [19] and are well suited for different basic studies. For instance, the Lotka-Volterra system [27, 28], the Higgins-Selkov-oscillator [29–31], and the "Brusselator" [32] have been studied extensively. Some years ago we identified the smallest chemical system with Hopf bifurcation [33]. Minimal MAK systems are summarized in Table 1. Recently, different 'smallest' or 'minimal' bistable systems for cell polarity [34] and G protein signalling [35] have been presented, as well as 'the smallest multistationary mass-preserving chemical reaction network' [36]. However, these systems are still too large to represent a minimal bistable system according to definition (1). Many different reaction topologies with 3 and 4 molecules have also been analysed computationally for the possibility of bistability [37]. Although this type of bistability detection may miss some bistable systems, the authors found nevertheless many topologies with switching behaviour (10% of tested configurations). The identified 'minimal' system contained 3 variables (5 reactants, 2 conservation relations) and 6 reactions. The bistable one-dimensional Schloegl system [38] contains trimolecular reactions and can therefore not represent a realistic elementary chemical system. Elementary chemical reactions are at most bimolecular.

Table 1 Distinguished minimal MAK systems

System	Reaction scheme	MAK model (ODEs)	Ref.
Minimal bi-stable MAK system	$S + 2X \xrightarrow{1} 3X$ $3X \xrightarrow{2} 2X + P$ $X \xrightarrow{3} P$	$\dot{x} = k_1 x^2 - k_2 x^3 - k_3 x$	[38]
Minimal bi-stable chemi-cal system	$S + Y \xrightarrow{1} 2X$ $2X \xrightarrow{2} X + Y$ $X + Y \xrightarrow{3} Y + P$ $X \xrightarrow{4} P$	$\dot{x} = 2k_1 y - k_2 x^2 - k_3 xy - k_4 x$ $\dot{y} = k_2 x^2 - k_1 y$	This paper
Minimal oscillating MAK system	$S + X \xrightarrow{1} 2X$ $X + Y \xrightarrow{2} 2Y$ $Y \xrightarrow{3} P$	$\dot{x} = k_1 x - k_2 xy$ $\dot{y} = k_2 xy - k_3 y$	[27, 28]
Minimal MAK system with limit cycle	$S \xrightarrow{1} X$ $X + 2Y \xrightarrow{2} 3Y$ $Y \xrightarrow{3} P$	$\dot{x} = v_1 - k_2 xy^2$ $\dot{y} = k_2 xy^2 - k_3 y$	[29–31]
Minimal chemical sys-tem with limit cycle	$S + X \xrightarrow{1} 2X$ $X + Y \xrightarrow{2} Y + P$ $Y \xrightarrow{3} P$ $X \xrightarrow{4} Z$ $Z \xrightarrow{5} Y$	$\dot{x} = (k_1 - k_4)x - k_2 xy$ $\dot{y} = k_5 z - k_3 y$ $\dot{z} = k_4 x - k_5 z$	[33]

S and P denote constant substrates and products.

Here we present and discuss the smallest bistable chemical reaction system. Application of our previously presented Instability Causing Structure Analysis (ICSA [39]) leads to additional insight into system functioning.

RESULTS

The smallest bistable chemical reaction system

We define the smallest chemical system (contains only mono- and bimolecular reactions, reversible reactions are considered as two irreversible ones) by the following criteria in decreasing order of importance:

1. Minimal number of reactants
2. Minimal number of reactions
3. Minimal number of terms in the ODEs

(1)

According to this definition, the following bistable system is unique (Methods section contains the proof for this statement).

$$S + Y \xrightarrow{\ 1\ } 2X$$

$$2X \xrightarrow{\ 2\ } X + Y$$

$$X + Y \xrightarrow{\ 3\ } Y + P$$

$$X \xrightarrow{\ 4\ } P \tag{2a}$$

Assuming spatially homogeneous conditions, the system can be described by the two-component mass-action kinetic ODE system (S is incorporated into k_1):

$$\dot{x} = 2k_1 y - k_2 x^2 - k_3 xy - k_4 x$$
$$\dot{y} = k_2 x^2 - k_1 y \tag{2b}$$

Due to its simplicity, the mathematical analysis of the system is simple as well. The system has two elementary flux modes [26], following directly from the two nullspace vectors of the corresponding stoichiometric matrix

$$\mathbf{S} = \begin{pmatrix} 2 & -1 & -1 & -1 \\ -1 & 1 & 0 & 0 \end{pmatrix}. \tag{2c}$$

One mode uses reactions 1-3, and the other reactions 1,2, and 4. Bistability can, of course, only arise if all reactions are active. Introducing dimensionless quantities ($x/c \rightarrow x, y/c \rightarrow y, k_1/(k_2 c) \rightarrow k_1, k_3/k_2 \rightarrow k_3, k_4/(k_2 c) \rightarrow k_4, tk_2 c \rightarrow t$), we set $k_2 = 1$, without restriction of generality ($k_1, k_3, k_4 > 0$). The system has three steady states: $\bar{x}_1 = \bar{y}_1 = 0, \bar{x}_2 = (k_1 - \sqrt{k_1 D})/(2k_3), \bar{y}_2 = \bar{x}_2^2/k_1$ $\bar{x}_3 = (k_1 + \sqrt{k_1 D})/(2k_3), \bar{y}_3 = \bar{x}_3^2/k_1$ with the discriminant $D = k_1 - 4k_3 k_4$. A saddle-node bifurcation occurs at $D = 0$, the three steady states are real if $D > 0$. The second and third steady state is always positive.

Generally, in two-component systems a steady state is locally stable if the trace tr and determinant det of the Jacobian at this point are negative and positive, respectively (node if $4det < tr^2$, focus otherwise). If the corresponding determinant is negative, the steady state is a saddle point. It can be seen from the Jacobian $\begin{pmatrix} -k_4 - 2x - k_3 y & 2k_1 - k_3 x \\ 2x & -k_1 \end{pmatrix}$ that its trace is always negative (phase flow of system (2) is confined to the positive part of the phase space). This excludes Hopf bifurcations (arising at $tr = 0$) and it means the system is dissipative, i.e. phase-space contracting all over the phase space (trace = two-dimensional Lyapunov exponent). The determinants of the Jacobian at the three steady states read $k_1 k_4$, $k_1/(2k_3)(D - \sqrt{k_1 D})$, and $k_1/(2k_3)(D + \sqrt{k_1 D})$, respectively. Therefore, the first and third steady state are always locally stable, the second is always locally unstable, a saddle-point (determinant of Jacobian at second steady state always negative, cf. point 3 in Methods). Simple calculation shows further that $4det < tr^2$ at the first and third steady state, so these are always stable nodes.

For $k_1 = 8$, $k_3 = 1$, $k_4 = 1.5$ the second and third steady state are $\bar{x}_2 = 2, \bar{y}_2 = 1/2$ and $\bar{x}_3 = 6, \bar{y}_3 = 9/2$ respectively. Figure1 shows a corresponding signal-response curve, also called bifurcation diagram [6]. The signal is the concentration of the constant outer substance S (assuming for the bimolecular rate constant $k_{bi} = 1$, the concentration S is identical to the apparent rate constant $k_1 = k_{bi} S$), the response is the steady state concentration of an internal reactant, here X. The saddle-node bifurcation occurs at $S = 3/4$. Beyond that point the system has two stable steady states. It follows from $\bar{x}_2 = (k_1 - \sqrt{k_1 D})/(2k_3)$ that $\bar{x}_2 \xrightarrow[k_1 \to \infty]{} 0$ (for fixed other parameters), so the real dynamic behaviour of the system is that of a toggle-switch (Figure 1): for sufficiently large k_1 small fluctuations in the concentrations would drive the system to the positive steady state (the 'on' state).

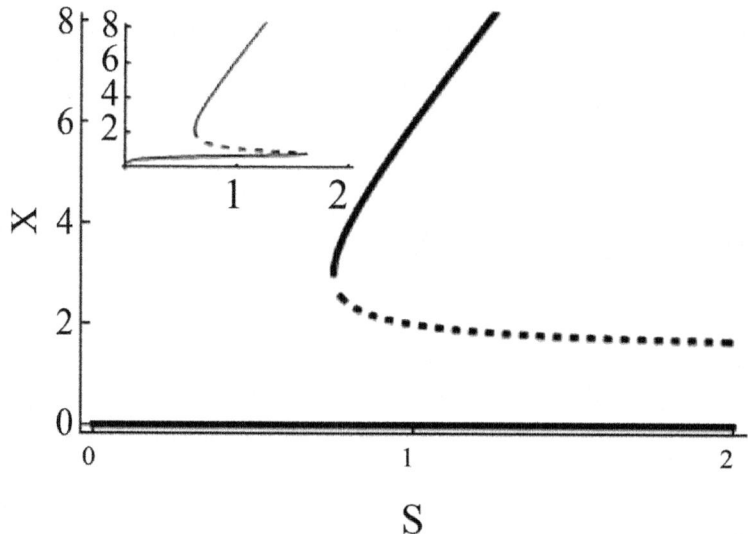

Figure 1: Signal-response curve (bifurcation diagram) of system (2) for the parameters $k_1 = 8$, $k_2 = 1$, $k_3 = 1$, $k_4 = 1.5$. Solid lines indicate locally stable steady states, the dashed line locally unstable steady states. The inset shows the signal-response curve if an additional small constant influx into X (here 0.6) is assumed (enabling a positive 'off' state, leaving the 'on' state and bifurcation point nearly unchanged). This is the classical toggle switch (terminology of Tyson et al. (6), others use the term toggle switch to describe a double negative (i.e. positive) feedback loop (4)) picture enabling the hysteresis cycle: starting with low values and increasing the signal continuously increases the response, until the saddle-node bifurcation at about S = 1.7 is reached. Further increase of the signal leads to a sudden jump of the response to the upper steady state. Decreasing the signal now leads to a continuous decrease of the response, the systems stays in the upper steady state until the left bifurcation point is reached where the response jumps back to the lower steady state.

Figure 2 shows rate curves of system (2). It can be seen that the three crossings of production and degradation rate (i.e. the three steady states) are due to the different contributions of the three degradation terms. This implies a simple general procedure for designing bi- or multistable systems: a bistable system can be created with one function for production and three different functions for degradation, e.g. a linear, a quadratic, and a cubic one as in our simple example system. Accordingly, tristable systems require 5 different functions to enable 5 crossings (three stable and two unstable steady states, cf. point 3 in Methods), and so forth for more steady states. This observation helps constructing minimal and/or realistic models of more complicated multistable systems. It can also be a starting point for the design of real bistable systems, for instance in synthetic biology. Note that all enzyme kinetic rate laws can be modelled with polynomial ODEs [40].

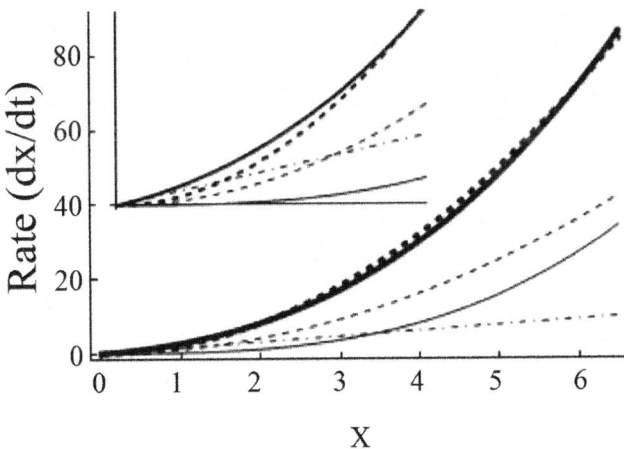

Figure 2: Rate curves [6] of system (2) for the parameters $k_1 = 8$, $k_2 = 1$, $k_3 = 1$, $k_4 = 1.5$. The thick solid line is the rate of the removal of reactant X (sum of the negative terms in $\dot{x}$) and the thick dashed line the rate of production (positive term in $\dot{x}$). The three crossings indicate the three steady states $(\bar{x}_1, \bar{x}_2, \bar{x}_3) = (0, 2, 6)$. The thin lines show the contributions of the three degradation terms separately: quadratic term $k_2 x^2$ dashed, the effectively cubic term $k_3 xy$ solid, and the linear term $k_4 x$ dotdashed. The inset shows a zoomed version for $x < 2.1$.

The Instability Causing Structure Analysis (ICSA) of system (2)

Recently we presented a new method for topological network analysis of dynamical systems, the *Instability Causing Structure Analysis* (ICSA [39]). Standard stoichiometric network analyses (such as elementary flux mode calculations [26]) are based on the assumption of steady states and lead to linear constraints in flux space. ICSA is a nonlinear network analysis. It is based on the assumption of locally stable steady states. The additional demand for local stability yields additional nonlinear constraints for the steady state flux space [39].

ICSA needs no knowledge about kinetic details. It can be applied to (bio) chemical systems where just the stoichiometric matrix is known (or even just the signs of its elements) or to signal transduction/gene regulatory networks represented by interaction graphs [18] (also called incidence graph [21] or causal influence graph [41]). ICSA leads to additional insight into system functioning by identifying all contained feedback loops and all the corresponding instability causing structures (ICS). An ICS is either a single feedback loop or a special combination of feedback loops [39]. ICSA yields a necessary condition for local instability of any steady state of the system: if there is no ICS all potential steady states are locally stable. For two-dimensional systems this also implies that the system has just one steady state (cf. point 3 in Methods).

We apply ICSA for additional analysis of system (2). The stoichiometric matrix $\mathbf{S}$ is given in (2c). Multiplication of $\mathbf{S}$ with the reaction velocity substrates ector (contains the substrates for each reaction) $(v_1 (y) v_2 (x) v_3 (x,$

$$\begin{pmatrix} 2v_1(y) - v_2(x) - v_3(x,y) - v_4(x) \\ -v_1(y) + v_2(x) \end{pmatrix}$$

$y) v_4 (x))^T$ leads to . Differentiation yields the general Jacobian $\mathbf{J}_G$ (39) of system (2):

$$\mathbf{J_G} = \begin{pmatrix} -v_{2x} - v_{3x} - v_{4x} & 2v_{1y} - v_{3y} \\ v_{2x} & -v_{1y} \end{pmatrix},$$

$$(3)$$

where the indices x and y denote the corresponding partial differentiation. The off-diagonal elements represent the fundamental activating and inhibiting interactions in the system: the positive v_{2x} in $\mathbf{J}_{G21}$ shows that x activates y by the second reaction, and equivalently for the two terms in $\mathbf{J}_{G12}$: $v_{1y} \rightarrow$ y activates x by the first reaction, - $v_{3y} \rightarrow$ y inhibits x by the third reaction. Figure 3 shows the corresponding interaction (incidence) graph summarizing these interactions. Interaction graphs can often be found in the biological literature and corresponding databases (KEGG [42]; BIOBASE [43,44]; Dynamic Signaling Maps http://vivo.library. cornell.edu/lifesci/individual/ vivo/individual5093. ICSA [39] was developed for structural analyses of (bio) chemical systems (KEGG [42]; BRENDA [45]) AND such interaction graphs.

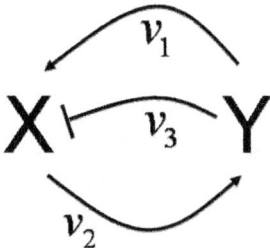

Figure 3: Interaction graph of system (2). It follows directly from the off-diagonal elements of the general Jacobian (3). The positive feedback loop is the only instability causing structure (ICS) in the system, allowing for a locally unstable steady state (presupposition for bistability).

The system contains one positive and one negative feedback loop. The positive loop is the necessary structural condition for bistability [18]. The second often discussed ingredient for bistability is any mechanism for filtering out small stimuli to make the off-state stable. Ferrell and Xiong [20] discussed different such mechanisms, as ultrasensitivity and back reaction saturation. Our system has the simplest mechanism to stabilize the off-state: the monomolecular efflux reaction 4. Analysis shows that without this reaction the second (unstable) steady state merges with the zero off-state making it (weakly) unstable. In fact, Figure 2 (especially the inset) shows that it is the linear degradation term that makes the overall degradation rate higher then production for low X concentrations.

System (2) is the minimal chemical reaction system with bistability. Therefore, any ingredient is essential. That also means, without the negative loop (without reaction 3 being catalyzed by Y) this system cannot be bistable. In fact, the negative feedback prevents explosion of the system: without reaction 3 the system has just one locally stable off-state and one unstable positive steady state. Figure 2 shows that without the cubic term there are just two crossings of the rate curves.

Summarizing, system (2) contains the three different necessary conditions for bistability: (i) a positive feedback loop, (ii) a mechanism for filtering out small stimuli, and (iii) a mechanism for preventing explosion. Interestingly, the third condition lacks discussion in the literature [19, 20, 22]. Moreover, these three conditions can be related to the three degradation terms in the balance equation of X and so to the last three of the four reactions in (2). The first reaction represents the production of the system, it is the only input. The second reaction closes the positive feedback cycle. Reactions three and four are degradation reactions. The third is the main degradation for higher concentrations, it prevents explosion. The fourth reaction filters out the small stimuli.

The first step in the general ICSA procedure is the analysis of feedback cycles resulting from the off-diagonal terms of the general Jacobian. The second step is the identification of the topological structures that actually cause instability. As mentioned above, standard local stability analysis of two-component systems needs consideration of the Jacobian's trace and determinant *det*. In ICSA we study the general Jacobian. It follows from (3) that the trace is always negative and $det = v_{2x} v_{3y} + v_{1y} (v_{3x} + v_{4x} - v_{2x})$. The only negative term in *det* comprises reactions 1 and 2, so this positive feedback cycle is the only ICS of the system. The negative feedback cycle cannot cause any instability here.

DISCUSSION

The minimal bistable chemical reaction system (2) is based on definition (1). This definition is based on a chemical/physical point of view, but other definitions might also be possible. The definition for the smallest chemical reaction system with Hopf bifurcation [33], for instance, was more mathematically motivated: here minimal number of quadratic terms in the ODEs had a higher priority than minimal number of reactions. So far it has not been clear whether any at most bimolecular 3-variable system with only four (irreversible) reactions and Hopf bifurcation exists. Meanwhile we found a corresponding system (

$$\dot{x} = -2k_2x^2 + k_3yz, \quad \dot{y} = k_1y - k_3yz, \quad \dot{z} = k_2x^2 - 2k_4z^2$$

) which has a supercritical Hopf bifurcation (e.g. at $k_1 = k_2 = 1,$ $k_3 = 2^{5/2}\sqrt{k_4} + 4k_4$). This might be the minimal chemical reaction system with sustained oscillations according to definition (1).

Interestingly, system (2) is similar to a previously presented "minimal reaction network" [12] modelling activation of the epidermal growth factor receptor (EGFR). Our $\dot{x}$ equation resembles the balance equation of the phosphorylated receptor tyrosine kinase (this superfamily contains EGFR) and $\dot{y}$ is similar to the differential equation for active protein tyrosine phosphatase.

Bistable systems play important roles also beyond biology. They are usually depicted by the mechanical example of a ball rolling into two different valley basins. Bistable chemical systems, in particular, have been studied extensively to analyse relaxation kinetics [46], non-equilibrium thermodynamics [47], stochastic resonance [48], as well as climate change [49].

Positive feedback is clearly associated with bi/multistability. Negative feedback, in contrast, is often discussed in the context of oscillations [6], and we recently conjectured that this is indeed a necessary condition for (sustained) oscillations [39]. However, this contradicts statements as "sustained oscillations can occur in models based on positive or negative feedback" [50]. Obviously, different understandings of feedback exist. We suggest to use the general Jacobian $\mathbf{J}_G$ (for system (2) it is given in (3)) for a simple definition of feedback: if the $\mathbf{J}_G$ terms close any cycles, then feedback exists (e.g. the positive and negative feedback cycles of system (2)), otherwise not. Analysis of $\mathbf{J}_G$ guarantees a unique identification of all in a system contained feedback cycles. Goldbeter [50] mentioned different examples where the oscillations should be based on a positive feedback, such as glycolytic, Ca^{2+}, and cAMP oscillations. However, a more detailed analysis of these systems shows that a negative feedback (according to our definition) is always contained (results unpublished). An example is the simplest model for glycolytic oscillations, the Higgins-Selkov oscillator ([29–31], Table 1): the corresponding general Jacobian is $\mathbf{J}_G = \begin{pmatrix} -v_{2x} & -v_{2y} \\ v_{2x} & v_{2y} - v_{3y} \end{pmatrix}$. Its trace and determinant are $v_{2y} - v_{2x} - v_{3y}$ and $v_{2x}v_{3y}$, respectively. Obviously, v_{2y} in $\mathbf{J}_{G22}$ is the only instability causing term (a positive feedback), so the second reaction is the only ICS in the system. However, inspection of the off-diagonal elements of $\mathbf{J}_G$ reveals a negative feedback as well: the larger x, the larger becomes y, but the larger y, the smaller becomes x. The same is realised in the bistable system (2), where the positive feedback is the only ICS and another negative feedback is contained. These examples show how the analysis of the general Jacobian helps clarifying the discussion of feedback loops.

We have shown that a mechanism for preventing explosions is a third necessary condition for bistability (complementing the previously discussed two other conditions positive feedback and filtering out of small stimuli). In system (2) this is achieved by a negative feedback. Other bistable systems contain negative feedbacks as well (e.g. ERK pathway [2]), so we hypothesize that this is indeed a typical feature of bistable systems. Interestingly, also oscillating systems typically contain (besides the necessary negative feedback) a positive feedback (for better tunable frequency, evolvability and robustness [51]). Thus, oscillating and bistable systems are practically based on the same set of feedback cycles.

CONCLUSION

Bi/multistability and oscillations are the two most important dynamic phenomena in biology. Limit cycle oscillations are associated with biological clocks and cell signalling [52], and spatial oscillations with proper cell division [53]. The fundamental importance of bistability is discussed in the introduction. Some years ago we presented the smallest chemical system with limit cycles [33]. Here we have derived the smallest chemical system with bistability (2).

Minimal systems are well suited for basic studies and for teaching purposes. This explains the great success of, for instance, the Lotka-Volterra [27, 28] and the Higgins-Selkov system [30]. We have demonstrated for the minimal chemical reaction system with Hopf bifurcation [33] that it is accessible for detailed mathematic-analytical examination [54] and a good example system for thermodynamic considerations [55]. We hope that also the minimal chemical system with bistability will serve for such purposes in the future.

METHODS

System (2) is the smallest bistable chemical reaction system according to definition (1) - proof (inductive proof systematically considering all possibilities):

1. The Schloegl system (Table 1) is the smallest bistable one-variable (1d) system: 1d systems need an unstable steady state to separate the attractor regions of two stable steady states, so we need at least three steady states to realize a bistable system. The simplest function (which is realizable as a MAK system) $f(x)$ with three zeros is the cubic polynomial. To realize a stable 'on' state the sign of the cubic term needs to be "-" ($\dot{x} = f(x) = \dots - ax^3, a > 0$). For three different non-negative steady states we also need a positive quadratic and a negative linear term: $f(x) = -a\,x^3 + b\,x^2 - c\,x + d$ (a, b, c>0) possesses two positive extrema as can simply be seen considering $f'(x) = 0$. Thus, the minimal bistable 1d system reads $\dot{x} = -ax^3 + bx^2 - cx$ (a, b, c > 0) . The minimal corresponding MAK system is the Schloegl system shown in Table 1 (a reversible monomolecular efflux reaction (d>0) allows for two positive

stable steady states). It follows that a chemically realistic bistable system with only mono- and bimolecular reactions needs at least two variables.

2. Using our general quasi-steady-state-approximation procedure [56], we have previously shown that any irreversible trimolecular reaction can be understood as limit case of a reversible bimolecular reaction and another irreversible bimolecular reaction by introducing one additional intermediate [57]. Transforming the Schloegl system accordingly proves that at most bimolecular bistable 2d systems exist, i.e. the number of variable reactants in the minimal bistable chemical system is fixed to two (cf. definition (1)).

3. The lemma of the index sum [58] states that the sum of indices of all steady states within a two-dimensional confined set (closed region in phase space where all trajectories point inwards [59]) equals one. The index values of a node, a focus, and a saddle are +1, +1, and -1, respectively (for stable and unstable nodes and foci [58]). Our minimal system should therefore contain one unstable (saddle-point) and two stable steady states (we are only considering non-exploding systems, such that a confined set could simply be constructed, trajectories point inward at the boundary of the positive orthant anyway). To get three steady states we need at least a cubic steady state equation $f(x) = 0$, i.e. an x^3 term (higher order polynomials would require more bimolecular reactions). In 2d MAK systems this can only be realized by one ODE with an xy term and the other with x^2 and y terms. Inserting the corresponding steady state expression $y = x^2...$, into the other ODE's xy term gives the cubic term. A direct x^3 term is forbidden in bimolecular systems. The symmetric case y^2 and x needs no extra consideration. These terms already correspond to at least three reactions. One can show that such three reactions are not sufficient to give a bistable system (the explicit proof is not necessary, because it turns out indirectly from the following analysis). So the minimal bistable system has at least four reactions.

4. To realize a $-x^3$ term in the corresponding steady state equation, one either needs the terms y and x^2 with different signs and -xy in the other ODE, or y and x^2 with the same signs and +xy in the other ODE. But the same signs variant cannot work: $\dot{x} = -x^2 - y$ is not chemical, $\dot{x} = x^2 + y$ would contain at least one trimolecular reaction, $\dot{y} = -x^2 - y$ is not chemical, and $\dot{y} = x^2 + y$ is also impossible. The latter applies because: (i) the $\dot{y}$ equation must also contain a term -xy (the corresponding reaction cannot be bimolecular

otherwise: a term $+xy$ in $\dot{x}$ implies the bimolecular reaction X+Y->2X, implying the term $-xy$ in $\dot{y}$), (ii) detailed analysis of the system $\dot{x} = -(2)k_1x^2 + k_2xy, \quad \dot{y} = (2)k_1x^2 - k_2xy + k_3y$ shows that the cubic term in the steady state equation always vanishes, independent of any other added mono- and/or bimolecular reactions. So the y and x^2 terms must have different signs in one ODE.

5. There are two variants for the different-sign-case: 1. $\dot{x} = y - x^2, \dot{y} = -xy$
2. $\dot{x} = -x^2 - xy, \dot{y} = x^2 - y$ (note that the x^2 term has to appear also in the $\dot{x}$ equation to be interpretable as bimolecular reaction). We show that no bistable system with only four reactions exists that corresponds to the first case: to realize a positive quadratic term in the cubic polynomial of the steady state equation (cf. discussion concerning the Schloegl system in 1.) $\dot{y} = 0$ we have three options: (i) $\dot{y} = -xy + x^2$, (ii) $\dot{y} = -xy + y$, (iii) $\dot{x} = y - x^2 + x, \dot{y} = -xy$. The second case could only be realized by 4 different reactions and would need a fifth reaction to yield a linear term in the cubic polynomial (cf. discussion concerning the Schloegl system in 1.). The same holds for the third case, even if another x^2 term (still realizable by four reactions) would be added in the $\dot{y}$ equation. The first case is realizable by three reactions. However, it is easy to see that a linear term in the steady state equation $\dot{y} = 0$ needs at least two additional reactions.

6. However, the second system $\dot{x} = -x^2 - xy, \dot{y} = x^2 - y$ can yield a bistable at most bimolecular system with only 4 reactions: The system $\dot{x} = -x^2 - xy, \dot{y} = x^2 - y$ is realizable by three reactions, we add y as the positive term in $\dot{x}$ (still three reactions) and a fourth reaction, a simple efflux of x to ensure a linear term in the cubic polynomial (to yield three steady states): $\dot{x} = -x^2 - xy + y - x, \dot{y} = x^2 - y$. Analysis shows that the reaction from y to x needs to be bimolecular (reaction 1 in (2a)) to enable three nonnegative steady states. In fact, this is the only input into the system. Note that this system is quite unique, there is no really different bistable chemical system with only four reactions: the positive term in $\dot{x}$ has to be y to enable a positive quadratic term in the steady state equation $\dot{x} = 0$. System (2) is unique with respect to definition (1). The very similar system where reaction $2X$

$\rightarrow X + Y$ is replaced with $2X \rightarrow 2Y$ is mathematically equivalent (factor 2 can be incorporated into the kinetic constant). $\square$

The slightly modified system replacing reaction $X + Y \rightarrow Y + P$ with $X + Y \rightarrow P$ is bistable as well (it has also just four reactions, but is not as minimal as system (2) for the third criterion of definition (1). Modifying further by replacing $2X \rightarrow X + Y$ with $2X \rightarrow Y$ and $S + Y \rightarrow 2X$ with $S + Y \rightarrow 3X$ gives another bistable system resembling a system which can be derived from the Schloegl-system using our previously discussed general transformation rules [57].

DECLARATIONS

Acknowledgements

I thank Klaus R. Schneider for referring to the lemma of the index sum. The work was supported by a BBSRC Core Strategic Grant for IFR.

REFERENCES

1. Eissing T, Conzelmann H, Gilles ED, Allgoewer F, Bullinger E, Scheurich P: Bistability analyses of a caspase activation model for receptor-induced apoptosis. *J Biol Chem* 2004, 279:36892–36897.

2. Kim D, Rath O, Kolch W, Cho K-H: A hidden oncogenic positive feedback loop caused by crosstalk between Wnt and ERK pathways. *Oncogene* 2007, 26:4571–4579.

3. Kellershohn N, Laurent M: Prion diseases: dynamics of the infection and properties of the bistable transition. *Biophys J* 2001,81:2517–2529.

4. Veening J-W, Smiths WK, Kuipers OP: Bistability, epigenetics, and bethedging in bacteria. *Annual Rev Microbiol* 2008, 62:193–210.

5. Yao G, Lee TJ, Mori S, Nevins JR, You L: A bistable Rb-E2F switch underlies the restriction point. *Nature Cell Biol* 2008, 10:476–482.

6. Tyson JJ, Chen KC, Novak B: Sniffers, buzzers, toggles and blinkers: dynamics of regulatory and signalling pathways in the cell. *Curr Opin Cell Biol* 2003, 15:221–231.

7. Ferrell JE, Machleder EM: The biochemical basis of an all-or-none cell fate switch in Xenopus oocytes. *Science* 1998, 280:895–898.

8. Xiong W, Ferrell JE Jr: A positive-feedback-based bistable 'memory module' that governs a cell fate decision. *Nature* 2003, 426:460–465.

9. Ozbudak EM, Thattai M, Lim HN, Shraiman BI, Van Oudenaarden A: Multistability in the lactose utilization network of Escherichia coli. *Nature* 2004, 427:737–740.

10. Paliwal S, Iglesias PA, Campbell K, Hilioti Z, Groisman A, Levchenko A: MAPK-mediated bimodal gene expression and adaptive gradient sensing in yeast. *Nature* 2007, 446:46–51.

11. Miller P, Zhabotinsky AM, Lisman JE, Wang X-J: The stability of a stochastic CaMKII switch: dependence on the number of enzyme molecules and protein turnover. *PLoS Biol* 2005, 3:e107.

12. Reynolds AR, Tischer C, Verveer PJ, Rocks O, Bastiaens PIH: EGFR activation coupled to inhibition of tyrosine phosphatases causes lateral signal propagation. *Nature Cell Biol* 2003, 5:447–453.

13. Bhalla US, Ram PT, Iyengar R: MAP kinase phosphotase as a locus of flexibility in a mitogen-activated protein kinase signalling network. *Science* 2002, 297:1018–1023.

14. Markevich NI, Hoek JB, Kholodenko BN: Signalling switches and bistability arising from multisite phosphorylation in protein kinase cascades. *J Cell Biol* 2004, 164:353–359.

15. Clarke BL: Stability of complex reaction networks. *Adv Chem Phys* 1980, 43:1–216.

16. Thomas R: The role of feedback circuits: Positive feedbacl circuits are a necessary condition for positive real eigenvalues of the Jacobian matrix. *Ber Bunsenges Phys Chem* 1994, 98:1148–1151.

17. Cinquin O, Demongeot J: Positive and negative feedback: Striking a balance between necessary antagonists. *J Theor Biol* 2002,216:229–241.

18. Soulé C: Graphic requirements for multistationarity. *ComPlexUs* 2003, 1:123–133.

19. Ferrell JE Jr: Self-perpetuating states in signal transduction: positive feedback, double-negative feedback and bistability. *Curr Opin Chem Biol* 2002, 6:140–148.

20. Ferrell JE Jr, Xiong W: Bistability in cell signalling: How to make continuous processes discontinuous, and reversible processes irreversible. *Chaos* 2001, 11:227–236.

21. Angeli D, Ferrell JE Jr, Sontag ED: Detection of multistability, bifurcations, and hysteresis in a large class of biological positive-feedback systems. *Proc Natl Acad Sci USA* 2004, 101:1822–1827.

22. Eissing T, Waldherr S, Allgoewer F, Scheurich P, Bullinger E: Steady state and (bi-)stability evaluation of simple protease signalling networks. *Biosystems* 2007, 90:591–601.

23. Craciun G, Tang Y, Feinberg M: Understanding bistability in complex enzyme-driven reaction networks. *Proc Natl Acad Sci USA* 2006,103:8697–8702.

24. Conradi C, Flockerzi D, Raisch J: Multistationarity in the activation of a MAPK: Parametrizing the relevant region in parameter space.*Math Biosci* 2008, 211:105–131.

25. Conradi C, Flockerzi D, Raisch J, Stelling J: Subnetwork analysis reveals dynamic features of complex (bio)chemical networks. *Proc Natl Acad Sci USA* 2007, 104:19175–19180.

26. Schuster S, von Kamp A, Pachkov M: Understanding the roadmap of metabolism by pathway analysis. *Methods Mol Biol* 2007,358:199–226.

27. Lotka AJ: Undamped oscillations derived from the law of mass action. *J Am Chem Soc* 1920, 42:1595–1599.

28. Volterra V: Fluctuations in the abundance of a species considered mathematically. *Nature* 1926, 118:558–560.

29. Higgins J: The theory of oscillating reactions. *Ind Eng Chem* 1967, 59:18–62.

30. Selkov EE: Self-oscillations in glycolysis. 1. A simple kinetic model. *Eur J Biochem* 1968, 4:79–86.

31. Schnakenberg J: Simple chemical reaction systems with limit cycle behaviour. *J Theor Biol* 1979, 81:389–400.

32. Prigogine I, Lefever R: Symmetry breaking instabilities in dissipative systems. II. *J Chem Phys* 1968, 48:1695–1700.

33. Wilhelm T, Heinrich R: Smallest chemical reaction system with Hopf bifurcation. *J Math Chem* 1995, 17:1–14.

34. Mori Y, Jilkine A, Edelstein-Keshet L: Wave-pinning and cell polarity from a bistable reaction-diffusion system. *Biophys J* 2008,94:3684–3697.

35. Csercik D, Hangos KM, Nagy GM: A simple reaction kinetic model of rapid (G protein dependent) and slow (β-Arrestin dependent) transmission. *J Theor Biol* 2008, 255:119–128.

36. Shiu A: The smallest multistationary mass-preserving chemical reaction network. *Lecture Notes Comp Sci* 2008, 5147:172–184.

37. Ramakrishnan N, Bhalla US: Memory switches in chemical reaction space. *PloS Comput Biol* 2008, 4:e1000122.

38. Schloegl F: Chemical reaction models for non-equilibrium phase transitions. *Z Physik* 1972, 253:147–161.

39. Wilhelm T: Analysis of structures causing instabilities. *Phys Rev E* 2007, 76:011911.

40. Wilhelm T, Hoffmann-Klipp E, Heinrich R: An evolutionary approach to enzyme kinetics: optimization of ordered mechanisms. *Bull Math Biol* 1994, 56:65–106.

41. Klamt S, Saez-Rodriguez J, Lindquist JA, Simeoni L, Gilles ED: A methodology for the structural and functional analysis of signalling and regulatory networks. *BMC Bioinformatics* 2006, 7:56.

42. Kanehisa M, Goto S, Hattori M, Aoki-Kinoshita KF, Itoh M, Kawashima S, Katayama T, Araki M, Hirakawa M: From genomics to chemical genomics: new developments in KEGG. *Nucleic Acids Res* 2006, 34:D354–357.

43. Krull M, Pistor S, Voss N, Kel A, Reuter I, Kronenberg D, Michael H, Schwarzer K, Potapov A, Choi C, Kel-Margoulis O, Wingender E:TRANSPATH: an information resource for storing and visualizing signalling pathways and their pathological aberrations. *Nucleic Acids Res* 2006, 34:D546–551.

44. Matys V, Kel-Margoulis OV, Fricke E, Liebich I, Land S, Barre-Dirrie A, Reuter I, Chekmenev D, Krull M, Hornischer K, Voss N, Stegmaier P, Lewicki-Potapov B, Saxel H, Kel AE, Wingender E: TRANSFAC and its module TRANSCompel: transcriptional gene regulation in eukaryotes. *Nucleic Acids Res* 2006, 34:D108–110.

45. Schomburg I, Chang A, Ebeling C, Gremse M, Heldt C, Huhn G, Schomburg D: BRENDA, the enzyme database: updates and major new developments. *Nucleic Acids Res* 2004, 32:D431–433.

46. Laplante JP, Borckmans P, Dewel G, Gimenez M, Micheau JC: Relaxation kinetics near the hysteresis limit of a bistable chemical system: The chlorite-iodide reaction in a CSTR. *J Phys Chem* 1987, 91:3401–3405.

47. Dutt AK: Non-equilibrium thermodynamics of a model bistable chemical system. *Chem Phys Lett* 2000, 322:73–77.

48. Strizhak PE, Demjanchyk I, Fecher F, Schneider FW, Muenster AF: Stochastic Resonance in a bistable chemical system: The oxidation of ascorbic acid by oxygen catalyzed by Copper(II) Ions. *Angew Chem Int Ed* 2000, 39:4573–4576.

49. Goldblatt C, Lenton TM, Watson AJ: Bistability of atmospheric oxygen and the Great Oxidation. *Nature* 2006, 443:683–686.

50. Goldbeter A: Computational approaches to cellular rhythms. *Nature* 2002, 420:238–245.

51. Tsai TY-C, Choi YS, Ma W, Pomerening JR, Tang C, Ferrel JE Jr: Robust, tunable biological oscillations from interlinked positive and negative feedback loops. *Science* 2008, 321:126–129.

52. Goldbeter A: *Biochemical oscillations and cellular rhythms: The molecular bases of periodic and chaotic behaviour* Cambridge Univ Press, Cambridge 2008.

53. Kruse K, Howard M, Margolin W: An experimentalist's guide to computational modelling of the Min system. *Mol Microbiol* 2007,63:1279–1284.

54. Wilhelm T, Heinrich R: Mathematical analysis of the smallest chemical reaction system with Hopf bifurcation. *J Math Chem* 1996,19:111–130

55. Wilhelm T, Schuster S, Heinrich R: Kinetic and thermodynamic analyses of the reversible version of the smallest chemical reaction system with Hopf bifurcation. *Nonlinear World* 1997, 4:295–321.

56. Schneider KR, Wilhelm T: Model reduction by extended quasi-steady-state approximation. *J Math Biol* 2000, 40:443–450.

57. Wilhelm T: Chemical systems consisting only of elementary steps - a paradigma for nonlinear behaviour. *J Math Chem* 2000, 27:71–88.

58. Arnold VI: *Ordinary differential equations* MIT Press, Cambridge 1980.

59. Murray JD: *Mathematical Biology* Springer, Berlin 1993.

CITATION

CHAPTER 1

Yang R, Han Y, Ye Y, Liu Y, Jiang Z, Gui Y, et al. (2011) Chemical Synthesis of Bacteriophage G4. PLoS ONE 6(11): e27062. doi:10.1371/journal.pone.0027062

CHAPTER 2

Zhu Q, Qin T, Jiang Y-Y, Ji C, Kong D-X, Ma B-G, et al. (2011) Chemical Basis of Metabolic Network Organization. PLoS Comput Biol 7(10): e1002214. doi:10.1371/journal.pcbi.1002214

CHAPTER 3

Kazuyoshi Tatsumi, Shunsuke Muto, Kazutaka Ikeda and Shin-Ichi Orimo, Chemical Bonding of AlH_3 Hydride by $Al-L^{2,3}$ Electron Energy-Loss Spectra and First-Principles Calculations, doi:10.3390/ma5040566.

CHAPTER 4

P. K. Chattaraj and B. Maiti, Chemical Reactivity Dynamics and Quantum Chaos in Highly Excited Hydrogen Atoms in an External Field: A Quantum Potential Approach, doi:10.3390/i3040338ss

CHAPTER 5

P. Geerlings and F. De Proft, Chemical Reactivity as Described by Quantum Chemical Methods, doi: 10.3390/i3040276

CHAPTER 6

Stefano Zambelli (2012). Chemical Kinetics, an Historical Introduction, Chemical Kinetics, Dr Vivek Patel (Ed.), ISBN: 978-953-51-0132-1, InTech, DOI: 10.5772/37081.

CHAPTER 7

Robin A. Cox, A Greatly Under-Appreciated Fundamental Principle of Physical Organic Chemistry, doi: 10.3390/ijms12128316

CHAPTER 8

Bapuji Pullepu, P. Sambath, and K. K. Viswanathan, "Effects of Chemical Reactions on Unsteady Free Convective and Mass Transfer Flow from a Vertical Cone with Heat Generation/Absorption in the Presence of VWT/VWC," Mathematical Problems in Engineering, vol. 2014, Article ID 849570, 20 pages, 2014. doi:10.1155/2014/849570

CHAPTER 9

Ravi, S., Prakash, J., Gottam, V. and Sibyala, V. (2015) Effects of Thermal Radiation and Radiation Absorption on Flow Past an Impulsively Started Infinite Vertical Plate with Newtonian Heating and Chemical Reaction. Open Journal of Fluid Dynamics, 5, 364-379. doi: 10.4236/ojfd.2015.54036.

CHAPTER 10

Yuyi Jiang, Zhiqing Shao, and Yi Guo, "A DAG Scheduling Scheme on Heterogeneous Computing Systems Using Tuple-Based Chemical Reaction Optimization," The Scientific World Journal, vol. 2014, Article ID 404375, 23 pages, 2014. doi:10.1155/2014/404375

CHAPTER 11

Thomas Wilhelm , The smallest chemical reaction system with bistability, DOI: 10.1186/1752-0509-3-90

INDEX